Biomimetic Architecture and Its Role in Developing Sustainable, Regenerative, and Livable Cities

Mohsen Aboulnaga • Samaa E. Helmy

Biomimetic Architecture and Its Role in Developing Sustainable, Regenerative, and Livable Cities

Global Perspectives and Approaches in the Age of COVID-19

Springer

Mohsen Aboulnaga
Department of Architecture
Faculty of Engineering
Cairo University
Giza, Greater Cairo, Egypt

Samaa E. Helmy
Department of Architecture
Faculty of Engineering
Cairo University
Giza, Greater Cairo, Egypt

ISBN 978-3-031-08291-7 ISBN 978-3-031-08292-4 (eBook)
https://doi.org/10.1007/978-3-031-08292-4

This Springer imprint is published by the registered company Springer Nature Switzerland AG
The registered company address is: Gewerbestrasse 11, 6330 Cham, Switzerland

Preface

Amid climate change and global warming's severe manifestation and impacts, world leaders convened at the COP 26 UN Climate Change Conference, (COP26) in Glasgow, UK, from October 31 to November 13, 2021, to tackle and discuss the climate crisis. Before COP26, the leaders of G7 met in Rome, Italy, to deal with and discuss the skyrocketing of prices for energy globally. These meetings highlighted two key global prime challenges and crises: climate change and energy, which are closely interconnected. The COP26 aimed at attaining climate neutrality by 2050 and allocating funding packages for developing countries to enable them to meet climate mitigation and adaption. One of the main topics discussed during COP26 was the role of nature and its benefits as a means to confront climate change risks.

The Glasgow Climate Pact (GCP), which was derived from the COP26 meeting, stated that the serious concern of climate and weather extremes and their adverse impacts on people and nature will continue to increase with additional increments of rising temperature. In terms of adaptation, the outcomes of COP26 emphasized the importance of scaling up action and support, including a) finance, b) capacity building and c) technology transfer, to improve adaptive capacity, strengthen resilience, and lessen vulnerability to climate change in the context of the best available science, taking into consideration the priorities and needs of developing countries. Regarding mitigation, GCP reasserted the long-term global goal to hold the increase in global average temperature to well below 2 °C above pre-industrial levels, in addition to recognizing that limiting global warming to 1.5 °C entails rapid, deep, and sustained reduction in global greenhouse gas (GHG) emissions. The latter includes the mitigation of global carbon dioxide emissions by 45% by 2030 relative to the 2010 level and to net zero by 2050, as well as deep reduction in other greenhouse gases. Among COP26 – GCP decisions are also to emphasize the significance of protecting, conserving, and restoring nature and ecosystems, including protecting the biodiversity while ensuring social and environmental safeguards. In a nutshell, the Glasgow Climate Pact outlines eight main folds: science and urgency, adaptation, adaptation finance, mitigation, finance, technology transfer, and capacity-building for mitigation and adaptation; loss and damage; implementation; and collaboration.[1]

In view of COP26 outcomes/pact and the prolonged coronavirus (COVID-19) pandemic, can biomimicry and biomimetic architecture inspired by nature be an ingredient of the trail towards achieving climate neutrality by 2050?

Tracey Ryan – Aurecon's New Zealand Managing Director and FIDIC Chair of Sustainable Development Committee – highlighted that "Net Zero 2025 is the right thing to do." She added that biomimicry and the need for a holistic approach to circularity, decarbonization, and resilience are becoming crucial.[2] The legendary and famous architect Frank Lloyd Wright said it is important to *study nature, adore nature, and stay close to nature, it will never fail you*. It is truly said since biomimicry – a new path to attaining sustainability. Thus, biomimicry is an apt approach to confront challenges facing humanity and cities with such huge resource consumption, air pollution, climate change risks, water scarcity, and above all the COVID-19 pandemic. Biomimicry in architecture can also be the right path to drive climate action and foster green and sustainable cities, and yet achieve sustainable development and the SDGs. In its recent report on mapping nature-based solutions and natural climate solutions, the World's Business Council for Sustainable Development (WBCSD) stated that accelerating business solutions for climate and nature is vital.[3]

There is mounting evidence that human overexploitation of nature is a key factor causing the outbreak and spread of new diseases such as COVID-19. Thus, protecting nature and integrating architecture into nature will preserve the global biodiversity and ecosystem. While climate generates the conditions for nature to flourish, nature controls climate via the carbon and water cycles.

Therefore, there is a great deal to learn, discover, and gain from nature, which is seemingly forgotten. Amid the global efforts to confront climate change, biomimicry stands as a motivating approach that has influential impacts on solving multifaceted human problems through imitating the elements, systems, and paradigms of nature. The term "biomimicry" is derived from the Greek word "bios." The latter term means life and "mimesis" implies imitate. Therefore, mimicking nature in architecture can be a substantial helix of inspiration to attain sustainability of the built environment. However, biomimicry is not a new idea, and it goes back to an American inventor and academic – Otto Schmitt, who first coined the world itself as early as 1982.

In fact, cities globally are encountering huge challenges. With the number of megacities to reach 43 by 2030, this puts colossal pressure on natural resources in

---

[1] UNFCCC. (2021). *Decision -/CP.26 Glasgow Climate Pact* (pp. 1–8). Glasgow: United Nations Framework Convention on Climate change. Retrieved 9 December 2021, from https://unfccc.int/sites/default/files/resource/cop26_auv_2f_cover_decision.pdf

[2] Pearson, G. (2021). *"Net zero 2025 is the right thing to do," says Aurecon's Tracy Ryan – Infrastructure Global.* Infrastructure Global. Retrieved 9 December 2021, from https://infra-global/net-zero-2025-is-the-thing-to-do-says-aurecons-tracy-ryan

[3] Docs.wbcsd.org. (2021).Retrieved 10 December 2021, from https://docs.wbcsd.org/2021/12/WBCSD-Accelrating-Buisness-Solutions-for-Climate-and-Nature.pdf

terms of energy, water, and material consumption. Moreover, the rapid population increase worldwide adds other dimensions to these immense challenges. With the global climate crisis and severity of the COVID-19 pandemic, which hit the world in the year 2019 and is still at large today, are megacities resilient enough to withstand the devastating impact of the ongoing SARS-CoV-2 pandemic coupled with climate change? Statistics published globally indicated that cities' infrastructure and sectors are not resilient. For example, as of December 6, 2021, the total confirmed cases amount to 265.8 million, and the total deaths reached 5.2 million and more to be recorded in the next months. Thus, there is an urgent need to build and transform cities to be resilient by focusing more on nature and exploiting biomimicry in architecture and urban spaces in the context of livability and sustainability.

The relationship with nature transformed into a more adaptive form through organic architecture in an attempt to integrate the spaces into a coherent whole, and the adaptation between the site and the built environment. For creating green, sustainable, and resilient cities in the age of the COVID-19 pandemic and attaining climate neutrality, biomimicry in architecture is key in this equation. Thus, there is a need to accelerate climate action, including (a) enhancing air quality, (b) protecting biodiversity, (c) ensuring clean energy transition, (d) fostering circular economic system and e) engaging local stakeholders. All of the above can be achieved through the implementation of nature-based development, low-emission development, circular development, and resilient development as well as equitable and people-command development.

Noticing the City Climate Finance Gap Fund, which wires cities in emerging and developing countries, to grasp and apprehend their climate ambition, turning low-carbon, climate-resilient concepts into strategies, policies, and finalizing projects, biomimicry in architecture could be a core and revolving point in reaching such goals, especially amid the COVID-19 crisis.

With more than 1430 cities in 210 countries around the world (e.g., the USA, India, Brazil Russia, and the UK) distressed and impacted by the coronavirus (COVID-19) and over 95% of the total cases being in urban areas and megacities, finding a solution through biomimicry is urgent and essential. Most of the people at risk are those living in informal settlements and slums in teeming and imperfect housing, estimated at more than one billion dwellers. Thus, biomimicry lingers as relevant to the health of our planet and cities as it was before the COVID-19 pandemic where nature has produced the ultimate medicines and design solutions. A wide range of multinational corporations and institutions now adjust their strategies and embrace biomimicry in their mission to efficiency, effectiveness, and sustainability. Related to climate change, organizations created applications and solutions to fight CC impacts, such as carbon mix (a new type of concrete which is inspired by coral), vertically aligned wind turbines (arrangement in the formation as seen in schools of fish), BioWAVE (which converts the energy of ocean waves into usable energy by imitating the movement of kelp under water to generate energy), and AirCarbon (which furnishes a distinctive way out of the current problem of plastics by producing plastics that are derived from greenhouse gases akin to methane instead of petroleum – the traditional source of material).[4]

Nature is a well of knowledge during the COVID-19 crisis. According to the Organization of Economic Co-operation and Development (OECD), the total number of unemployed workers in the OECD regions fell to 38.3 million, still 2.8 million higher than the pre-pandemic level.[5] Hence, taking inspiration from nature in promoting healthy and green buildings and cities by virtue of biomimicry and biomimetic architecture can foster the transition toward green and livable cities by creating new job opportunities. Therefore, crises can present opportunities for cities and their economic recovery and growth. For instance, evidence has indicated that the quality of air has improved in cities across the world since transportation, road congestion, and heavy traffic were lessened, more people were working from home, and economic activities were decreased. This had a direct impact on people's health but on the other hand, created a challenge to existing well-ventilated buildings or air-tight buildings and spaces. In this context, biomimicry in architecture can play a major role in creating livable spaces and healthy buildings. With the high cost of climate change impacts globally, specifically the devastating floods in Europe (Belgium, Germany, The Netherlands, and Spain) and other parts of the global, city leaders must utilize the current crisis to plan and manage how net zero carbon aspiration can be achieved and how cities can become more resilient and sustainable in environmental, economic, and social stipulations. After all, Professor Tim Dixon – Chair of Sustainable Future in the Built Environment, University of Reading, UK, and the Oxford COVID-19 Evidence Service Team, University of Oxford – in his essay "What impacts are emerging from COVID-19 for urban future?" indicated that 1496 cities in 30 countries had affirmed climate emergencies before COVID-19 affected people's lives. In the UK, for example, cities like Oxford and Reading have not only set out their vision for 2050 but also have climate change strategies in place or under revision.[6] Ultimately, net zero carbon and managing "'transition" remains a huge challenge for cities and regions alike, but coupling these goals to innovating thinking for the future is primarily essential if cities worldwide are to become resilient enough and meet COP26 outcomes and the Glasgow Climate Pact.

Post COP26, the prime global warming is to be curbed to 1.5 °C, and all cities need to be net zero by 2050. The very concern is how to reach climate neutrality and curb the severity of climate change, and attain Sustainable Development Goals (SDGs), which have been hindered globally by COVID-19 crises. As the supply chain and food crises led to the increase in inflation worldwide, the authors intended

---

[4] Fransen, B. (2020). *Learning from Mother Nature – Biomimicry – EcoMatcher*. EcoMatcher. Retrieved 9 December 2021, from https://www.ecomatcher.com/learning-from-mother-nature-biomimicry

[5] *Post-Covid-19 Recovery Plan Key to Bolstering Growth and Public Finance*. oecd.org. (2021). Retrieved 10 December 2021, from https://www.oecd.org/post-covid-19-recovery-plan-key-to-bolstering-growth-and-public-finance

[6] Dixon, T, (2021). *What impacts are emerging from Covid-19 for Urban Futures? – The Centre for Evidence-Based Medicine.* The Centre for Evidence-Based Medicine. Retrieved 10 December 2021, from https://www.cebm.net/covid-19/what-impacts-are-emerging-from-covid-19-for-urban-futures

to write this book due to the importance of the topic globally in terms of nature, biodiversity, and ecosystem. Going back to nature is a key issue that was discussed by leading international institutions during COP26, for instance, the United Nations, The World Bank, ICLEI – Local Governments for Sustainability, and OCED as well as the Biomimicry Institute, International Institute for Sustainable Development (IISD), C40 Cities Climate Leadership Group, and many more.

In articulating this book, the authors contemplated this topic since it is in the forefront and tackled globally, especially with reference to urban climate and challenges facing megacities and regions post COP26. Thus, we developed this book to focus mainly on biomimicry and biomimetic architecture as a means to green megacities and to fully comprehend biomimicry's approach through learning its background and history as well as to provide the origins of biomimetic architecture and its meaning, and how to integrate such an approach as a way of thinking to achieve climate and carbon neutrality. We also highlight the biomimicry role in the early conceptual design, overlapping between architecture and nature by virtue of taking nature as an inspiring model, measure, and mentor for the biomimicry design spiral. In addition, the authors emphasize the importance of biomimetic architecture amid the COVID-19 pandemic and climate crisis.

It was essential to highlight why the book focuses on biomimicry and biomimetic architecture's definitions, and the significance of biomimicry in the age of COVID-19 in terms of attitudes, connectivity to nature, and pro-environmental behaviors during COVID-19, yet learning through neighborhood's access to green spaces amid COVID-19. It is imperative to state that nature, a great source of inspiration to biomimicry, is a daily lifestyle these days, and the need for new approaches to attain livability, sustainable development and, SDGs in cities including biomimicry and energy efficiency, and circular economy and built environment are vital. There is a key question that this book is trying to answer. Can biomimicry aid in developing urban spaces and creating healthier buildings indoors? The book also highlights the interwoven relationship between biomimicry and urban design as well as biomimicry and climate change adaptation, in addition to biomimicry and post-COVID regulations, and showcases innovation in biomimicry and biomimetic architecture.

One of the challenges deduced from the literature in this book is that the topic occupies a large scale; therefore, we dived in to intensely read the literature, and research deeply to address such topic briefly, coherently, and holistically. This was really a time-demanding chore, and the COVID-19 pandemic lockdown delayed the process of writing this book, but the authors managed to overcome such a challenge in a timely manner. Another challenge derived from the literature is that people living next to coastal areas reported more than half of the world population, thus putting a severe strain on natural ecosystems and resources in addition to climate change (extreme threats including strong storms, devastating floods, and heat waves). Also, biodiversity loss and climate change – strongly interlinked – are major challenges of a similar significance and urgency to achieve SDGs. Obviously, nature and biodiversity have many possible solutions and inspiration to future pandemics and existing viruses. The green and inclusive recovery to COVID-19 has been focusing largely on climate change (CC), but less attention on biodiversity. Therefore, in any post-COVID regulation, biodiversity loss, and climate action should be mutually addressed as part of wider efforts to attain a green, sustainable, and resilient recovery in addition to attain the Glasgow Climate Pact and Climate neutrality by 2050.

Another defy is that global urban areas are the most affected regions when it comes to the spread of COVID-19, as seen in many cities in the USA, Europe, Africa, and Asia. What are the set policies, strategies, and action plans needed to curb the pandemic that hindered the progress effort done before 2019 in terms of the Sustainable Development Goals (SDGs)? This has also instigated the authors to write this book. Additionally, the current policies and mechanisms need to address the strategic actions to meet low-carbon communities, and yet integrate net zero process with the SDGs, specifically Goal 4 – Quality Education; Goal 7 – Affordable and Clean Energy; Goal 9 – Industry, Innovation and Infrastructure; Goal 11 – Sustainable Cities and Communities; Goal 12 – Responsible Consumption and Production, as well as Goal 13 – Climate Action; Goal 14 – Life on Land; and Goal 17 – Partnerships for the Goals.

This book furnishes the information needed to learn and comprehend biomimicry thinking, types, levels, and approaches as well as the six sustainability elements to achieve biomimetic concept for creating biomimicry in architecture. The aim is to highlight the importance of nature as a source of inspiration in order to withstand the severe impacts of climate change as well as natural disasters in megacities, in addition to the impacts of the COVID-19 crisis in more than 1400 cities worldwide.

The book also portrays a comprehensive overview on innovative approaches and sustainable biomimetic architecture examples in Asia, Europe, Africa, and the USA to assist the reader in facilitating and comprehending the content. The book is structured into two main parts. The first part presents some of the key issues related to biomimicry and architecture. Part I portrays a full review of biomimicry and nature – milieu, history, and approaches as well as design methods and process. It highlights the evolution of mankind and its relationship with nature biomimicry, and early conceptual design, as well as overlap between architecture and nature by taking nature as a model, measure, and mentor. Part I also furnishes the readers with a full comprehension of biomimicry through different aspects and underlines the drivers of the entire unsustainable practice of modern civilization that lies in the dualistic detachment of nature and culture. Additionally, it presents a review on the influence of biomimicry on architectural design and the built environment through several points. It also highlights and discusses the cause of biomimicry thinking, biomimicry in architecture, approaches, and levels of biomimicry as well as the levels of biomimetics information and the application types of biomimicry in architecture, and yet focuses on inspiration levels in nature and showcases many iconic global exemplary buildings. Moreover, it portrays the biomimicry design lens and nature's design principles and how biomimicry attains sustainability. While the second part of this book encompasses four chapters dedicated to highlighting the key issues of biomimicry in relation to sustainability and global applications of biomimicry approaches and levels in architecture, it also presents an overview and discusses climate change efforts, and actions concerning mitigation and adaptation in three countries (Egypt, Italy, and Germany). Part II also depicts biomimicry in criticism including the argument and defense as well as the direction towards a sustainable future. Moreover, it presents an overview of the COVID-19 pandemic and addresses how biomimicry innovation and post-COVID regulations can curb the devastating impacts of such a crisis. At the end of this book, an important question is highlighted – "Can biomimicry innovation assist in limiting the spread of COVID-19?"

With contributions from the lead author, an international and national expert who has more than 35 years of experience in higher education, government, senior management, and consultancy in strategy and policy related to sustainable urban development, sustainable energy policies, and climate change mitigation and adaptation, this book portrays an important and indispensable knowledge resource on the topic of **biomimetic architecture and its role in developing sustainable, regenerative, and livable cities.**

The *authors* strongly believe that the book will be a valuable resource to governments, policy makers, and professionals, along with research centers, libraries, and academicians as well as researchers and students who are interested in this field, in addition to industry stakeholders involved in the development of cities and buildings, for attaining climate neutrality by 2050 and offsetting climate change and achieving climate neutrality by adopting biomimicry approaches in architecture and getting inspired from nature, especially amid the COVID-19 pandemic and global crises.

You are welcome to contact the lead author through the e-mail address – maboulnaga@eng.cu.edu.eg – to share your thoughts and comments upon reading this book or afterward. I assure you to get back to you in a speedy manner.

Giza, Greater Cairo, Egypt Mohsen Aboulnaga

Acknowledgments

We would like to express our sincere thanks to everyone who has contributed tirelessly and significantly to the completion of this book. We also appreciate the thoughts, research work, and wealth of materials put in the development and articulation of this book volume.

We would like to express our sincere appreciation to Springer Nature's editorial board and production team for their valuable advice, incessant support, and patience during the development of this book. We would like to thank Mr. Michael McCabe, Senior Editor – Applied Sciences; Mr. Brian Halm, Project Coordinator – Book Production; and Zoe Kennedy, Editorial Assistant. Additionally, we appreciate the support and coordination of Mr. Shabib Sheikh, book manager, for his cooperation and endless assistance during the process and production of this book.

The authors would also like to convey their appreciation to Santiago Calatrava Global, specifically Architect Reem Nassour at Calatrava International LLC in Dubai for her effort in reviewing the related text and the images of the UAE Pavilion. In this regard, we earnestly thank Mr. Oliver Schuh and Ms. Barbara Burg, Palladium Photodesign, Köln, Germany, for their kindness in granting the permission of two images of the UAE Pavilion at EXPO 2020 Dubai. We also express our sincere thanks to The Japan Pavilion Press Team for reviewing the related text in Chap. 1. Special thanks to the designer – Simmetrico Network, Italy – and engineering company – SCE Project Milan – who were in charge of the structural, architectural, and MEP engineering work, as well as the project managers, especially Architect Maurizio Burragato and Mr. Roberta Sorrentino, Communication Department, SCE Project, for their full support and providing the photos and related permissions of the Azerbaijan Pavilion at EXPO 2020, which appear in Chap. 1. Additionally, we thank Mr. Musthafa Ebrahim Khumanpur, Managing Director – Photography, ESC Project, for the photos of the Azerbaijan Pavilion EXPO 2020 Dubai.

Thanks also go to Ms. Emily Binet Royall, MIT'16 City Planning, USA, for her kind support and granting the permission of an image depicted in Chap. 2.

Also, we truly extend our heartfelt appreciation to Zaha Hadid Architects in London, namely Ms. Malin and Mr. Henry Sequeira, press team, for granting the permission of London Aquatics Centre photos. We also express our sincere appreciation to Architect Jasper Jägers and Mrs. Judith Jägers, AEMSEN Rotterdam, the Netherlands, for providing the images of two of their sustainable and iconic projects and granting the permission. Special thanks go to Dr. Stefano Boeri and Stefano Boeri Architetti, Milan, particularly Ms. Siriana Jinag Guoyin and Ms. Fiamma Colette Invernizzi, communication team, for reviewing the text and figures in addition to providing permission for five images illustrated in Chap. 3.

Additionally, we convey our thanks to Kengo Kuma and Associates, Tokyo, Japan, particularly Mariko Inaba, press team, for reviewing as well as providing the drawings and images of their mixed-use building in Odawara-shi, Japan, portrayed in Chap. 4. Thanks also due to Matthew Dean Parkes and Tam Nguyen, Atlas Industries, for granting the permission. We also express our true appreciation to Professor Juan Herreros, founder of Estudio Herreros in Madrid and author of the Hispasat Headquarters in Spain, its responsible of media Architect Miguel del la Ossa, and photographer José Hevia for providing and granting the permission for the six images of the building presented in Chap. 4. Special thanks to William McDonough + Partners, an architectural firm in Virginia, USA, for their support and providing the permission of two images appearing in Chap. 4 of this book.

Moreover, we also extend our sincere thanks to the World Meteorological Organization (WMO) in Geneva, Switzerland, for granting the permission for a photo in Chap. 5, especially Ms. Sylvie Castonguay, Chair of Publication, and Ms. Clare Nullis, Officer – Strategic Communication Office.

Furthermore, the authors extend their thanks to Ms. Yasmina Ragab, Ms. Sherifa El-Haggan, and Mr. Mohamed Shahwna for their effort and time to proofread the book chapters. We also thank Architect Salma Abdelkader for her support and effort in enhancing some images in book and developing some of the picture permission forms.

Finally, we sincerely appreciate the effort of Amina El-Haggan, Aya Ghobasy, Rana El-Bakry, and Sherry El-Ghoraiby for capturing the images of the UAE Pavilion, The Terra – The Sustainability Pavilion, and The Mobility and Continuity Pavilion as well as Slovenia Pavilion and Japan Pavilion at Expo 2020 Dubai, which are depicted in Chap. 1 of this book.

Authors would also like to thank Arch. Haidy Mousa for her effort and assistance in pursuing the electronic corrections to the book proofs.

Last but not least, thanks are extended to all those who made this book process a success.

Mohsen Aboulnaga
Samaa E. Helmy

Contents

List of Figures

List of Tables

List of Boxes

About the Authors

Mohsen Aboulnaga has more than 36 years of experience in higher education, government, senior management, and consultancy in sustainable development, sustainable urban development, sustainable energy, energy policy development, and green cities as well as climate change mitigation and adaptation. Dr. Aboulnaga is Professor of Sustainable Built Environment in the Faculty of Engineering, Cairo University, and a government strategy and policy adviser. He is also an associate and senior consultant in strategy and sustainable development at Oil and Gas Skills (OGS), the human development arm of the Ministry of Petroleum (2014–2019), and an associate expert at the Regional Centre for Renewable Energy and Energy Efficiency (RCREEE), which he represented at a high-level meeting at the European Commission – Joint Research Center (EC-JRC) in Ispra, Italy, to review the Guidebook for South-ENPI. Professor Aboulnaga is an expert on sustainable urban development, sustainable development, energy efficiency, green building, and green and smart cities. His areas of competency include strategy planning, policy development, low-carbon society, and sustainable building as well as renewable energy. Other areas of expertise include executive capacity building and coaching training on strategic planning and leadership and academic activities, as well as developing vision, mission, strategic goals, and action plan of programs and coordination. Between 2015 and 2018, Professor Aboulnaga was appointed by the European Union and the Ministry of Foreign Affairs (MoFAs) as a senior adviser and national expert on the "Clean Energy Savings Mediterranean Cities – CES-MED" project, a project funded by the European Union (2013–2018). In October 2015, Professor Aboulnaga was appointed senior expert by the Human Dynamics Consortium – KG to develop the CES-MED National Recommendation Report (NRR) on Sustainable Urban and Energy Savings Actions (SUESA) for Egypt, as part of the CES-MED project. In May 2016, he has been appointed by the EU and Ministry of Foreign Affairs, the National Expert of CES-MED project in

Egypt – Associated to Transitions - Sustainability Consultant – Paris, France and developed the Sustainable Energy and Climate Action Plans (SECAPs) for the Governorate of Red Sea and the Governorate of Luxor – Egypt. Professor Aboulnaga has been an invited expert and has participated in the 5th Steering Committee Meeting, as part of the framework of the European Union funded projects "Cleaner and Energy Saving Mediterranean Cities" (CES-MED) and "Sustainable Urban Demonstration Energy Projects" (SUDEP), and the high-level conference on the Role of Local Authorities in Sustainable Energy Al-Skhirat, Morocco, 4–7 May, 2015.

Furthermore, Professor Aboulnaga is a senior expert on sustainable urban development and green cities with the Union for The Mediterranean (UfM) in Barcelona. In 2020, Professor Aboulnaga organized with the Technical University of Lisbon, Portugal; the University of Florence, Italy; and the University of Bath, UK, the international design competition "HOME-KIT Sustainable Modular Houses for Slums' Dwellers amidst COVID-19," which concluded on July 30, 2021, in Lisbon, Portugal (https://wrec2020.tecnico.ulisboa.pt/~wrec2020.daemon/homekit.html).

Professor Aboulnaga has developed many concept notes for leading international entities such as ENI ZOHR Sustainability Initiative in 2019, European Union (EU) funded projects in 2018–2019, and Sustainable Urban Development for The Union for the Mediterranean (UfM) in 2017.

Professor Aboulnaga is a former strategy and policy adviser of environment and infrastructure at the Prime Minister's Office of the UAE (February 2009–April 2010), responsible for implementing and monitoring the Federal Government Strategy (cycle 2008–2010). Prior to that, he held a position in the capacity of strategy and policy adviser at the Executive Council, Government of Dubai (May 2007–January 2009), where he developed many strategic projects, namely Dubai Government Green Building Policy 2007, Dubai Heat Island Effect, Orthophoto Flyover project in collaboration with LBNL in the USA, Detailed Policy for a Sustainable Dubai Urban Development Framework 2020, and Green Building Guidelines for the UAE Ministry of Public Works and LAS in January 2009. Professor Aboulnaga founded the Emirates Green Building Council in 2006 and was elected the first chairman of the Council and Board of Directors.

Professor Aboulnaga is an honorary fellow of the Institute of Green Professionals, USA. He has been an associate partner with the European Sustainable Development Network (ESDN) in Austria since 2010 and a registered expert with UNEP and ESCWA since 1997. In December 2013, he participated in the European Commission – Joint Research Centre (EC-JRC) expert meeting in Ispra, Italy, to review the Sustainable Energy Guidebook for ENPI South countries.

Professor Aboulnaga is an author of 4 books published by Springer and more than 50 chapters in books published by Springer and Elsevier. He has been an invited keynote speaker and speaker at more than 400 international and regional conferences, seminars, workshops, and training programs, and has more than 200 international publications and presentations to his credit. In addition, he is an external examiner of the PhD examination panel at the University of Sheffield in England, UK, and has taught at the University of Leeds, UK, and at other universities. Currently, he is teaching at Cairo University.

Samaa E. Helmy has seven years of experience in higher education, sustainable urban development, and climate change. Ms. Helmy is currently a teaching assistant in the Architectural Engineering and Technology Programme (AET) at Cairo University and a teaching assistant in the Department of Architecture at Misr University for Science and Technology (MUST) since 2017. She received a 3-week summer school training on "Sustainability and Energy Conservation in Architecture" at the University of Lincoln, UK, in 2012. In addition, she is has been assisting Professor Mohsen Aboulnaga in many projects and courses, namely smart building, since 2016.

Ms. Helmy is a trained and qualified architect in Egypt and is skilled in sustainable urban development, strategic environment assessment, and energy efficiency in low-income dwellings. Her research areas of interest are sustainability, sustainable development, and sustainable built environment, including energy efficiency and green cities in developing countries. Her areas of competency include sustainable communities, biomimicry architecture, and climate change vulnerability and adaptation. Other areas of expertise include informal areas upgrading related to cultural heritage and low-carbon society. She spoke and participated in an the International Sustainable Ecological Engineering Design for Society (SEEDS) 2019 conference, and published a paper entitled. "Future Cities: The Role of Biomimicry Architecture in Improving Livability in Megacities and Mitigating Climate Change Risks" in the SEEDS proceedings.

She also a published a chapter in the book *Sustainable Ecological Engineering Design* Selected Proceedings from the International Conference of Sustainable Ecological Engineering Design for Society (SEEDS) 2019 (Springer, 2020). She obtained her Bachelor of Science (BSc) in the Architectural Engineering and Technology Programme, Faculty of Engineering, Cairo University, in 2014, and a Master of Science (MSc) in sustainable environmental design from the same university in 2018. Finally, Mrs. Helmy registered for the degree of Doctor of Philosophy (PhD) in March 2019 and currently is conducting her research work for it.

Chapter 1
Biomimetic Architecture: The Path to Drive Climate Action, Foster Sustainable Cities, and Attain the SDGs, and the Coverage

Study nature, love nature, and stay close to nature. It will never fail you. (Frank Lloyd Wright)

1.1 Introduction

Biomimicry, as a new path toward attaining sustainability, is becoming an essential approach more than ever, especially amid many challenges facing humanity between 2019 and 2022, and beyond. Huge resource consumption coupled with global energy skyrocketing prices due to the war between Russia and Ukraine, climate change dire risks, and the Coronavirus (COVID-19) pandemic are among the challenges facing cities across the world. Biomimicry in architecture can also be the right path to drive climate action and foster green and sustainable cities, yet achieve sustainable development and the SDGs. In this context, mimicking nature into architecture can be a great source of inspiration to achieve sustainability (Fig. 1.1). Cactus plants, for example, store water and even survive without water in desert through the developed waxy skin, spiky spines, and the reserved opening and closing of the stomata to lower the water loss rate by limiting the airflow near the cactus as shown in Figs. 1.2 and 1.3 [1]. Buildings can adopt the same by minimizing water use and losses through the 3-R (Reduce, recycle, and reuse).

1.2 Why Focus on Biomimicry in Architecture Amid the COVID-19 Pandemic and Crisis

There are enough reasons why all stakeholders in urban development should focus on biomimetic architecture amid the severe impacts of climate change and the Coronavirus (COVID-19) pandemic. In fact, European cities are moving toward climate neutrality participation, indicators, and setting new benchmarking [2]. This means lessening the GHG emission, mainly the CO_2, to almost zero. The term "carbon neutral" is increasingly used to refer to "net-zero emissions." In a simple

M. Aboulnaga, S. E. Helmy, *Biomimetic Architecture and Its Role in Developing Sustainable, Regenerative, and Livable Cities*,
https://doi.org/10.1007/978-3-031-08292-4_1

Fig. 1.1 Nature and biology inspire biomimetic architecture. (Image credit: Mohsen Aboulnaga)

description, carbon neutral means the sum of all greenhouse gases (GHGs) – that is put into and taken out of the atmosphere – should balance out to zero. But Patrick Lau stated that in practice this is proven to be a confusing and divisive concept [3].

Between 2020 and the present, the Coronavirus (COVID-19) pandemic caused and is still causing serious and catastrophic impacts on many sectors globally such as economic, health, education, aviation, and industry as well as the social sector. Nature can be the platform to learn from amid COVID-19 crisis (Figs. 1.4 and 1.5). To illustrate the importance of biomimicry during COVID-19, the UNDP, through its Accelerator Labs and in partnership with the Biomimicry Institute in the United States and in collaboration with a number of Ukrainian NGOs, invited innovators, engineers, designers, and inventors to take part in the Biomimicry Global Challenge (BGC) – an intentional competition to support biomimicry-based solutions [4].

Biomimicry in architecture can play a major role in creating green cities in the age of the COVID-19 pandemic and supporting the reach toward climate neutrality. By virtue of the GreenClimateCities (GCC) program, the ICLEI – Sustainability for Local Governments provides a guideline for local and regional governments to address climate change [5]. Nonetheless, the Climate Neutrality Framework (CNF) – as shown in Fig. 1.6 – is part of the GCC program that assists cities, towns and regions in setting and advancing toward climate neutrality [6]. In this context, the Global Covenant of Mayors for Climate and Energy (GCoM), where ICLEI is a founding member, gathered over 11,500 cities and local governments steadfast to develop low-emission and resilient communities [7]. It is clear from Fig. 1.6 that Nature-based Development is the first among ICLEI five pathways to achieve CNF.

Fig. 1.2 Nature-inspiring biomimicry in architecture. (Image credit and source: Lee Aboulnaga)

With the City Climate Finance Gap Fund (Gap Fund) that supports cities in emerging and developing countries realize their climate ambition, turning low-carbon, climate-resilient concepts into strategies and policies as well as finalizing projects, biomimicry in architecture could be a revolving point in reaching such goals, especially amid COVID-19 crisis [8]. Biomimetic architecture could also be in the heart of ICLEI Global Research Strategy (ICLEI-GRS) that aims at furnishing cities and towns with the needed information, policy recommendations, and tools to assist them get through the challenges facing sustainability [9]. Authors believe that such ICLEI-GRS platform could be a perfect hub to support cities advancing research for biomimetic buildings to achieve climate neutrality. Box 1.1 presents the definition of climate neutrality [6].

Can urban spaces with natural landscape be the solution to lessen the impact of the Coronavirus (COVID-19) in the open air or spaces with clean air and high level

(b) (b)

Fig. 1.3 Nature-inspiring biomimicry in architecture. (**a**) Cactus plants with prickly spines growing in clusters. (**b**) Cactus flowers ending the biological process. (Images' credit and source: Mohsen Aboulnaga)

Fig. 1.4 Nature as a platform to learn from during the COVID-19 pandemic. (Image credit: Mohsen Aboulnaga)

Fig. 1.5 Nature as knowledge and inspiration for biomimicry and confronting COVID-19 crisis. (**a**) Italian Alpine grass (Prato Alpino Italia). (Image credit and source: Silvia Marcellino, https://commons.m.wikimedia.org/wiki/Category:Plants#/media/File%3Aprato_alpino_italia.png). (**b**) *Hibiscus Tiliaceus* plant from nature. (Image credit & source: Wafid Kholishatul, https://commons.m.wikimedia.org/wiki/Category:Plants#/media/File%3ADaun_Waru_(Hibiscus_tiliaceus_di_pagi_hari. jpg)

of oxygen during the day as illustrated in Fig. 1.7? Since large open spaces with social distances can reduce the spread of the virus, it has provided a chance to enhance walkability and pedestrians' experience coupled with social distances cities (Figs. 1.7 and 1.8). In this line, Fritjof Capra highlighted the importance of biomimicry to address the global biological crisis as shown in Box 1.2.

In the advancing sustainability, ICLEI Global Research Strategy (ICLEI-GRS) is striving to implement its mandate through various activities in multiple working

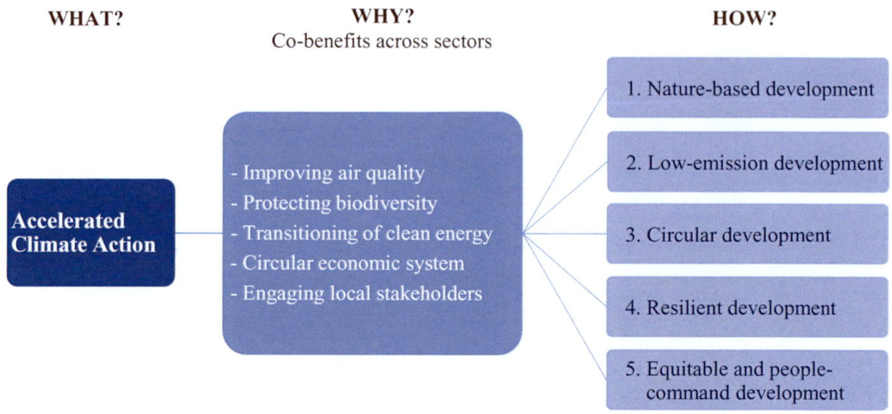

Fig. 1.6 GreenClimateCities – Climate Neutrality Framework. (Image source: Developed by authors after ICLEI World Congress 2021–2022: The Road to Malmö)

Box 1.1: Definition of Climate Neutrality

In the context of local and regional governments, climate neutrality is defined as:

> The targeted reduction of greenhouse gas (GHG) emissions and GHG avoidance in own operations and across the community in all sectors to an absolute net-zero emission level at the latest by 2050. (ICLEI – Local governments for sustainability)

areas [9]. Figure 1.9 illustrates these eight activities. In this regard, nature-based biomimetic architecture could be in the core of ICLEI activities and the global research strategy as well.

It is worth mentioning some of the important facts about the impact of the Coronavirus (COVID-19) pandemic in the urban world. For instance, more than 1430 cities in 210 countries around the world are affected by COVID-19 and over 95% of the total cases are in urban areas and megacities. Most of the people at risk are those living in informal settlements and slums in overcrowded and inadequate

Box 1.2: Importance of Biomimicry

> At this point in our global biological crisis, the survival of humanity will require a fundamental shift in our attitude toward nature: from finding out how we can dominate and manipulate nature to how we can learn from it. In this brilliant and hopeful book, Jay Harman shows us how far the new field of biomimicry has already progressed toward this goal. (Fritjof Capra)

Fig. 1.7 Walkability in Vienna's urban spaces with nature (landscape) as a solution in the age of COVID-19. (Image credit: Mohsen Aboulnaga)

Fig. 1.8 Walkability in urban spaces (trees and water) in Prague as a solution during COVID-19. (Image credit: Mohsen Aboulnaga)

housing, estimated at more than 1 billion dwellers [10]. Figure 1.10 shows some facts on the impact of the COVID-19 pandemic in the urban world.

Biomimicry remains as relevant to the health of our planet and cities as it was before the COVID-19 pandemic. Nature has generated the ultimate medicines and design solutions, for example, a chemical amalgam found in the dogfish shark that safeguards against viral infection. This could be utilized in the reaction to COVID-19 pandemic [4].

Nature can be a source of knowledge during the COVID-19 crisis. This is what has been learned from nature, the detection of viruses is difficult and time consuming. Living cells, nonetheless, are capable of noticing even faint trace of molecules with their sensitive biochemical receptors. Mimicking the way receptors behave, and letting them spur a macroscopically visible color change, has allowed the

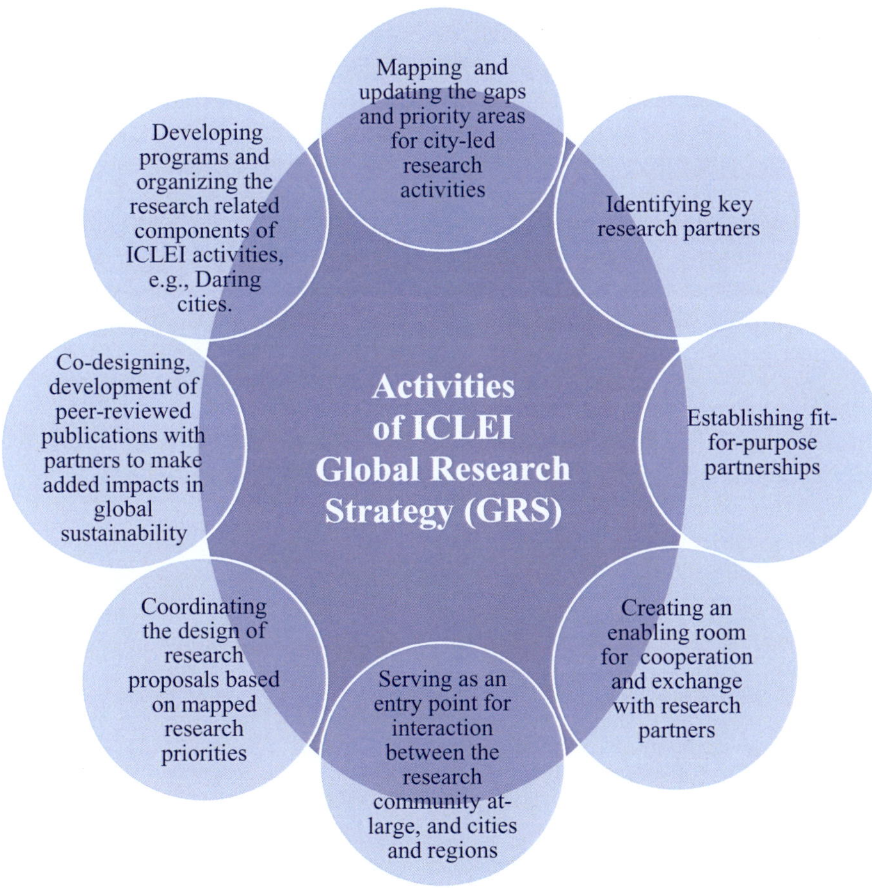

Fig. 1.9 Activities of ICLEI Global Research Strategy – GRS. (Image source: Developed by authors after ICLEI, https://www.iclei.org/en/Global_Research_Strategy.html)

Fig. 1.10 Facts about the Coronavirus (COVID-19) pandemic impacts on the urban world. (Image source: Developed by authors after FES Regional Climate and Energy Project MENA)

instantaneous recognition of molecules related with pathogens and other chemicals [4]. Also, in the healthcare sector, one of the greatest shortages is the availability of personal protective products equipment (PPEs). Hence, there is a high need for better material to develop such lifesaving yields. By looking at nature, we can mimic the mean nature generates such materials, e.g., taking carbon dioxide (CO_2) from air and transforming it into biomass similar to the way a tree does. Inventors developed polymers that have up to 50% of CO_2 by mass that assist in locking up this harmful greenhouse gas GHG [4]. For surfaces, where the COVID-19 virus survives for many hours or more on metal (3 days), there is an urgent necessity to regularly disinfect these surfaces. A type of Red Seaweed can emit an antibacterial film that shields it from contaminations [4]. Mimicking this performance to develop similar antivirus films could result in an innovative approach to safeguard buildings' users in cities.

1.3 Biomimicry and Biomimetic Architecture Definitions

It is imperative to digest the term "biomimicry"before reading this chapter and the book's coverage. Biomimicry presents an immense potential for urban planning and cities following the analogy of city lungs (parks including trees) and human lungs that demonstrate such ability, while the basic principles of sustainability appear to be "returning back to the roots of nature."

1.3.1 Biomimicry Definition

Biomimicry is defined as "a new science that examines nature's models and then emulates these forms, process systems and strategies to solve the human problems through sustainably." This definition considers nature as a "Model." When considering nature as a "Measure," biomimicry is explained as "Biomimicry exploits an ecological standard to judge the sustainability of our innovation." Nevertheless, if nature is considered as a "Mentor," biomimicry is a new way of viewing and valuing nature"[11]. In addition, biomimicry is a type of bioinspired design, but not all bio-inspired designs reflect biomimicry [12]. When nature's strategies are interpreted into design, the science of practice mainly entails three elements: emulate, ethos, and (re)connect. Figure 1.11 illustrates these elements in detail. In contrast, biomimicry in architecture is defined as "Multi-disciplinary scientific approach to sustainable design that goes beyond using nature as inspiration for aesthetic but rather deeply studying and adapting construction principles that are found in natural environments and species" (Box 1.3) [12].

Box 1.3: Biomimetic Architecture Definitions

Biomimicry is a practice that learns from and mimics the strategies found in nature to solve human design and challenges – and to find hope along the way. (Biomimicry Institute [12])

Multi-disciplinary scientific approach to sustainable design that goes beyond using nature as inspiration for aesthetic but rather deeply studying and adapting construction principles that are found in natural environments and species. [12]

1.3.2 Types of Biomimicry

Biomimicry has many types to manifest. According to Janine Benyus, the biomimicry types are classified into three main types. The first type is copying form and shape; the second type is copying a process, like photosynthesis in a leaf; and the third type is mimicking an ecosystem's level, like building a nature-inspired city [12, 13]. Box 1.4 and Fig. 1.12 present these three types of biomimicry. "Biomimicry" as a terminology was first recognized in the scientific literature in 1962. The term is

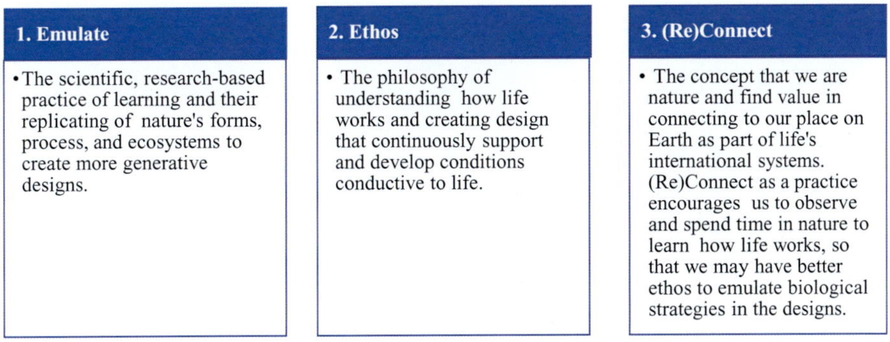

Fig. 1.11 Elements of nature strategies in design. (Source: Developed by authors after Biomimicry Institute)

utilized especially in the middle of materials researchers in the 1980s. There has been a huge impetus in the current decade, by renowned Professor of Biology Janine Benyus, American biomechanics researcher Steven Vogel, and Professor of Biomimetics Julian Vincent – known as one of the most imaginative and thorough researchers, as well as an inspirational teacher and knowledge-based writer, in the area of bioinspired materials. The three scholars broadly composed such bough of knowledge (Box 1.5).

1.4 Biomimicry's Significance in the Age of COVID-19

The Coronavirus (COVID-19) pandemic imposed drastic adjustments and diverse changes to our patterns of lives (work, education and mobility, and others). COVID-19 has brought up a new "trend" and changed our norms, perceptions, lifestyle, and priorities. Creating virus resilience and flexibility was not an easy task and experience, whether at the family scale, neighborhoods and communities, and governments alike. This resilience has led to the fact that fundamental elements of

Box 1.4: Types of Biomimicry

There are types of biomimicry – one is copying form and shape, another is copying a process, like photosynthesis in a leaf and the third is mimicking at an ecosystem's level, like building a nature inspired city. (Janine Benyus)

Box 1.5: Natural World

The more our world functions like the natural world, the more likely we are to endure on this home that is ours, but not ours alone.

When we look at what is truly sustainable, the only real model that has worked over long periods of time is the natural world. (Janine Benyus)

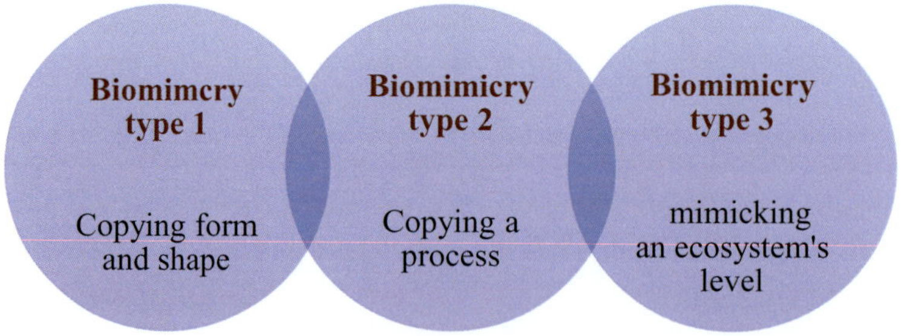

Fig. 1.12 Biomimicry types' classification. (Source: Developed by authors after Janine Benyus)

architecture design have been changed and altered to meet the Coronavirus precautions and social distances.

The bamboo tree is a good example of biomimicry since it has optional fiber distribution for lightness and toughness. Fibers surrounding the heart-shape opening in the stem have a denser distribution toward the outer part. Hence, it creates strength [14]. Figure 1.13 shows bamboo trees in Arashiyama forest, Kyoto, Japan. The spatial distribution of fibers in the hollow bamboo cylinders is optimized to reinforce flexural rigidity. Such a new finding sheds light on biomimetic approaches in the development of materials and biomimetic architecture. Being light and tough, bamboo is broadly utilized as a natural material in Asian countries, particularly in Japan and Thailand. However, its lightness leaves it vulnerable to strong crosswinds [14]. Box 1.6 presents the statement of Professor Motohiro Sato at Hokkaido University [14, 15]. Figure 1.14 also illustrates small trees' stems growing in water

Fig. 1.13 Bamboo trees inspiring biomimetic architecture. (Image source: Mohsen Aboulnaga)

Fig. 1.14 Nature and ecosystems inspire biomimicry. (Image source: Mohsen Aboulnaga)

and the ecosystem of Arashiyama forest at the fringes of Kyoto in Japan, where nature inspires biomimicry.

Box 1.6: Nature – Bamboo Model to Inspire Biomimicry
Our study could help develop advanced materials by mimicking the bamboo model for its lightness and toughness. (Motohira Sato, Hokkaido University, Japan [15])

As the world is slowly reopening in July 2021 and the lockdown measures were variably lifted, people globally are adapting to new veracity. Evaluating our environment can provide solutions to confront the severe impacts of the COVID-19 crisis and can foresee a relatively unknown future. Nonetheless, logical analogies can be developed based on the current situation to inform better inventions exploiting nature. The issue of nature and biomimicry's approach could have an influence on adjusting to COVID-19 crisis. Future adaptation could be through inspirations and looking to nature and finding solutions to the current persisting problems. In this line, architects globally furnish their skills and knowledge to use in the fight against the coronavirus and have created innovative solutions that might stick around us for a while and perhaps be the acclimated lifestyle on the planet.

1.4.1 Attitudes, Connectivity to Nature, and Proenvironmental Behaviors During COVID-19

During COVID-19 lockdown, adults' behaviors and attitudes toward nature have specifically changed to be connected more with nature. The University of Cumbria in Carlisle, UK, conducted a study in 2020 to explore experience and associate changes in attitudes to nature, connection to nature, and proenvironmental behaviors; about 704 people participated representing all adult age groups across England, Scotland, and Wales; and the study was carried out from 29 May to 14 June 2020 [16]. The findings of the survey are summarized as follows:

- Large increase in desire to spend time in nature among adults in the UK.
- 72% and 60% of women and men, respectively, stated that they are likely to spend time in nature in the future.
- Lockdown witnesses all age groups spending more time in nature daily.
- 66% of the surveyed parents/guardians stated that their children are spending more time than usual in nature during the lockdown (only 16% report less time than normal).
- Listening to birdsong was the most common way adults noticed nature during the lockdown (94% of respondents), followed by watching wildlife (87%), and taking time to notice bees and butterflies specifically (83%).
- Over a third of respondents had intentionally studied nature in detail during the UK lockdown, e.g., by learning to identify trees or birds.
- 77% of respondents have captured a photograph or video of nature during lockdown.
- Large increase in the UK adults speaking about nature with friends and relatives throughout the lockdown (two-thirds of respondents report doing this quite often).
- The results suggest experience during lockdown may be related to increased nature connectedness among adults population and 95% agreed with the statement "I find beauty in nature" [16].

Figure 1.15 illustrates these findings in brief and depicts UK adults' attitude and behaviors toward nature throughout the lockdown (May – June 2020). Such findings indicate the significance of nature in changing peoples' attitude toward nature and proenvironmental behaviors; biomimicry is yet to be inspired with nature.

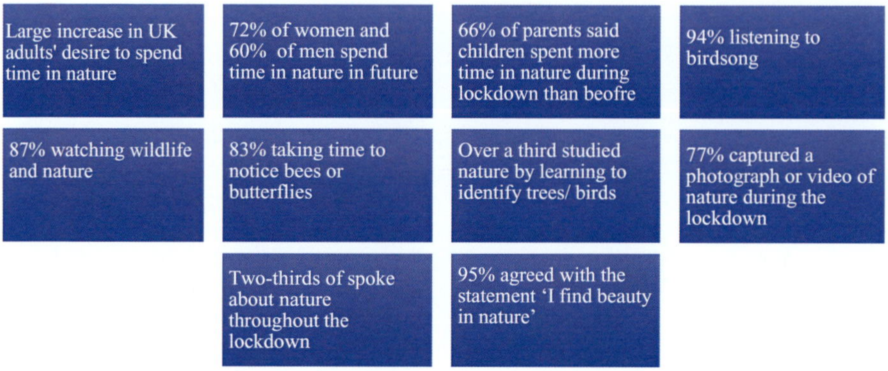

Fig. 1.15 Changes in UK adults' attitudes and connectivity to nature and proenvironmental behaviors. (Source: Developed by authors after Tania Lemmey, University of Cumbria, 2020)

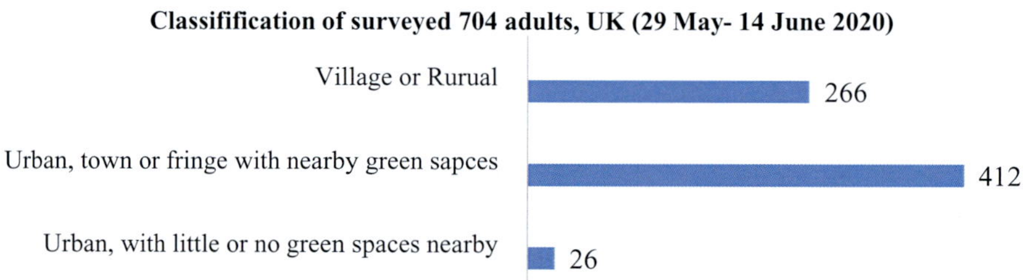

Fig. 1.16 Description of the neighborhood and local green spaces, UK. (Image source: Developed by authors after Tania Lemmey, University of Cumbria, 2020)

1.4.2 Biomimicry Learning Through Neighborhood Access to Green Spaces Amid COVID-19

The distribution of surveyed people in the study carried out by Cumbria University shows that 59% of the respondents lived in urban, town, or outskirts with nearby green spaces whereas those living in village and rural residents form 38%, but about only 4% of the respondents lived in urban areas with little or no green spaces nearby [16]. Figure 1.16 presents these percentages. Nevertheless, a survey conducted by YouGov for the RSPB revealed that only 51% of people in households in England reported living within a 10-miunte walking distance of any publicly accessible natural green space [17]. It is clear from Fig. 1.16 that the majority are living in urban towns with nearby green spaces. Another important factor is the daily time spent in nature. Figure 1.17 shows the daily time spent in nature during and before COVID-19 lockdown mid-2020 in the UK. People who spent more than 1.5 hours per day in nature increased from 27% before lockdown to 45% during lockdown. Comparing

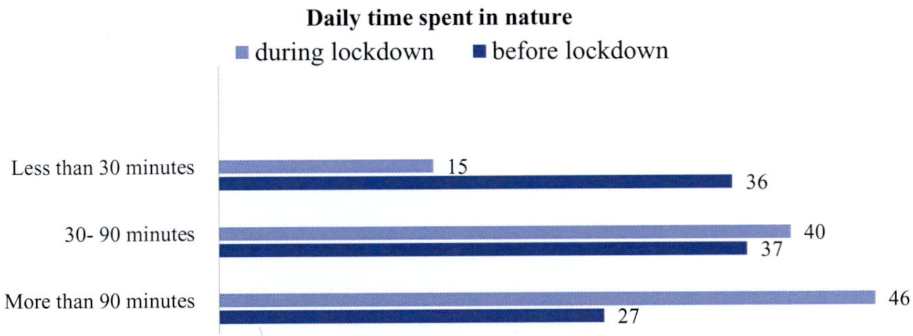

Fig. 1.17 Pattern of time people spent daily in Nature in the UK during and before 2020 lockdown. (Image source: Developed by authors after Tania Lemmey, University of Cumbria 2020)

this with an urban seaside part of Naples, Italy, that has no or very little green spaces (Fig. 1.18) compared to the UK case highlights the importance of nature (urban green space) during COVID-19. Hence, nature is a great source of inspiration for biomimicry and biomimetic architecture.

1.5 Nature, a Great Source of Biomimicry Inspiration

Inspiration from nature can be a great source of biomimicry. This can be noticed in the front car of the Shinkansen Bullet train in Tokyo, Japan, as illustrated Fig. 1.19c. This train is the fastest in the world (320 km/h and it can reach 500 km/h) using the Maglev technology. The innovation was inspired by the water bird Pelecaniformes, which is also known by other names (totippalmates or Steganopodes). Figure 1.19.a shows the long beak of the Pelecaniformes bird while Fig. 1.19.b presents the birds in the sky traveling in high speed over water where the sharp end of the beak enables it to penetrate and pierce the coming high-speed wind. Such exemplary model of mimicking the Pelecaniformes bird is considered as the first type of biomimicry shown in Fig. 1.19. Figures 1.20 and 1.21 also present Heydar Aliyev Cultural Center in Baku, Azerbaijan, where late Zaha Hadid mimicked reptiles and waves together.

1.6 Needs for New Approaches to Attain Livability, Sustainable Development, and SDGs

Climate action by virtue of biomimicry innovation for attaining sustainable development (SD) and sustainable development goals (SDGs) is getting attention by many intentional scholars and organizations. Rebecca Carlson recently highlighted

Fig. 1.18 View of seaside urban area with less green spaces, Naples, Italy, before COVID-19. (Image credit: Mohsen Aboulnaga)

Fig. 1.19 Biomimicry example of the Shinkansen bullet train in Japan inspired by the Pelecaniformes water bird. (a) Sharp and long beak of Pelecaniformes water bird. (b) The Pelecaniformes bird flying in strong high-speed winds by virtue of its beak. (c) The Shinkansen bullet train in Tokyo, Japan. (Images' source: (a) https://commons.m.wikimedia.org/wiki/Commons:Featured_pictures/Animals/Birds/Pelecaniformes#/media/File:Eeastern_great_egret_(Close-up_of_the_head_area)_at_Tennöji_Park_in_Osaka,_November_2016_1410.jpg; (b) https://commons.m.wikimedia.org/wiki/Commons:Featured_pictures/Animals/Birds/Pelecaniformes#/media/File:Eeastern_great_egret_in_flight_at_Tennöji_Park_in_Osaka,_November_2016_1479.jpg; and (c) Mohsen Aboulnaga)

the issue of Biomimicry Global Design Challenge and SDGs. She stated in the following opening statement [18]:

> The tagline for this year's Biomimicry Global Design Challenge, "Design° for people + Planet," ask participants to look at what degree of change they will make for a particular SDG, using nature as their mentor.

To attain livability and SDGs, innovative approaches to biomimicry could be a significant tool to drive the change toward this aspect and address cities' challenges. Biomimicry is a practice that inspires from nature and mimics the strategies and lessons found in biology, through the exploitation of nature, as a model, mentor, and measuring tool for sustainability and SD as well as SDGs, yet solves human design challenges by developing analogous solutions from nature [18].

As indicated by Sustainability Brands, the alignment between the SDGs and Biomimicry Global Design Challenge (BGDC) is expressed as all innovation

Fig. 1.20 Mimicking a reptile – Heydar Aliyev Cultural Center in Baku, Azerbaijan, where late Zaha Hadid mimicked reptiles and waves together. (Image source: https://ru.m.wikipedia.org/wiki/%D0%A4%B0%D0%B9%BB:Heydar_Aliyev_Center.JPG)

Fig. 1.21 A view of Heydar Aliyev Cultural Center in Baku, Azerbaijan, where the design mimics water vaves. (Image source: https://en.m.wikimedia.org/wiki/File:Heydar_Aliyev_Cultural_Center.jpg)

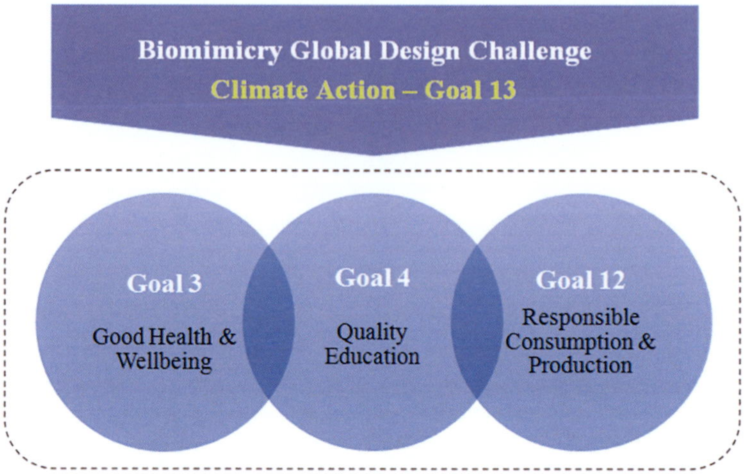

Fig. 1.22 Biomimicry's link to the SDGs. (Source: Developed by authors after M Lopez)

points to Climate Action – Goal 13, distinctively addressing Climate Action; to all goals interconnected within the climate crisis, for instance, Goal 3: Good Health and Wellbeing, Goal 12: Responsible Consumption and Production, and Goal 4: Quality Education, in addition, Goal 14: Life Below Water that calls for the conservation and sustainable use of oceans, seas, and marine resources for sustainable development. By-and-large, every SDG, in some form or shape, can be attributed to furthering regenerative ecosystems that all species do [19, 20]. As part of BGDC, Biomimicry San Diego established a "Biomimicry Launchpad" to foster learning and practicing of biomimicry and accelerates its development [20]. Figure 1.22 illustrates such connection between biomimicry and SDGs. Also, by emulating the way organisms like manta rays and basking sharks filter food from water, Floating Coconet is able to collect and direct free-flowing plastics, small and large in size, to help fight the growing of plastic pollutions challenge in oceans [18].

Redacción Mapfre indicated that the branch of science (bio meaning "life" and mimesis meaning "to imitate") studies nature as a source of inspiration for innovative technologies with a view to solving human problems – and it is gaining traction with regard to biodiversity – compatible architecture, and construction [21]. Box 1.7 highlights António Gaudi on nature as source of art for livability, while Box 1.8 summarizes how nature is a source of inspiration for biomimicry. In fact, biomimicry is known as biomimetics or biomimetism, which refers to the process of studying, observing, understanding, and applying natural solutions to human problems in accordance to biological principles and by exploiting all kinds of biomaterials [21].

Box 1.7: Nature as an Art for Livability

Nothing is art if it does not come from nature. (António Gaudi)

Box 1.8: Nature as a Source of Inspiration for Biomimicry

The branch of science (bio meaning "life" and mimesis meaning "to imitate") studies nature as a source of inspiration for innovative technologies with a view to solving human problems – and it is gaining traction with regard to biodiversity – compatible architecture and construction. (Redacción Mapfre)

1.6.1 Biomimicry and Energy Efficiency

In a published paper in the Risk Management magazine by MAPFARE Global Risks, which refers to biomimicry as a mean for enhancing energy efficiency, this will be achieved by designing buildings similar to the structures found in natural organisms (Fig. 1.23). Nevertheless, Marlén López – Director of Biomimetics Laboratory – highlights that one of the prime challenges will be fulfilling the SDGs in an endeavor to lessen the environmental impacts while reacting to the social and energy defies facing the global [21].

In this sense, developing plenty of efficient skeletons (structures), producing materials, and developing zero-consumption systems, as well as resource management, monitoring, and controlling thermal comfort – all under one roof (envelop) – besides generating energy for buildings are considered part of such process.

1.6.2 Circular and Biomimicry Built Environment: CBBE and SDGs

As the built environment is increasingly responsible for the global environmental problems such as vast amount of natural resources (material, water, and energy use) and waste and their related GHG emissions, attributed to the habitats and humans have developed around themselves; it is obviously becoming vital that a change in the consumption rates is obligatory more than ever to curb such emissions and the way in which the built environment is designed and developed if further damages to climate and ecosystems are to be evaded.

Biomimicry is characterized as the design and creation of materials, structures, and systems that are modeled on biological process [22]. Biomimicry designs are frequently very innovative and can furnish sustainable, eco-friendly, and intuitive solutions to confront challenges, particularly in the built environment. In the meantime, initiating circular design components guarantees the resilience and sustainability of the built environment.

Fig. 1.23 Design buildings like structures of natural organisms. (Image credit and source: Mohsen Aboulnaga)

In this line, the United Nations Economic and Social Commission for Asia and the Pacific (UNESCAP) has established the Circular and Biomimicry Built Environment Group (CBBEG) and opened on the SDG Community of Practice – a collaborative hub and gateway for enthusiasts of the CBBEG under the SDGs – with the 17 SDGs as a road map [21]. Implementing these sustainable principles is vital in the post-COVID and recovery, and reconstruction, once development returns back to normal rates. In this context, authors foresee nature as a source of sustainability and SDGs attainment especially amid the impact of the COVID-19 crisis, which hindered the progress gained for SDGs before the year 2020, including most of the SDGs.

According to Oslo's manifesto, biomimicry is proven to be adopting sustainable solutions inspired by nature [23]. Based on biomimicry, it is clear that biomimicry exits in all what are surrounding us in the environment, ecosystems, and nature. Box 1.9 highlights Janine Benyus view on sustainability and the natural world, while Box 1.10 presents biomimicry as a value of nature.

By examining the configurations of urban settings in various cities such as Naples, Italy – Prague, Czech Republic – Kingston, Southwest of London, the UK – Arashiyama – Kyoto, Japan (Figs. 1.24, 1.25, 1.26, 1.27, 1.28 and 1.29), it is clear that greenery and water are the most common features of the urban areas in these

Box 1.9: Nature as a Source of Sustainability and SDGs Attainment

When we look at what is truly sustainable, the only real model that has worked over long periods of time, is the natural world. (Janine Benyus)

Box 1.10: Biomimicry and Valuing Nature

Biomimicry is about valuing nature for what we can learn, not what we can extract, harvest, or domesticate. In the process, we learn about ourselves, our purpose, and our connection to each other and our home on earth. (Biomimicry Institute)

cities, but in Naples, it shows only sea water as the common feature in addition to buildings. Hence, which configuration may be better for people during COVID? Table 1.1 lists the assessment results of the best configurations of these settings amid the COVID-19 crisis

In evaluating the urban settings in aforementioned cities and to link biomimicry approaches – as a design process – that are classified under two approaches: (a) design looking to biology; and b) biology influencing design [24–26], a scale from 1 to 5 was utilized to evaluate these urban settings in the time of COVID-19 (Table 1.1). Figure 1.30 shows the biomimicry's approaches [27]. To connect nature to architecture, it is imperative to comprehensively understand and digest the intersecting lines between them. There is an overlap between nature and architecture, which is known as the 3-Ms' relationship, including three layers (Model layer, Measure layer, and Mentor layer) as illustrated in Fig. 1.31.

Fig. 1.24 View of the heart of Naples, Italy, before COVID-19. (Image credit and source: Mohsen Aboulnaga)

Fig. 1.25 View of the heart of Prague, Czech Republic, before COVID-19. (Image credit and source: Mohsen Aboulnaga)

Fig. 1.26 View of a river in Kingston South East of London, UK, before COVID-19. (Image credit and source: Mohsen Aboulnaga)

Fig. 1.27 Arashiyama on the fringes of Kyoto, Japan, before COVID-19 where mountains are covered with greenery and water is running beneath them. (Image credit and source: Mohsen Aboulnaga)

Fig. 1.28 View of trees on the mountains in Arashiyama forest, Kyoto, Japan, before COVID-19. (Image credit: Mohsen Aboulnaga)

Fig. 1.29 View of the bamboo trees in Arashiyama in Kyoto, Japan, before COVID-19. (Image credit and source: Mohsen Aboulnaga)

Table 1.1 Assessment of the best configurations of these settings in the age of COVID-19 and post-COVID

City/ country	Image	Sustainability	Openness	Green areas	Water	Walkability	Clean air/ O2	Livability
Naples, Italy		XXX	X	X	XX	XX	XXX	XX
Prague, Czech Republic		XXXX	XXXX	XX	XXXX	XXXX	XXXX	XXXX
Kingston, Southwest London, UK		XXX	XXXX	XXX	XXXX	XXXX	XXXX	XXXX
Arashiyama, Kyoto, Japan		XXX	XXXX	XXX	XXXX	XX	XXXX	XXXX

Arashiyama, Kyoto, Japan		XXXX	XXX	XXXXX	XX	XXXX	XXXX	XXXX
Arashiyama, Kyoto, Japan		XXXX	XXX	XXXXX	XX	XXX	XXX	XXX

Key: 5 = Excellent, 4 = Very good, 3 = Good, 2 = Average, 1 = Below average

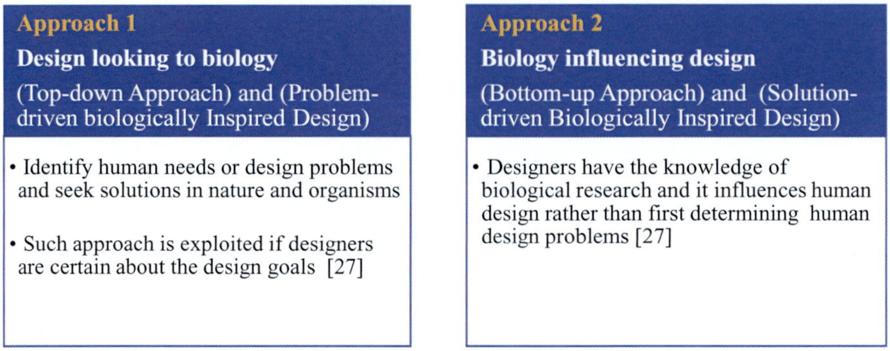

Fig. 1.30 Biomimicry approaches for evaluating urban setting in cities. (Image source: Developed by authors after Blizzard and Klotz [27])

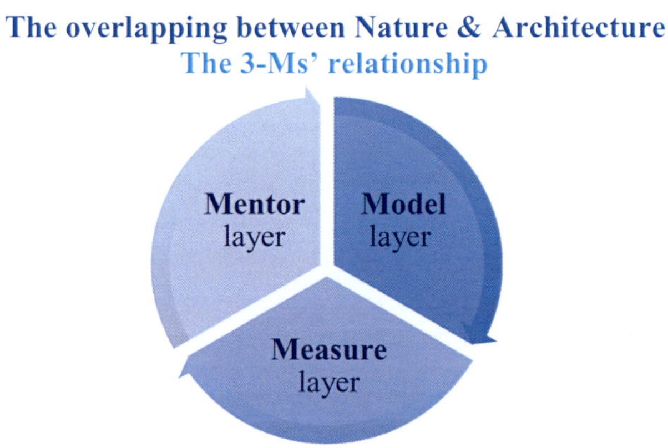

Fig. 1.31 The layers connecting architecture to nature. (Image source: Developed by authors after Janine Benyus)

1.7 Can Biomimicry Aid in Developing Urban Spaces and Healthier Buildings' Indoors?

In the recent years, the use and application of biomimicry have stretched gradually into various fields such as architectural, urban design, and scaling up from materials to a sizable scale [24]. In fact, nature has stimulated built space since antiquity, when natural proportions were on loan for esthetic reasons [28]. Mimicking biological morphology is considered one of many traditional applications of biomimicry in the architectural domain [29], and the subjects of this mimicry are not only single organisms or organism as such, but also the products of their biological behavior, such as nests [30–32].

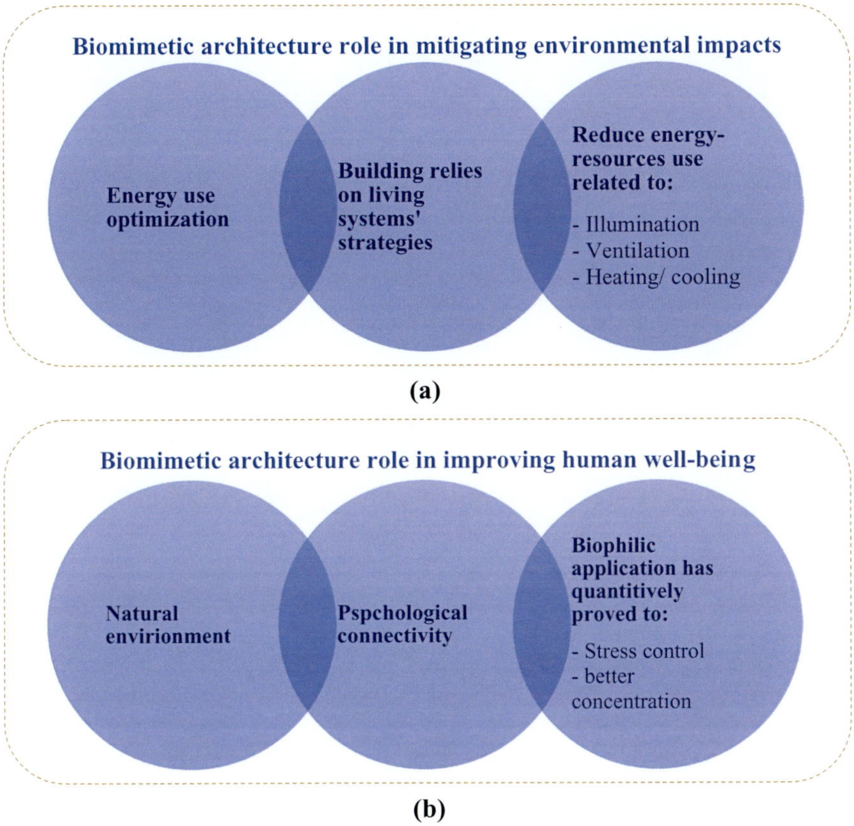

Fig. 1.32 Biomimicry applications aim at mitigating environmental impacts and enhancing human well-being. (**a**) Biomimicry's applications to mitigate environmental impacts. (**b**) Biomimicry's applications to enhance well-being impacts. (Image source: developed by authors after Uchiyama et al. [24])

Figure 1.32a presents biomimicry's applications of targeting the mitigation environmental impacts. A vivid example of biomimicry in architecture is the art nouveau style by Architect Buckminister Fuller, in which biological structures were often mimicked [33]. Biomimicry applications in architecture and healthier buildings aim regularly at reducing environmental impacts and enhancing human well-being through many elements as illustrated in Fig. 1.32b.

Integration biomimetic patterns into the structure and decoration of buildings and investigation of visual patterns in biomimicry can greatly improve psychologically well-being [28, 34]. Also, biophilic design (Box 1.11) as one of biomimicry subcategories can be comprehended as a biomimetic application of targeting human well-being [35, 36]. Such concept concentrates on the psychological connectivity between humans and natural environment in an effort to foster a sense of belonging [37]. The biophilic applications have garnered several quantifiable pieces of evidence in humans, in areas such as stress control and better concentration [38]. In

Fig. 1.33 Examples of biophilic design and biomimetic buildings. (**a**) One Central Park, Broadway, Sydney. (**b**) UT Dallas students service building, Texas. (**c**) Santé building in nature, SE Univ., Nanjing. (**d**) Offices looking out over nature, Florence. (Images' source: (**a**) https://commons.m.wikimedia.org/wiki/File:(1)Central_Building_Broadway_Sydney-1.jpg (**b**) https://commons.m.wikimedia.org/wiki/File:UT_Dallas_Students_Service_Building.JPG. (Images' credit and source: (**c**) and (**d**) Mohsen Aboulnaga)

demonstrating such idea, it is about the utilization of natural lightning, organic morphologies, and views of natural spaces [39]. Figures 1.33, 1.34, and 1.35 present examples of biophilic design in various global buildings.

Box 1.11: Biophilic Design

Biophilic design is the practice of connecting people and nature within our built environment and communities. (Biophilic Design Initiative, International Living Future Institute [35])

Box 1.12: Value of Nature in Cities

In every walk with nature, one receives far more than one seeks. (John Muir, 1877)

Fig. 1.34 Global and local examples of biophilic design and biomimetic buildings. (**a**) Houses look out over nature, Kingston, UK. (**b**) A recreational building in Baku, Azerbaijan. (**c**) An office building facing nature, Barcelona. (**d**) Baku Int'l airport faces nature, Azerbaijan. (**e**) EWHA University built in nature, Seoul, SK. (**f**) An INU building facing nature, Incheon, SK. (**g**) Biophilic design of Westin hotel, New Cairo. (**h**) Nature surrounding the hotel, Cairo, Egypt. (Images' Credit: Mohsen Aboulnaga)

Fig. 1.35 Global examples of biophilic design for biomimetic buildings. (**a**) Biophilic design of Bosco Vertical, Milan. (**b**) Biophilic design of a shopping mall, Bangkok. (**c**) Restaurant looking out over nature, Bangkok. (**d**) Office buildings facing nature in Brussels. (**e**) Biophilic design of Russian embassy, Berlin. (**f**) Biophilic design of residential bldg., London. (Image source: (**a**) https://commons.m.wikimedia.org/wiki/File:Bosco_Verticale_(205282347).jpeg; (**b–f**) Mohsen Aboulnaga)

1.7.1 *Biomimicry and Urban Design*

There is a strong rapport between architecture and environment. We cannot create a good building without the woven relationship with the natural systems and the surrounding environment [40]. In nature, trees talk to each other! Can buildings communicate with each other? Biomimicry can play a vital role in attaining such idea, especially in the digital world. In addition, nature and forests provide remedy for people visiting those places or moving from one building to another via nature (Box 1.12 and Figs. 1.36, 1.37, 1.38, 1.39, 1.40 and 1.41). Can buildings and urban spaces furnish remedy for humans and their well-being in the age of COVID? This is a question to be investigated by architects, designers, and urban designers, not only to

Fig. 1.36 Biophilic design, Khoo Teck Puat hospital in Singapore. (Image source: https://en.m.wikimedia.org/wiki/Khoo_Teck_Paut_Hospital#/media/File%3ALink_bridge_at_KTPH.jpg)

Fig. 1.37 Bosco Vertical towers as part of an urban space in Milan, Italy. (Image source: https://de.m.wikipedia.org/wiki/Datei:Bosco_Verticale_Milano.jpg)

Fig. 1.38 Urban space biophilic design, Incheon National University in Songdo – Incheon, South Korea. (Image credit and source: Mohsen Aboulnaga)

Fig. 1.39 Nature (trees & green) in the Champ de Mars park adjacent to Eiffel Tower and overlooking the École Militarie in Paris, France. (Image credit and source: Mohsen Aboulnaga)

(a) (b)

Fig. 1.40 Biophilic design inside NOI Center in Bolzano, Italy. (**a**) The interior of NOI building, a biophilic design and greenery. (**b**) The atrium of NOI building. (Image credit and source: Mohsen Aboulnaga)

Fig. 1.41 Biophilic design of the exterior of one of Paris' buildings, France. (Image credit and source: Mohsen Aboulnaga) It is imperative that architects and urban designers develop buildings and urban spaces to attain eco-friendly and sustainable measures, meaning to ensure resources management and energy efficiency, yet mitigating the overall impact of the built environment on human health and nature. In order to achieve this goal, several considerations should be tackled and examined. The main considerations are listed in Fig. 1.42

Fig. 1.42 Consideration for eco-friendly buildings. (Image source: Developed by authors after IEREK, https://www.ierek.com/news/inde.php/2016/04/13/the-relationship-between-architecture-and-environment)

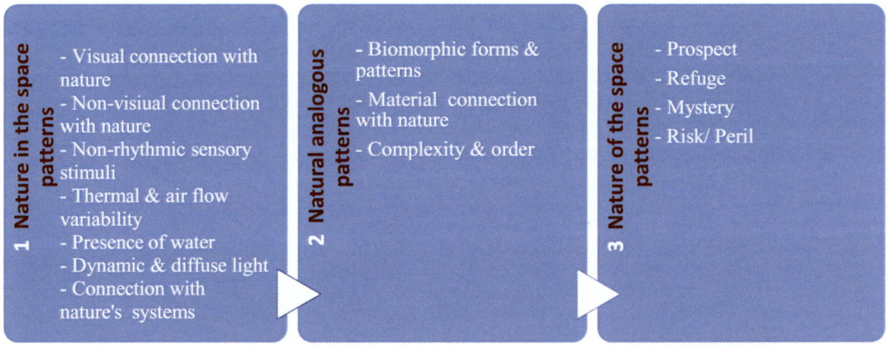

Fig. 1.43 The 14 patterns of biophilic design of spaces. (Image source: Developed by authors after TERRAPIN Bright Green, https://www.terrapinbrightgreen.com/reports/14-patterns)

achieve sustainable cities and buildings, but also to attain healthier, livable, and resilient cities, communities, and buildings' interior.

Ensuring the above is not an easy process as they require good will and knowledge to pursue and implement them in design. Therefore, 14 patterns of biophilic design for improving health and well-being in the built environment were developed to foster the sustainability of urban spaces. This addresses three principles: (a) nature in the space patterns; (b) natural analogous patterns; and (c) nature of the space patterns [41]. Figure 1.43 illustrates the patterns for each principle, where the nature in the space is addressed by seven patterns; natural analogous is covered by three patterns, and nature of the space is presented by four patterns.

1.8 Biomimicry and Climate Change Adaptation

Can biomimicry play a vital role in the adaption to climate change? Can biomimicry be a catalyst in the adaptation process? In fact, this a good notion to be addressed and examined since climate change is severalty manifested in many parts of the world with destructive impacts in Europe (Germany, Belgium, and The Netherlands in August 2021), and Asia (Vietnam, and Sultanate Oman in October 2021). Thus, biomimicry and biomimetic architecture can play an important part in climate change mitigation and adaptation.

With the high cost of climate change (CC) impacts and damages (severe storms, floods, heat waves, and wildfire) amounting to US$ 210 billion in 2020 [42], actions are urgently needed by all of us. The UK is ahead of the COP26 conference due in November 2021 in Glasgow, as it is focusing on "ADAPTATION." Climate change adaptation (CCA) is urgently needed to protect communities and natural habitats currently living with devastating and damaging extreme events and weather intensified by climate change [43]. In Europe, the cost of CC could reach 4% of GDP as per a study published in 2014 [44], but by 2050 climate change will cost the world US$ 7.9 trillion as per a published report in November 2019 [45].

According to the World Bank, natural disasters cost nearly US$18 billion yearly in low- and middle-income countries through devastation and destruction of power-generation and transport infrastructure alone [46]. Natural disasters due to CC also trigger vast disruptions for households and firms costing a minimum of US$ 390 billion per year [46].

The Biomimicry Institute stated that people living next to coastal areas are estimated to be above 50% of the world population. Such concentration of people is putting severe pressure on natural ecosystems and resources. Climate change (CC) is even causing more extreme threats in the form of sturdy storms, destructive floods, heat waves, and sea-level rise. Therefore, coastlines globally need a huge amount of finance for development and maintenance [47].

In reality, natural ecosystems are remarkable at absorbing carbon dioxide (CO_2) and transforming it into the building blocks of life. The exploitation of biomimicry, to craft from and for our ecosystems, adds values to humans. This is due to the fact that the designs we make encourage healthier habitats; hence, the ecosystems and infrastructure can perform collectively to create climate change resilience [47, 48].

Actually, biomimicry opens the gates for designers, influencers, and industries to connect development and sustainability in urban spaces and cities, which are immensely affected by climate change risks. In fact, biomimicry permits policy makers, engineers, architects, and landscape architects as well as contractors to integrate eco-engineering technologies while building in harmony with nature. As a matter of fact, urban waterfronts can mitigate their ecological footprint, while becoming an interface for the public to intermingle with marine life [48]. In addition, cities can lower the effects of storms on their waterfronts though offering prospects for recreation and experience livability.

Nevertheless, biomimicry in architecture can be one of the main players in the efforts needed to mitigate the harmful impacts of climate change including the decrease of CO_2 emissions, to near zero-carbon emissions in the next two decades. The new book "The Green New Deal and Beyond – Ending *the* Climate Emergency While We Still Can" by Stan Cox argues on the ability of Green New Deal to assist in saving the ecological and economic collapse [49–51]. Thus, biomimetic architecture can assist in the implementation of the legislation's vision to culminate "green" energy capacity and its promise to uphold and accelerate economic growth through clean energy manifesto.

Despite the fact that adaptation of finance flows has increased by 35% in recent years, such increase is still falling short of the needed target, particularly in developing countries, to curb climate change risks and impacts [46]. Today's sustainable development trends could rapidly become tomorrow's global net-zero transformation, and all countries – including the poorest and most vulnerable – can tremendously gain from such transition [46]. In this path, many examples of biomimicry in architecture exhibit how inspiration from nature can reach near zero-carbon emissions.

1.9 Biomimicry and Post-COVID Regulations

The Coronavirus (COVID-19) pandemic has changed the world in every avenue of life. With the social distance preventive measures and regulations, it seems that the COVID age needs a new set of regulations on the city level, urban level, and buildings' level. Thus, biomimicry and biomimetic architecture could be the pathway to the future.

Making the built environment resilient and sustainable is essential for both outdoor and indoor spaces like houses, public buildings, especially government and educational buildings, and workplaces (due to the large number of occupants and visitors in these types of buildings). Hence, redesigning buildings and urban spaces for well-being, public health, and livability in the living spaces for post-COVID should be considered [52]. The transformation of cities and buildings into bioinspired systems could benefit the health of the built environment.

The Policy Brief, published by the Organization of Economic Cooperation and Development (OECD), highlights the vital role of biodiversity for human life and the significance of incorporating biodiversity considerations into the recovery of the COVID-19 crisis. It is imperative to learn that the biodiversity loss is a key driver of emerging infectious diseases and causing a variety of other mounting risks to societies and global economy [53]. The economy and human well-being depend on biodiversity such as food, clean water, flood protection, erosion control, inspiration for innovation, and many more. In fact, more than 50% of the World's GDP is fairly or highly dependent on biodiversity [53]. Therefore, upholding and protecting biodiversity is vital and in evading the decline of biodiversity. Such decline poses significant risks and impacts to society especially during the COVID-19 crisis and beyond.

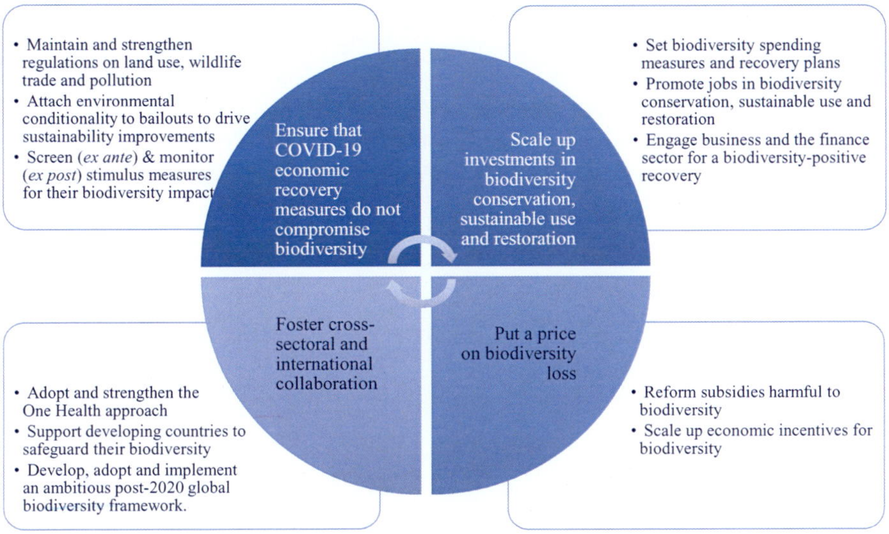

Fig. 1.44 Post-COVID regulations for biodiversity and well-being – OECD. (Image source: Developed by authors after OCED Policy Brief, https://www.oecd.org/coronavirus/policy-responses/biodiversity-and-the-economic-response-to-covid-19-ensuring-a-green-and-resilient-recovery-d98b5a09)

As part of post-COVID regulations, many countries and governments integrated biodiversity measures in their COVID-19 policy acts. For instance, these measures included changes in the regulations of wildlife trade to protect human health and programs that concentrate on ecosystem restoration, sustainable forest management, and invasive species control [53]. Moreover, OECD put forward steps for governments to consider for integrating biodiversity measures in the COVID-19 recovery plans [53]. Figure 1.44 presents these steps in brief.

For post-COVID regulations, it is clear that nature and biodiversity have many possible solutions and inspirations to future pandemics and existing viruses. The green and inclusive recovery to COVID-19 has been focusing largely on climate change (CC), but less attention has been given toward biodiversity. In fact, biodiversity loss and climate change are major challenges of a similar magnitude and urgency, yet strongly interlinked. Therefore, in any post-COVID regulation, biodiversity loss and CC should be mutually addressed as part of wider efforts to attain a green, sustainable, and resilient recovery. Also, post-COVID regulations should address the new zero-energy building (nZEB) standards to be enforced from 2021 to achieve climate-neutrality standards. In addition, the existing buildings stock should also be upgraded by 2050 to enhance energy efficiency by effectively using renewable energy to meet the criteria for a climate-neutral building. Hence, the energy requirements of existing buildings will be developed gradually by 2030 to take into account the economic aspects [54, 55].

It has been proven that biomimicry expresses an amazing capability to mitigate human-made GHG emissions, while adjusting to environmental changes and effects by looking at living organisms and ecosystems. Nevertheless, there are more venues to discover in the applications of biomimicry in the built environment. Exploiting biomimetic architecture in urban built environment can rely on nature to comprehend better mimicking. It has been shown that biological system biomimicry is a technique for managing the plentiful methods of building sustainable structures.

1.10 Innovation in Biomimicry and Biomimetic Architecture

Innovation in biomimicry and biomimetic architecture is crafted in many examples that were developed across the globe. Figure 1.45 depicts a summary of biomimetic, iconic, and low-carbon or net-zero buildings constructed in Paris, France; Shanghai, China; and Dubai, UAE.

The first innovative design was created by the French architect Jean Nouvel, who designed the famous building – The Arab World Institute (Institut du Monde Arabe) in Paris, France, in 1987. This building was inspired from the human eye iris, as illustrated in Fig. 1.46. Such biomimicry innovation in architecture is depicted through the moving Brise-soleil façades, which are created like the lens of a human eye (camera) to control the amount of daylight and sunlight entering the building as well as limiting glare. For each square in the façade, some are opened more than others while the center of each square is formed like an aperture blade system of an eye's lens [56]. In this innovative and iconic building, there are 30,000 photoelectric-sensor apertures that open up and close like the iris in an eye according to where the sun is at a specific time as illustrated in Fig. 1.47 [57, 58]; as the adaptive façade of the envelope mimics the iris of the eye, accordingly, it changes daylight and sunlight levels indoor (Fig. 1.48). I have personally gained the experience of witnessing such biomimetic architecture in operation during my visit to the building in Paris in the year 2010.

In the age of COVID-19, this building would be considered a very good example since large areas of the floors have sunlight impinging on the indoor spaces and enabling occupants to have excellent environmental indoor quality (EIQ) as well as a clean and hygienic building.

The Spanish Pavilion in Shanghai Expo 2010 depicts another example of biomimicry architecture, designed by Miralles Tagliabue. The structure of the building is inspired from the skin of a fish, and it aims to be sustainable, environmentally friendly, and innovative and technological (Fig. 1.49). The wrapped fabric emulates fish skin by making the steel in a wicker form – a technique for crafting a product that is woven from a variety of pliable plant materials used in such process (Fig.1.50) [59]. Figure 1.51 shows the wicker-woven fabric of the pavilion at night. Furthermore, Shanghai Expo 2010 vividly reflected innovation and biomimetic architectural models. Emulating nature, in the form of sand dunes, is manifested clearly in the Emirates pavilion where the changing light and colors of the desert were evoked in the building's reflecting skin of gold-colored stainless steel [60] as depicted in Fig. 1.52.

Fig. 1.45 Innovative and iconic biomimetic architecture mimicking nature in Europe and Asia. (**a**) Arab World Institute in Paris, France 1987. (**b**) Spanish Pavilion in Shanghai Expo, 2010, China 2010. (**c**) Emirates Pavilion in Shanghai Expo, 2010, China 2010. (**d**) UAE Pavilion Expo 2020 Dubai, UAE 2021–2022. (**e**) Slovenia Pavilion Expo 2020 Dubai, UAE 2021–2022. (**f**) Sultanate Oman Pavilion, Expo 2020 Dubai, UAE 2021–2022. (**g**) Azerabijian Pavilion, Expo 2020 Dubai, UAE 2021–2022. (**h**) Terra -The Sustainability Pavilion, Expo 2020 Dubai, UAE 2021–2022. (**i**) The Mobility Pavilion, Expo 2020 Dubai, UAE 2021–2022. (**j**) Japan Pavilion, Expo 2020 Dubai, UAE 2021–2022. (Images source: (**a**) Mohsen Aboulnaga; (**d**) (**h**) (**j**) Mohsen Aboulnaga and Amina El-Haggan; and (**e**) (**i**) Mohsen Aboulnaga and Aya Ghobashy (**b**) https://en.m.wikipedia.org/wiki/File:Spain_ Pavilion_of_Expo_2010.jpg (**c**) https://commons.m.wikimedia.org/wiki/File:United_Arab_ Emirates_Pavilion,_daytime.jpg (**f**) Government of Sultanate Oman, https://www.arabnews.com/node/1975621/middle-east (**g**) SCE Project Milan, https://www.sceproject.it/featured/azerbaijian-pavilion-dubai-expo-2020)

Recently, biomimicry and biomimetic architectural innovation have been strongly exhibited globally. One of the current walking hubs for learning and knowledge is Expo 2020 Dubai, through innovative iconic examples of buildings inspired from nature. One of the most vivid examples of biomimitec architecture is the UAE Pavilion, which was inspired from the desert nature biodiversity – the falcon. The

Fig. 1.46 Close-up view of the moving Brise-soleil façade showing the windows' iris controlling daylight, sunlight, and glare. (Image credit and source: Mohsen Aboulnaga)

UAE Pavilion, which occupies over 15,000 square meters was vividly created and designed by the Spanish and Swiss famous architect – Santiago Calatrava as illustrated in Fig. 1.53 [61]. This iconic pavilion form has 28 floating movable wings that mimic the desert falcon's wings and cover the building's roof. Each wing has PV panels to generate clean energy from renewable sources as well as to protect the building from rain and sandstorms as shown in Fig. 1.54 [62]. Box 1.13 presents Santiago Calatrava's concept of environmental sustainability while Figs. 1.55, 1.56, 1.57 and 1.58 also illustrate various views and details of the UAE Pavilion at Expo 2020.

Box 1.13: Environmental Sustainability Practice

We are committed to ensure that the leading practices of environmental sustainability are embedded at the heart of the design process. (Santiago Calatrava)

Fig. 1.47 The Institute of Arab World – Paris, main façade inspired from the human eye where 30,000 photoelectric-sensor apertures open up and close based on weather conditions. (Image credit and source: Mohsen Aboulnaga)

(a)

Fig. 1.48 View of the façade of the Institute of Arab World controlling daylight and glare. (**a**) Close-up view of the main façade illustrating some of photoelectric-sensor apertures adapting to control daylight, sunlight, and glare. (**b**) Side view of modular apertures. (**c**) The back façade in shadow. (Image credit and source: (**a–c**) Mohsen Aboulnaga)

(a) (b)

Fig. 1.48 (continued)

Fig. 1.49 General view of the structure of the Spanish Pavilion in Shanghai Expo 2010, exhibiting biomimetic and high-tech architecture. (Image source: https://es.m.wikipedia.org/wiki/Pabellón_de_España_(Shanghài)#/media/archivo:Spain_Pavilion_of_Expo_2010_2.jpg)

Fig. 1.50 Close-up view of the structure and wicker membranes of the fabric of the Spanish Pavilion in Shanghai Expo 2010. (Image source: https://en.m.wikipedia.org/wiki/File:Spain_Pavilion_of_Expo_2010.jpg)

Fig. 1.51 The Spanish Pavilion in Shanghai Expo 2010 at night. (Image source: https://es.m.wikipedia.org/wiki/Pabellón_de_España_(Shanghài)#/media/archivo:Spain_Pavilion_of_Expo_2010_2.jpg)

(a) **(b)**

Fig. 1.52 The Emirates Pavilion in Shanghai Expo, 2010. (**a**) Mimicking sand dunes in pavilion form. (**b**) The roof's gold-colored stainless steel. (Images source: (**a**) https://commons.m.wikimedia.org/wiki/File:United_Arab_Emirates_Pavilion,_daytime.jpg (**b**) https://www.fosterandpartners.com/projects/uae-pavilion-shanghai-expo-2010)

The Pavilion of Slovenia at Expo 2020 Dubai is another model of a real green oasis. It emulates the nature of the country, which is at an important European crossroad of three natural elements: forest, water, and wood (Fig. 1.59) [63]. The pavilion, which was designed by architect Robert Klum, has a 6-meter-high green wall covering an area of 465m^2 with plants representing the forests of Slovenia [64] as shown in Fig. 1.59a, b.

Another innovative example of biomimicry in architecture that mimics nature has been crafted in the Pavilion of the Sultanate of Oman at Expo 2020, as

Fig. 1.53 Bird's-eye view of the iconic UAE Pavilion Expo 2020 Dubai mimicking nature in the form of a falcon. (Image and source: Oliver Schuh + Barbara Burg, Palladium Photodesign, Germany, www.palladium.de)

illustrated in Fig. 1.60. The building emulates nature by mimicking an authentic and old Omani's Frankincense tree growing in its valleys (Fig. 1.60c) [65].

The Azerbaijan pavilion is also a successful example of collaboration between designers, engineers, and suppliers that brings it to an outstanding integrated result. In accordance to SCE Projects, they state that:

> In order to reach a perfect integration between concept and construction, our job has been focused on identifying a construction strategy that guarantees the best possible result in terms of precision, quality, speed. The complex geometry of the roof, consisting of an ellipsoidal solid, was subject to a careful structural analysis and value-engineering as well as the footbridge which was improved achieving a complete suspension by steel cable to the wooden tree. The 900 square meters leaf was built with an insufflated ETFE membrane, the design was developed in almost two years; (https://www.macotechnology.com/portfolio/pavilion-azerbaijan-expo-dubai-2020/); all the material was fabricated and tested in Italy and delivered to Dubai to be installed in just three days. This result was the outcome of a team effort to design and coordinate all the involved actors: designers, contractors and suppliers.

Similarly, the Azerbaijan Pavilion Expo 2020 is inspired by nature in the form of a tree that engulfs the building (Fig. 1.61), attempting to generate a tree stem holding a reverse mushroom covered by a long curved leaf [66]. Furthermore, there is no doubt that the Azerbaijan Pavilion can be viewed as a unique and innovative model of biomimicry in architecture.

The creative network Simmetrico (Italy) conceived, designed, and built the Azerbaijan Pavilion; the SCE Project, in collaboration with Simmetrico Architettura

Fig. 1.54 View of one of the UAE Pavilion's entrances, the floating wings with PV panels, and dynamic covers. (Image and source: Oliver Schuh + Barbara Burg, Palladium Photodesign, Germany, www.palladium.de)

Fig. 1.55 General view of the UAE Pavilion showing the left side of the 28 floating wings with solar PV panels installed on them to generate clean energy. (Image credit and source: Mohsen Aboulnaga and Amina El-Haggan)

(b) (c)

Fig. 1.56 The floating wings generating clean energy from renewable sources for the building energy annual consumption. (**a**) Front façade of the UAE Pavilion's entrance and the solar PV panels appearing on the wings; (**b**) one of the wings shading for pedestrians; and (**c**) close-up view of the floating wings. (Images' credit and source: (**a**) Mohsen Aboulnaga and Amina El Haggan, (**b, c**) Mohsen Aboulnaga and Aya Ghobashy)

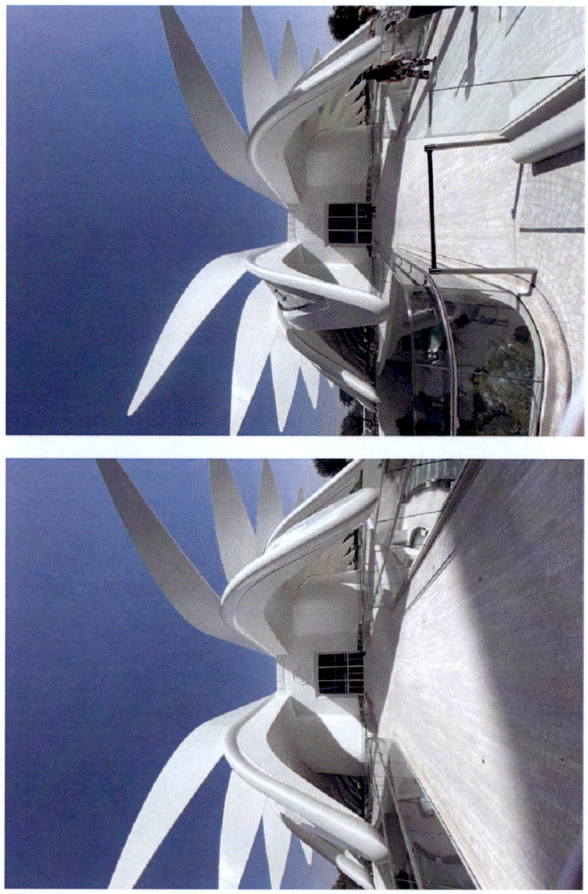

Fig. 1.57 One of the entrances of the UAE Pavilion with white floating rips to reflect solar radiation in summer. (Images' credit and source: Mohsen Aboulnaga and Rana El-Bakry)

(a) (b)

Fig. 1.58 The wings of the UAE Pavilion at night. (**a**) The kinetic rips to cover the floating wings; (**b**) smart rips to protect the PV panels from dust. (Images credit and source: Mohsen Aboulnaga and Amina El-Haggan)

and iDeas, is the engineering company that was in charge of the structural, architectural, and MEP engineering, as well as the project manager and construction management [67].

Figure 1.62 presents the complex geometry of the roof while Fig. 1.63 shows external views of the Azerbaijan pavilion and the suspended footbridge. Figure 1.64 also illustrates the 900 Sq.m leaf made of ETFE membrane that covers the wooden tree that mimics a mushroom with an insufflated at night. Also, the suspended visitors' footbridge leading to the pavilion is shown in Fig. 1.65, while Fig. 1.66 presents the pavilion's exterior and interior.

The main focus of Expo 2020 Dubai is directed toward sustainability, which is extensively manifested in one of the three attractive pavilions, the "Terra – The Sustainability Pavilion." The word "Terra" means earth or land in the Latin language, as well as the Mobility Pavilion. Figure 1.67 presents the main gate entrance to the "Terra – The Sustainability Pavilion" which is seen through the lattice metal structure gate welcoming the pavilion's visitors and leading to the solar complex, which consists of a large solar tree surrounded by 18 smaller energy trees. A visitor can see them from afar before entering the gate. The Terra – The Sustainability Pavilion, the largest space area in Expo 2020, mimics nature to create the complex through a set of solar trees as illustrated in Figs. 1.68, 1.69b, and 1.69c. The concept of the pavilion was inspired from the Dragon trees found in the Desert of Socotra Island in Yemen [68, 69] as shown in Fig. 1.69a.

The complex encompasses 19 energy trees (E-trees); each of the 18 E-tree canopies – 130 meters wide – emulates the sun flower (Figs. 1.69b, c). All E-trees rotate 180 degrees to increase the efficiency of the cells that generate 4 GWh of renewable energy as well as produce water from recycled rainwater [68]. There are 4912 ultra-efficient Monocrystalline photovoltaic panels, covering an area of more than 6000 Sq.m, which are installed on the 18 trees (each ranging from 15 to 18 m in diameter) and implanted in glass panels [68, 69]. The Terra – Sustainability Pavilion generates 28% of the energy needed to power the building [66] as shown in Fig. 1.70. The

(a)

(b)

Fig. 1.59 The Slovenia Pavilion, a real green oasis emulating nature and representing the country's forest, wood, and water. (**a**) Forest, the first element of nature, combined with wood as the second element of nature. (**b**) Water, the third element of nature, along with wood and green walls, a biophilic design. (Images' source and source: Mohsen Aboulnaga and Aya Ghobashy)

Fig. 1.60 Sultanate Oman's Pavilion Expo 2020, Dubai, emulating nature. (**a**) The building form reflecting a frankincense tree; (**b**) Façade of the Omani Pavilion mimicking nature; (**c**) Frankincense trees in Dhofar, Oman. (Image source: (**a**) (**b**) Sultanate Oman Government, https://www.arabnews.com/node/1975621/middle-east (**c**) https://commons.m.wikimedia.org/wiki/File:Frankincense_Tree_in_Dhofar_3.jpg)

combination of the cell and the glass casing permits the building to harness solar energy (Figs. 1.71 and 1.72), while granting shade and daylight for visitors. Each E-tree has a canopy serving as a large storm water harvesting and dew that replenishes the building's water systems [69] and harnesses energy (Figs. 1.64c, 1.70, 1.72, and 1.73). In addition, the pavilion after Expo will be used as a museum of sustainability. The Sustainability Pavilion is designed by the famous British firm Grimshaw Global. According to Grimshaw Global, key sustainability issues were centered on six main facts, including the following: (a) Project site; (b) Transect Zone/ Climate Zone; (c) Ecoregion; (d) Operation Energy/ Carbon; (e) Embodied Energy; and (f) Water [70]. In alignment with the SDGs, the pavilion attained five SDGs: goal 6 – Clean Water and Sanitation, goal 7 – Clean and Renewable Energy, goal 9 – Innovation and Infrastructure, goal 12 – Responsible Consumption and Production, and goal 13 – Climate Action in addition to goal 4 – Quality Education in the Legacy mode. Figure 1.74 shows the sustainability facts in detail, and key sustainability features are listed as follows [70]:

(a) **Canopy** – The 135-meter wide multifunction canopy encompasses 1055 solar PV panels covering an area of 8000 m^2.
(b) **The pavilion energy demand** – Generates its own power (4 GWh) that is made available by energy-saving techniques through the building design.

Fig. 1.61 The Azerbaijan Pavilion at Expo 2020 mimicking nature in the form a long leave and mushroom tree structure made from sustainable timber. (Image credit and source: Ph. Musthafa Ebrahim Khumanpur – SCE Project, https://www.sceproject.it/featured/azerbaijian-pavilion-dubai-expo-2020)

(c) **Buried accommodation** – The pavilion sits partially below ground, providing a thermal effect which is generally cooler than the ambient temperature in summer.
(d) **Daylight** – Captured where appropriate, and a range of light pipe and fiber–optic systems are incorporated to furnish daylight to deep spaces.
(e) **Canopy shading** – The shading provided by the canopy reduces energy use of the internal exhibition spaces by decreasing the solar radiation impinging on them.
(f) **Water** – Rainwater is harvested and percolated into landscape to recharge the ground water, which is then being extracted and treated for use as potable water within the building, corresponding to a net-zero water ambition.
(g) **Ecology** – Integrated into this landscape of native and adapted species are new crops that produce zero-km food and biofuels.
(h) **Energy Trees** – Eighteen E-Trees that support a smaller axially rotating PV array dish on long stem, furnishing 4000 m^2 of solar PV panels generating about 2.6 GWh of power yearly.
(i) **Comfort** – Earth below ground is fairly cooler due to its high thermal mass; this provides comfort during used periods.

(a)

(b)

Fig. 1.62 The complex geometry of the roof consisting of an ellipsoidal solid emulating a leaf covering a wooden mushroom. (**a**) Entrance of Azerbaijan Pavilion showing the 900 Sq.m-insufflated ETFE membrane leaf covering the structure; (**b**) Close-up view of an insufflated ETFE membrane leaf. (Images' credit and source: Ph. Musthafa Ebrahim Khumanpur – SCE Project, https://www.sceproject.it/featured/azerbaijian-pavilion-dubai-expo-2020)

Fig. 1.63 External view of the Azerbaijan pavilion portraying the complete suspension by steel cable over the mushroom wooden tree and visitors' suspended footbridge. (Image credit and source: Ph. Musthafa Ebrahim Khumanpur – SCE Project https://www.sceproject.it/featured/azerbaijian-pavilion-dubai-expo-2020)

(j) **Night cooling** – Low landscape to southeast of the site maximizes inflow of cool night air.
(k) **Canopy materials** – It is built from steel with 97% recycled content and manufactured at a site 15 minutes away from the pavilion.
(l) **Cement and embodied carbon reduction** – Strategies to mitigate the use of cement included constructing about 10,900 m^2 of upper floors with bubble decks that use around 25% less concrete and resulting less steel compared to solid concrete slabs.

(m) **Education** – The design of each exhibit space promotes learning by school groups and other arranged uses, specifically after Expo and in Legacy mode.

The second complex in Expo 2020 Dubai's three largest iconic signatures is The Mobility and Opportunity Pavilion – named Alif, which is inspired from nature and

(a)

Fig. 1.64 Night view of the ellipsoidal solid roof of the Azerbaijan Pavilion and the suspended footbridge. (**a**) Night view of the insufflated ETFE membrane leaf structure; (**b**) general view of the wooden mushroom tree structure at night; and (**c**) view of the external staircase and suspended wooden footbridge. (Images' credit and source: Ph. Musthafa Ebrahim Khumanpur – SCE Project, https://www.sceproject.it/featured/azerbaijian-pavilion-dubai-expo-2020)

(b)

Fig. 1.64 (continued)

designed by Sir Norman Foster and Partners. This Mobility complex is created, where the façades evoke movements and waves at the same time through their ripped and cured shapes as presented in Figs. 1.75 and 1.76. The ripped facades are made of highly reflective stainless-steel cladding that is inspired by Chrome fenders and air craft wings. The building is developed to LEED Gold standards, and it is

(c)

Fig. 1.64 (continued)

self-shading, where its metal cladding reflects heat in summer [71] as illustrated in Fig. 1.76a–c.

In addition to all the aforementioned examples of the innovation of biomimicry in architecture, comes the Japan Pavilion that emulates nature and Japanese heritage. The façades of the Japan Pavilion evoke a combination of the traditional Japanese asanoha pattern (hemp leaf pattern) and arabesque, representing the long-standing historical ties between Japan and the Middle East [72] as presented in Fig.1.77. The façades are also coupled with three-dimensional lattices that represent Japanese origami (Fig. 1.78). Details of the façade's 3-D lattices are shown in Figs. 1.79 and 1.80.

1.11 The Coverage

The main focus of this book, under its broad umbrella, is biomimicry and biomimicry in architecture as means to ensure human well-being, attain climate action, and create sustainable and green communities in order to provide healthier spaces in buildings and livable cities worldwide. This entails biomimicry role amid the COVID-19 pandemic and crisis. It also covers the definition of biomimicry and biomimicry in architecture as well as the biomimicry significance in the age of

Fig. 1.65 General view of the visitors' suspended footbridge leading to the Azerbaijan Pavilion at Expo 2020. (Image credit and source: Ph. Musthafa Ebrahim Khumanpur – SCE Project, https://www.sceproject.it/featured/azerbaijian-pavilion-dubai-expo-2020)

(a)

Fig. 1.66 External and interior views of The Azerbaijan Pavilion. (**a**) The external view of the revered mushroom wooden tree; (**b**) part of the mushroom wooden tree inside the pavilion; and (**c**) interior of the pavilion showing the horizontal and vertical wooden rips of mushroom tree structure from the inside. (Images' credit and source: Ph. Musthafa Ebrahim Khumanpur – SCE Project, https://www.sceproject.it/featured/azerbaijian-pavilion-dubai-expo-2020)

(b)

Fig. 1.66 (continued)

(c)

Fig. 1.66 (continued)

Fig. 1.67 The metal lattice gate to the Terra – The Sustainability Pavilion at Expo 2020 Dubai. (Image credit and source: Mohsen Aboulnaga and Amina El-Haggan)

COVID-19. In this regard, it is imperative to understand the attitudes, connectivity to nature, and proenvironmental behaviors during the COVID pandemic as well as learning through neighborhood access to green spaces amid COVID-19.

The book coverage highlights nature, as a great source for inspiration to biomimicry to solve the challenges encountered in the built environment (Fig. 1.81). It also demonstrates the need for new approaches to attain livability, sustainable development, and sustainable development goals (SDGs) by addressing the role of biomimicry and energy efficiency as well as the circular and biomimetic built environment and SDGs. The book addresses an important question "*Can biomimicry aid in developing urban spaces and healthier buildings' indoors? What is the role of*

Fig. 1.68 The Terra – The Sustainability Pavilion at Expo 2020 mimics nature in the form of the Dragon trees to create the solar complex. (Images' source: Mohsen Aboulnaga & Amina El-Haggan)

biomimicry in urban design?" In addition, the book illustrates the significant role of biomimicry in climate action and adaptation, as well as the part of biomimicry in the crafting of the Post-COVID regulations.

One of the challenges derived from the literature is that people living next to coastal areas constitute more than half of the world population, thus, posing a severe stress on natural ecosystems and resources in addition to climate change (extreme threats including strong storms, devastating floods, and heat waves). Also, biodiversity loss and climate change – strongly interlinked – are major challenges of similar significance and urgency in order to achieve the SDGs. It is clear that nature and biodiversity have many possible solutions and inspirations for future pandemics and existing viruses. The green and inclusive recovery of COVID-19 has focused largely on climate change (CC), but less attention has been paid toward biodiversity. Therefore, in any post-COVID regulation, biodiversity loss and CC should be mutually addressed as part of wider efforts to attain a green, sustainable, and resilient recovery.

Thus, the book *"Biomimetic Architecture Role and Innovation in Developing Sustainable, Regenerative, and Livable Cities – Global Perspective, Approach, and Solutions amid COVID-19 Pandemic*"is structured into two main parts that encompass a total of seven chapters focusing on issues of a vital topic "Biomimicry in

Fig. 1.69 The Terra – The Sustainability Pavilion generating renewable energy of 4 GWh annually. (**a**) Dragon trees in Socotra Island, Yemen, inspiring The Terra – Sustainability Pavilion design. (**b**) part of the 18 E-trees PV array canopies, each is supported on carbon fiber; and (**c**) dynamic "E-trees" canopies to track the sun by rotating 180 degrees throughout the day. (Images' credit and source: (**a**) Rod Waddington, https://commons.m.wikimedia.org/wiki/File:Dragon%27s_Blood_Tree,_Socotra_Is_(12473612124).jpg (**a**) Rod Waddington, https://commons.m.wikimedia.org/wiki/File:Dragon_Blood_Tree,_Socotra_Island_(10098980413).jpg (**a**) Andrey Kotov200514, https://upload.wikimedia.org/wikipedia/commons/e/e3/%D0%92%D0%B5%D0%BB%D0%B8%D0%BA%D1%96_%D0%B4%D0%B5%D1%80%D0%B5%D0%B2%D0%B0_%D0%B7_%D0%90%D1%84%D1%80%D0%B8%D0%BA%D0%B8.jpg (**b**, **c**) Mohsen Aboulnaga and Aya Ghobashy)

Fig. 1.70 General view of the Terra – Sustainability Pavilion to educate people and generate energy by the solar complex. (Image credit and source: Mohsen Aboulnaga and Aya Ghobashy)

Fig. 1.71 The shed fabric **covering the walking** passage around the 18 solar energy movable tree canopies. (Image credit and source: Mohsen Aboulnaga and Rana El-Bakry)

architecture and the built environment" which is gaining the attention of international organizations amid the COVID-19 crisis to achieve the sustainable development goals (SDGs). The book furnishes the information needed to understand biomimicry thinking, types, levels, and approaches as well as the six sustainability elements to achieve biomimetic concept for creating biomimicry in architecture in the context of sustainability.

In this coverage, it is imperative to narrate the term "biomimicry," the origins of biomimicry, and to depict the history of biomimicry in architecture with a focus on

Fig. 1.72 The main Energy tree "E-tree" solar canopy fixed on polycarbonate material stem, the Terra – The Sustainability Pavilion Expo 2020. (Image credit and source: Mohsen Aboulnaga and Rana El-Bakry)

Fig. 1.73 The energy tree "E-tree" solar canopy for harnessing the light from the sun as well as harvesting rain water. (Image credit and source: Mohsen Aboulnaga and Amina El-Haggan)

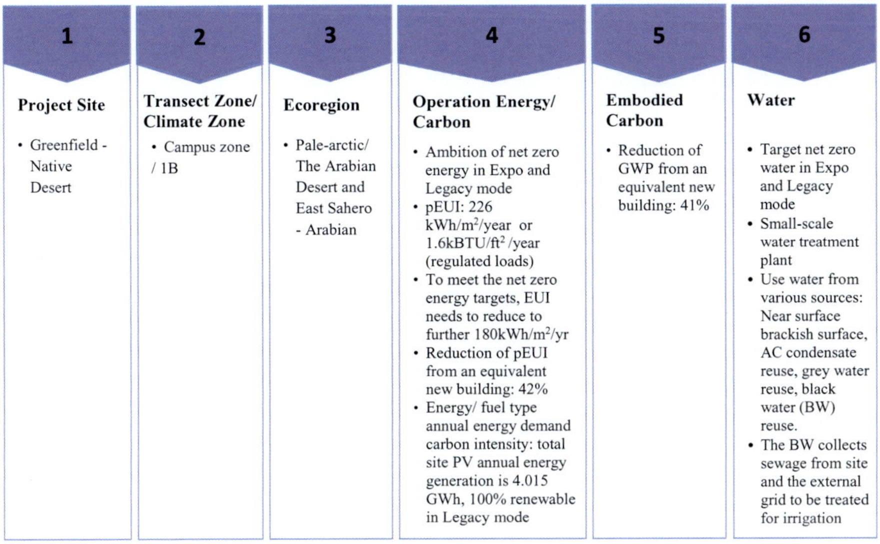

Fig. 1.74 Key sustainability facts achieved in the Terra – The Sustainability Pavilion Expo 2020 Dubai. (Image source: Developed by authors after Grimshaw Global, https://grimshaw.global/sustainability/expo-2020-sustainability-pavilion-case-study)

the evolution of mankind and its relationship with nature. The book also explores the way of integrating biomimicry as a way of thinking in the built environment. Biomimicry, early conceptual design, and the revival of nature in the architectural field as well as the overlap between architecture and nature by focusing on the matrix (nature as a model, nature as a measure, and nature as a mentor) are addressed. The coverage also includes biomimicry design methods and processes including the HELIX model of biomimicry as well as the relationship of biomimicry with the environment and sustainability. The biomimetics' design strategies and goals to achieve sustainable and resilient cities, yet the importance of biomimicry in architecture in the time of COVID-19, are also highlighted.

In addition, this book deals with the causes of biomimicry thinking and biomimicry in architecture, which are applied in many building designs in different parts of the world encompassing examples in the United States, Europe (Germany, United Kingdom, and Poland), Asia (China, Singapore, and Qatar), and Australasia (Australia and New Zealand) as well as Africa (Zimbabwe and Namibia). It also sheds light on biomimicry approaches and the types of biomimicry in architecture as an inspiration source. In terms of sustainability, biomimicry – as a tool for achieving sustainability – has been presented with an overview of the future of biomimicry. The book also concentrates on the role of robotic strategies in biomimetic architecture and innovation in biomimicry that were crafted in many examples worldwide including Jean Nouvel's iconic building, Institut du Monde Arabe in Paris, that was inspired from biology. Examples of innovation in biomimetic architecture inspired from nature include the Spanish Pavilion and Emirates Pavilion in

Fig. 1.75 The Mobility and Continuity Pavilion at Expo 2020 Dubai mimicking nature through the wave continuity and movements. (Image credit and source: Mohsen Aboulnaga and Sherry Elghoraiby)

(a)

(b)

Fig. 1.76 Alif – The Mobility Pavilion at Expo 2020 mimicking nature. (**a**) Front view of the Mobility Pavilion illustrating continuity and movements; (**b**) the ripped and curved shape of the façade mimicking movements of the Mobility Pavilion; and (**c**) side view of the flowing curves, evoking movements at the same time on the façades. (Images' credit and source: (**a**, **b**) Mohsen Aboulnaga and Aya Ghobashy (**c**) Mohsen Aboulnaga and Sherry Elghoraiby)

(c)

Fig. 1.76 (continued)

(a)

(b)

Fig. 1.77 The main façades Japan Pavilion combining the traditional Japanese asanoha pattern (hemp leaf pattern) and arabesque. (**a**) The Japan Pavilion during daylight; (**b**) the Japan Pavilion at night. (Images' credit and source: Mohsen Aboulnaga and Aya Ghobashy (**b**) Mohsen Aboulnaga and Amina El-Haggan)

Shanghai Expo 2010. Moreover, innovative biomimicry in architecture is at the center stage of Expo 2020 Dubai; many iconic examples inspired from nature were vividly exposed such as the UAE Pavilion by Architect Santiago Calatrava, The Terra – The Sustainability Pavilion, and The Mobility Pavilion by Foster + Partners

(a)

(b)

Fig. 1.78 The Japan Pavilion Expo 2020 coupled with three-dimensional lattices that represent Japanese origami. (**a**) The lattice of the pavilion façade illuminating and glittering at night; (**b**) the iconic façade with three-dimensional lattices reflected on dazzling water. (Images' credit and source: Mohsen Aboulnaga and Amina El-Haggan)

(a)

(b)

Fig. 1.79 Details of the Japan Pavilion façades depicting a traditional Japanese asanoha pattern. (**a**) Part of the façade mimicking traditional Japanese asanoha pattern (hemp leaf pattern); (**b**) close-up view of the 3-D lattices of the façade. (Images' credit and source: (**a**) Mohsen Aboulnaga and Amina El-Haggan; (**b**) Mohsen Aboulnaga and Rana El-Bakry)

Fig. 1.80 Details of the 3-D lattices in a harmonious setting to create shades and reflect sun light. (Image credit and source: Mohsen Aboulnaga and Rana El-Bakry)

as well as the Pavilions of Azerbaijan, Japan, Sultanate Oman, and Slovenia – all are inspired from nature as well.

More and above, the book presents biomimicry applications in the context of sustainability and architectural design examples in many parts of the world, e.g., the UK (Waterloo international terminal and the Swiss Re – Gherkin tower in London – the Eden project in Cornwall, England – Selfridges building in Birmingham);

Fig. 1.81 Nature – a source of inspiration for architecture. (Image source: Mohsen Aboulnaga)

Europe such as Germany (Munich Olympic Stadium), Poland (Tardigrada Botanica); Spain (Sagrada Familia Church in Barcelona – The Las Palmas' water theater – Tenerife Auditorium); Armenia (The all seasons tent tower in Yerevan); Singapore (Esplanade Theatre); Asia – China (Beijing Olympic Stadium and National Aquatic Center); Japan (Omotesando building, Tokyo); South Korea (Residential building in Songdo – Incheon); India (The lotus temple in Delhi – Lavasa Township) and Gulf (Aldar Headquarters, UAE, and The Sahar project – Giant cactus building in Doha); Australia (Council House 2); The United States (Ultima tower – Fennell residence – Portland, Oregon – Spiraling Chicago tower, Chicago); and Africa (Hydrological Centre for the University of Namibia – Eastgate Building in Harare, Zimbabwe).

The coverage also includes a comparative evaluation of biomimicry architecture models and sustainability. Moreover, it highlights the environmental, economic, and social benefits of sustainable architecture in the twenty-first century. Moreover, the coverage focuses on climate change (CC) in global context and vital signs as well as causes and effects of CC. It addresses the vital role of biomimicry in climate change mitigation and adaptation (CCMA). It highlights climate change in three selected countries: Egypt, Italy, and Germany. It also highlights Egypt's position related to CCMA activities. The book not only presents CC in Italy in terms of environmental action plans, adaptation strategies, their implementation, monitoring, and reporting as well as stakeholders' engagement and assessment, but also CC in Germany by highlighting policy and legal framework, adaptation strategies, actions, and assessment. In addition, the coverage discusses how biomimetic design could respond efficiently to CC and mitigate GHG emissions in the built environment (biomimicry for energy efficiency, segregation, and storage of carbon) and finally emphasizes the benefits of a biomimetic approach to architecture.

Furthermore, this book bestows biomimicry in criticism, argument, and defense, as well as a direction toward sustainability. It also presents the notion of biomimicry as it is not constrained by evolutionary instrumentalism, but it is an aid and does not restrict human thinking to build. The biomimetic architecture limitation is also addressed with focus on both strong concept and weak concept of biomimicry. It illustrates the future direction from bioinspired innovations to eco-mimetic cities.

There is an important question this book tries to address "Can biomimicry innovation assist in limiting the spread of COVID-19?" The coverage finally deals with biomimicry innovation and post-COVID regulations. It also highlights the challenges facing the world during COVID-19 as well as the innovation for COVID-19 and beyond, including materials' inspiration from nature and natural nano-structures mimicry. It focuses on healthier urban spaces and buildings' indoor spaces, to furnish guidelines, especially in the age of COVID-19. Moreover, it highlights the post-COVID regulations for creating sustainable, regenerative, and livable cities as shown in Figs. 1.82 and 1.83. The book closes by depicting the discussions and recommendations for the future of biomimetic architecture.

1.11.1 Part I

This volume is dedicated to present some of the key issues related to biomimicry and Nature as an approach to inspire the relationship between architecture and nature; mainly the origins of biomimicry in architecture are highlighted along with the review of sprawl factors, slums' challenges, and upgrading processes, while emphasizing the role of policies and governance toward achieving slums' resilience. In addition, it covers the exact meaning of biomimicry and its definitions; it also presents how important it is to integrate biomimicry approach as a way of thinking, yet the integration of the nature of surrounding environments as well as the biomimicry design methods and processes.

This part is outlined into two chapters (Chap. 2 is entitled: "Biomimicry and Nature – Milieu, history, approaches, and design methods and processes, and Chap. 3 addresses "Biomimicry Influence on Architecture and Design: Thinking, Approaches, Levels, Application Types, and Inspiration'"

Chapter 2 portrays a full review of biomimicry in architecture, its inspiration, and related information needed to comprehend the biomimicry approach through background and history. It also defines what biomimicry is and provides the origins of biomimicry in architecture and its meaning, and how to integrate this approach to become a way of thinking. The chapter highlights the evolution of mankind and its relationship with nature biomimicry, and early conceptual design, overlap between architecture and nature by taking the nature as a model, measure, and mentor of the

(a)

(b) (c)

Fig. 1.82 Nature (trees, green, and material) fort livability in urban spaces. (**a**) Trees, greenery, and material forming the urban spaces; (**b**) water and material – urban space's elements; and (**c**) water – nature element to reduce stress. (Images' credit and source: Mohsen Aboulnaga)

biomimicry design. In addition, the chapter presents the helix model of biomimicry and its relationship with the environment and sustainable architecture as well as integrating biomimicry approach as a way of thinking. Moreover, the chapter presents the importance of biomimicry and biomimetic architecture amid the Coronavirus (COVID-19) pandemic. Furthermore, the biomimicry design methods and processes are reviewed and presented. Finally, the chapter furnishes the readers with a full comprehension of biomimicry through different aspects and underlines the drivers of the entire unsustainable practice of modern civilization that lies in the dualistic detachment of nature and culture.

Fig. 1.83 Biomimicry influencing city's spaces. (Image credit and source: Mohsen Aboulnaga)

Chapter 3 presents a review of the needed information to comprehend the influence of biomimicry on the architectural design and the built environment through several points. It also highlights and discusses the causes of biomimicry thinking, biomimicry in architecture, and approaches and levels of biomimicry (organism, behavioral and ecosystem) as well as the levels of biomimetics information and the application types of biomimicry in architecture. This chapter also focuses on inspiration levels (visual inspiration nature, conceptual inspiration, and computational inspiration in nature) and showcases many iconic global exemplary buildings. It also discusses biomimicry design lens and nature's design principles and how biomimicry attains sustainability. Moreover, this chapter highlights the future of biomimicry and biomimicry conditions in Egypt. Finally, this chapter draws conclusions and lessons learned for a successful design process.

1.11.2 Part II

This volume encompasses five chapters dedicated to present some of the key issues of biomimicry in relation to sustainability and global applications of biomimicry approaches and levels in architecture, for climate change mitigation and adaptation with an overview of Egypt, Italy, and Germany's efforts and action in this regard. It also highlights biomimicry in criticism including the argument and defense as well as the direction toward a sustainable future. This part also illustrates an overview of the Coronavirus (COVID-19) pandemic and addresses how biomimicry innovation and Post-COVID regulations can curb the devastated impacts of such a crisis. This chapter presents an important question – "Can biomimicry innovation assist in limiting the spread of COVID-19"? Finally, the chapter highlights the discussion made by the authors regarding their thoughts "can nature and biomimicry be the solution for net zero cities" and presents the recommendations for the future of biomimetic architecture.

Chapter 4 portrays the sustainability and global applications of biomimicry approaches and levels in architecture. The objective of this chapter is to review global examples to draw lessons learned about biomimicry approaches. The chapter also presents a review of biomimicry applications in line with sustainability. It also depicts many examples of biomimetic buildings worldwide along with different levels of application, which are being investigated. Although biomimicry concepts are not widely recognized in the field of architecture, many countries across the globe follow this design approach. As the country's awareness regarding the environment, nature, and sustainability increases, the biomimetic buildings' design improves and flourishes. In addition, the ideas formed from the formal or structural functional point of view are presented. A high level of engineering knowledge is essential in order to imitate the designs of nature and apply them to architectural design. Moreover, different global examples of biomimetic architecture were reviewed, assessed, and evaluated. A comparative analysis and evaluation of these buildings were conducted and discussed. Finally, the chapter highlights the reasons and benefits of developing sustainable architecture.

Chapter 5 offers a full review of climate change through several main mitigation and adaptation actions. The objective of this chapter is to appraise and present the major role of biomimetic architecture in climate change mitigation and adaptation. Climate change in a global context and its vital signs, causes, and effects are presented and discussed. The role of biomimicry in climate change mitigation and adaptation and climate change in Egypt (Egyptian position related to climate change, needs for adaptation, and adaptation and mitigation activities), Italy (the environmental action plan and adaptation strategies in Italy including sustainable development and environment, environmental actions; and implementation means and monitoring, reporting, and evaluation as well as sectors and actions based on engaging stakeholders, and assessment), and Germany (policy and legal framework, assessments, and climate action plan 2050) are highlighted. Biomimicry, climate change, and how biomimetic design could respond efficiently to climate change and benefits of a biomimetic approach to architecture are discussed. At the end of this chapter, readers will learn about the role of biomimicry in architecture in mitigating and adapting to climate change. Finally, the chapter concludes on expressing the amazing capability of biomimicry to mitigate human-made GHG emissions, yet adjusting to the environmental changes and the effects by looking at living organisms and ecosystems, yet there is more to discover in the applications of biomimicry in the built environment.

Chapter 6 divulges biomimicry as one of the biologically sustainable approaches that assist in creating environmentally friendly architecture and cities. There are many critical thinking and arguments toward imitating nature's ideas, concepts, and behaviors from its organisms; on the other hand, there are many defenses toward this approach and many proofs that it is an excellent solution in different aspects. This chapter discusses the arguments toward biomimicry and naïve biomimetic architectural design examples. The chapter also presents the defense of biomimicry and debates that biomimicry is not constrained by evolutionary instrumentalism. In addition, the chapter highlights the importance of biomimicry as an aid in order to enhance the human ability to build innovative designs. Moreover, the limitations in biomimetic architecture are presented and discussed. Based on the comparison of the strength versus weakness of biomimicry, the strong concept of biomimicry and weak concept of biomimicry were highlighted and discussed with future directions toward sustainability. Finally, it portrays the notion "from bio-inspired innovations to eco-mimetic cities" and the conclusion's framework for the criticism against biomimicry.

Chapter 7 highlights an overview of the Coronavirus (COVID-19) pandemic and addresses how biomimicry innovation and Post-COVID regulations can curb the devastating impacts of such a crisis. This chapter presents an important question for all – "Can biomimicry innovation assist in limiting the spread of COVID-19"? The chapter tries to furnish some answers to this question and also sheds light on the scoping challenge that the world faces, and the context of COVID-19. The chapter additionally illustrates the innovation for COVID-19 and beyond to fight the COVID-19 pandemic through the exploitation of materials' inspiration from nature including natural nano-structures mimicry and nano-materials and nano-technology. The change in the face of life and the COVID-19 positive insights and opportunities are addressed. Moreover, this chapter presents the guidelines for healthier urban spaces and buildings' indoors in COVID-19, including seven folds (a) Urban spaces, (b) Healthy cities, (c) Green cities, and (d) Smart cities as well as All in cities (inclusiveness), (f) Sustainable cities, (g) Resilient cities, and (h) Rising cities. Furthermore, this chapter focuses on the urban context and pandemic role as well as the healthy building's spaces. Finally, it depicts a discussion on the post-COVID age and its conclusion.

Chapter 8 presents the discussions on how nature and biomimicry will be in the center stage and one of the solutions for achieving net-zero cities by 2050. The chapter also depicts the conclusions derived from the book and discusses the tendency of office buildings developed away from integrating them with nature and illustrating examples from Europe, Asia, and the MENA region. Finally, it draws the recommendations for the future of biomimetic architecture to create sustainable, regenerative, and livable cities in the age of COVID-19, and above all be resilient enough to adapt to climate change.

In this context, each country has its own location and contour lines on the world's map as well as its own weather and climate. Such a topographic setting provides a good source of inspiration for biomimicry and biomimicry in architecture. These points allow for dealing with challenges and setting priorities in the age of COVID-19. In retrospection, examples from nature (Fig. 1.84) can furnish excellent solutions, whether at the global level or local level, yet enable us to achieve better results. It is worth noting that adopting any approach, type, and level in biomimicry

Fig. 1.84 Ecosystem and nature inspiration for biomimicry in architecture. (Image credit and source: Mohsen Aboulnaga)

architecture models should be innovative and based on clear strategies and inclusive processes to bestow architects, engineers, and urban designers to attain climate neutrality, sustainability, and the SDGs.

1.12 A Final Note

Why do we urgently need sustainable and resilient cities, and adapt to climate change (CC) as well as confronting COVID-19? In fact, the Coronavirus (COVID-19) pandemic has shed light on the severe problems in megacities and urban slum areas worldwide due to their weak resilience and ability to cope with the severe impacts of CC and the COVID-19 crisis. In this context, biomimicry conveys a remarkable capability in mitigating human-related GHG emissions and adjusting to environmental changes and effects through looking at biodiversity, living organisms, ecosystems, and nature as illustrated in Figs. 1.85, 1.86, 1.87, 1.88, 1.89 and 1.90. In the age of COVID-19 crisis, buildings ought to have enough sunlight, daylight, fresh airflow, and cross-ventilation inside indoor spaces. Such provision of these parameters would enable occupants to experience satisfactory environmental indoor quality (EIQ), clean, and hygienic buildings.

Biomimicry can also play a vital role in climate change mitigation and adaptation, developing resilient and regenerative cities, and protecting communities and natural habitats that are living under extreme events and weather intensified by CC

Fig. 1.85 Nature to inspire biomimicry in architecture for future cities in the post-COVID age. (**a**) *Scilla siberica* plant to inspire biomimetic architecture; (**b**) *Dracophyllum recurvum* plant to inspire biomimetic architecture; (**c**) *Euphorbia ingens* plant with fruit to inspire biomimetic architecture; (**d**) *Dracophyllum menziesii* plant to inspire biomimetic architecture; (**e**) *Ludwigia sedioides* plant to inspire biomimetic architecture; and (**f**) *Lepidium oleraceum* plant to inspire biomimetic architecture. (Images' source: (**a**) https://commons.m.wikimedia.org/wiki/File:Scilla_siberica_21(2)._Balka_Vilna.jpg# (**b**) https://commons.m.wikimedia.org/wiki/Category:Plants#/media/File:Dracophyllum_recurvum.jpg (**c**) https://commons.m.wikimedia.org/wiki/File:Euphorbia_ingens_with_fruit_.jpg# (**d**) https://commons.m.wikimedia.org/wiki/File:Dracophyllum_menziesii.jpg (**e**) https://commons.m.wikimedia.org/wiki/Category:Plants#/media/File%3ALudwigia_Sedioides_close-up.jpg (**f**) https://commons.m.wikimedia.org/wiki/File:Lepidium_oleraceum.jpg#)

Fig. 1.86 Nature inspiring biomimicry in architecture for future cities in the post-COVID age. (**a**) *Corydalis* marshaling plant to inspire biomimetic architecture; (**b**) *Dracophyllum filifolium* plant to inspire biomimetic architecture; (**c**) *Miplanta* plant to inspire biomimetic architecture; and (**d**) *Corydalis paczoskii* plant to inspire biomimetic architecture. (Images' source: (**a**) https://commons.m.wikimedia.org/wiki/File:Corydalis_paczoskii_08(4)._Khortitsa.jpg# (**b**) https://commons.m.wikimedia.org/wiki/File:Dracophyllum_filiolium.JPG# (**c**) https://commons.m.wikimedia.org/wiki/Category:Plants#/media/File:Miplanta.jpg (**d**) https://commons.m.wikimedia.org/wiki/File:Medicago_Citrina_2c_JPG#)

Fig. 1.87 Paper mulberry plant from nature to inspire biomimicry in architecture for future cities in the post-COVID age. (Image source: https://commons.m.wikimedia.org/wiki/File:Paper_mulberry.jpg#)

across the globe. With the cost of CC impacts amounting to US$ 210 billion in 2020 and to be increased to US$ 7.9 trillion by 2050 [42, 45, 73, 74], urgent actions are essentially required by all governments and local authorities to CC mitigation and adaptation ahead of COP26 due to be held in Glasgow – Scotland, UK, with "ADAPTATION" slogan at the core of discussion [43].

In addition, biomimicry is becoming more important than ever due to the rapid increase of the world's population and the corresponding need to meet their basic needs and services like housing, health, and educational buildings, especially with more than 55% of world's population living in urban areas – which is estimated to rise to almost 70% by 2050 [75]. Thus, sustainability and biomimicry architecture applications contribute to solving such challenges. In fact, biomimicry is a key player in attaining the Sustainable Development Goals (SDGs), mainly goal 3 – Good Health, goal 4 – Quality Education, goal 7 – Clean Energy, goal 11 – Sustainable Cities and Communities, goal 12 – Responsible Consumption and Production, and goal 13 – Climate Action as well as goal 15 – Life on Land.

Box 1.14: Nature for Life

The value of nature in achieving the Sustainable Development Goals (SDGs). (Eva Gurria, Nature-Based Solutions Coordinator, Global Programme on Nature for Development, UNDP)

Fig. 1.88 *Fritillaria ruthenica* plants from Nature and ecosystem inspiring biomimicry for future cities in post-COVID. (Image source: https://commons.m.wikimedia.org/wiki/Category:Plants#/media/File%3AFritillaria_ruthenica._Verhnya_Khortitsa.jpg)

The UNDP and more than 40 partners created a virtual platform – "Nature for life Hub" –which coincided with the 75th UN General Assembly where global and local leaders share stories on the importance of nature for Sustainable Development. According to Eva Gurria, nature is becoming nature for life: The value of nature in achieving the SDGs (Box 1.14) [76]. For many reasons, now is the time to return to nature, as land use accounts for nearly 25% of global GHG emissions.

Based on UNDP, restoring, protecting, and sustainably managing the ecosystem globally could provide more than a third of climate mitigation solution. Also, nature is important to security, health, and well-being. It provides buffers to humans from natural disasters, averts the next pandemic, supplies water, food, and offers respite from an increasingly urban world. By 2050, up to 6.7 billion people could be facing some level of water scarcity [76]; thus, ensuring healthy biodiversity and ecosystems is essential for our health and well-being [77]. Figure 1.91 summarizes the importance of nature in main five points. According to the UNDP report in 2020, biodiversity is deteriorating at an unprecedented rate. Such a decline is threatening

Fig. 1.89 Nature inspires biomimicry in architecture for future cities in the post-COVID age. (Image source: https://commons.m.wikimedia.org/wiki/Category:Plants#/media/File%3ADraccophyllum_fiordense_foliage.jpg)

Fig. 1.90 *Fritillaria ruthenica* plant – biodiversity and ecosystem to inspire biomimicry in architecture for future cities in the post-COVID age. (Image source: https://commons.m.wikimedia.org/wiki/File:Fritillaria_ruthenica_3a._Zaporizhia.jpgs#)

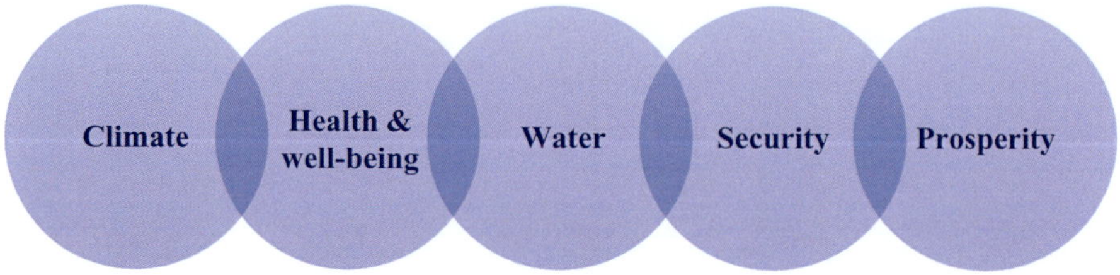

Fig. 1.91 Importance of nature to areas of global concern. (Source: Developed by authors after the UNDP)

the attainment of the SDGs particularly for the more than two billion people who mainly depend on nature for their livelihoods. With COVID-19 devastated impacts, the linkage between nature, people, and prosperity is becoming more important than ever.

Since human settlements depend largely on resources, benefits, and services from ecosystems, while being apt to deplete ecosystem health, there is an urgent need for a new urban design approach [78]. Eduardo Blanco et al., 2021, assessed the urban ecosystem – level biomimicry and regenerative design by linking ecosystem functioning and urban built environments, yet they concluded that it is vital to integrate understating of local ecological structures, processes, and functions within design phases [78]. Eduardo Blanco et al. also highlighted the gap between biomimetic teams and biological research and ecological practitioners [79]. They indicated that there is a major challenge in facilitating knowledge transfer from biological and ecological disciplines to architecture and urban design [79]. To bridge this gap, a rethinking approach is immensely needed especially to curb climate change impacts. Moreover, biomimicry tenders a potential to optimize the handling process of the retrofitting buildings with a sustainable development standpoint [80, 81].

In line with the ambitious target of COP 26 and the Glasgow Climate Pact to reach Net Zero by 2050, it is vital and agile to link ecosystem functioning and urban areas in cities through focusing on ecosystem-level biomimicry and regenerative approaches and resilient design. This should be vastly implemented to existing and new buildings to contribute to achieving Net Zero or at least to mitigating carbon emissions 45% by 2030 [43].

In closing, it is reported that half of global GDP (estimated US$44 trillion) is likely vulnerable to the loss of biodiversity [76]. With nature granting US$125 trillion in assets to humanity [76], protecting and restoring biodiversity and integrating the built environment with nature to promote biomimicry in architecture are vital to foster the recovery efforts from the COVID-19 pandemic and converge toward greening. It is obvious though that achieving global goals (ending poverty and advancing the SDGs) cannot be achieved unless healthy and functioning ecosystems for humanity are reached. In this context, Eva Gurria, Nature-Based Solutions Coordinator, Global Programme on Nature for Development, UNDP, says [76]:

> One health approach – a healthy planet is the foundation for health communities.

Therefore, it is fundamentally urgent that governments and all stakeholders focus on protecting biodiversity and ecosystem to improve the socioeconomy and sustainability of cities, yet develop inclusive future, regenerative, livable, and resilient cities that can adapt and cope with severe climate change events and the COVID-19 crisis with the aim to attain SDGs, mainly goal 3 – Good Health, goal 11 – Sustainable Cities and Communities, and goal 13 – Climate Action.

Finally, architects, engineers, urban designers, and planners should concentrate on **NATURE** to inspire innovative biomimetic solutions and foster biomimicry in architecture by integrating biophilic buildings and urban areas that focus on habitat provision strategies, encompassing urban green, gardens, green walls, and green roofs, all connected with ecological corridors to confront cities' colossal challenges and achieve sustainable, regenerative, and livable cities in the age of the COVID-19 crisis and post-COVID to ensure resilience.

References

1. Moller, K. (2013). *Plants of desert. ASU – ASK a biologist.* Retrieved September 23, 2021, from https://askbiologist.asu.edu/plants-desert
2. Brandt, N., Cambell, F., Deakin M., & Johansson S., et al. (2014). *European cities moving towards climate neutrality participation, indicators, and benchmarking.* Technical report. CLUE – Climate Neutral Districts in Europe. September 2014. https://doi.org/10.13140/2.1.3166.3369. https://www.researchgate.net/publication/266137403
3. Lau, P. (2021). *What is Carbon Neutrality, and can it really be achieved?* Giving Compass. Retrieved September 19, 2021, from https://givingcompass.org/article/what-is-carbon-neutrality-and-can-it-really-be-achieved/?gclid=EAIaIQobChMIjd-0u4yM8wIVVUeRBR0wDgq1EAAYBCAAEgIhv_D_BwE
4. Udovyk, O. (2021). *Biomimicry and COVID-19 response.* UNDP. Retrieved September 21, 2021, from https://www.ua.undp.org/content/ukraine/en/home/blog/2020/biomimicry-and-covid-19-response.html
5. *ICLEI. GreenClimateCities Program.* Iclei.org. (2021). Retrieved September 22, 2021, from https://www.iclei.org/en/GreenClimateCities.html
6. *ICLEI. The ICLEI Climate Neutrality Framework: Cities and regions on the path to net zero emissions and climate resilience.* Iclei.org. (2021). Retrieved September 22, 2021, from https://www.iclei.org/en/climate_neutrality.html
7. *Home – Global Covenant of Mayors.* Global Covenant of Mayors (2021). Retrieved September 24, 2021, from https://www.globalcovenantofmayors.org
8. *Turn resilient low-carbon ideas into strategies and finance-ready projects | CityGapFund.* Citygapfund.org. (2021). Retrieved September 24, 2021, from https://www.cityapfund.org
9. *ICLEI – Global Research Strategy.* Iclei.org. (2021). Retrieved September 24, 2021, from https://www.iclei.org/en/Global_Research_Strategy.html
10. Aboulnaga, M., Badran, M., & Barakat, M. (2021). *Resilience of informal areas in megacities: Magnitude, challenges, and policies* (1st ed., pp. 1–30). Springer.
11. Dicks, H. (2017). A new way of valuing nature: Articulating biomimicry and ecosystem services. *Environmental Ethics, 3*(39), 281–299. https://www.pdcnet.org/collection/show?id=enviroethics_2017-0039_0003_0281-0299&file_type=pdf
12. Benyus, J. (2021). *What isn't Biomimicry?* Biomimicry.org. Retrieved September 27, 2021, from https://biomimicry.org/what-is-biomimicry

13. Peake, S. (2021). *Renewable Energy –Ten short lessons* (pp. 87–92). In chapter 6: Plants and water are powerful, John Hopkins University Press, Baltimore, Maryland, USA.
14. *Bamboo biomimicry by staff*. Enginneering.con. (2017). Retrieved September 27, 2021, from https://www.engineering.com/story/bamboo-biomimicry
15. Sato, M., Inoue, A., & Shima, H. (2017). Bamboo-inspired optimal design for functionally graded hollow cylinders. *PLoS One, 12*(5), e0175029. https://doi.org/10.1371/journal.pone.0175029
16. Lemmey, T. (2020). *Connection with nature in the UK during the COVID-19 lockdown* (pp. 1–16). University of Cumbria. Retrieved September 26, 2021, from https://insight.cumbria.ac.uk/id/eprint/5639/1/Nature%20Connection%20and%20Covid%20OTL.pdf
17. RSPB and YouGov. (2020). *Recovering together*. Retrieved September 27, 2021, from https://www.rspb.org.uk/globalassets/downloads/recovery-together-report_nature-and-green-recovery_respbyougov_june-2020.pdf
18. Carlson, R. (2021). *Climate action through biomimicry: Innovation for the SDGs*. Sustainable Brands. Retrieved September 19, 2021, from https://sustaiablebrands.com/read/product-serve-design-innovation/climate-action-through-biomimicry-innovation-for-the-sdgs
19. *Biomimicry Global Design Challenge*. Biomimicry.org. (2021). Retrieved September 27, 2021, from https://biomimicry.org/globaldesignchallenge
20. *Biomimicry Design Challenge*. Biomimicry San Diego. (2021). Retrieved September 27, 2021, from https://www.biomimicrysandiego.org/learn-and-practice-biommimicry
21. *Announcement: Circular and Biomimicry Built Environment Group Opened on the SDG Community of Practice | SDG Help Desk*. Sdghelpdesk.unescap.org. (2021). Retrieved September 27, 2021, from https://sdghelpdesk.unescap.org/media-centre/announcement-circular-biomimicry-built-environment-group-opened-sdg-community-practice
22. Mapfre, R. (2021). *Biomimicry: Observing nature and emulating solutions in pursuit of sustainability*. Grupo MAFRE Corporativo – Acerca de MAPFRE. Retrieved September 28, 2021, from https://www.mapfre.com/en/insights/sustainability/biomimicry
23. *Biomimicry: Adopting Proven, Sustainable Solutions Inspired by Nature |Oslomanifesto*. Oslomanifesto| Design and Architecture for SDGs. (2021). Retrieved September 28, 2021, from https://oslomanifesto.org/biomimicry-adopting-proven-sustainable-solutions-by-nature
24. Uchiyama, Y., Blanco, E., & Kohsaka, R. (2020). Application of biomimicry to architectural and urban design: A review across scales. *Sustainability, 12*, 9813. https://doi.org/10.3390/su12239813
25. Shimomura, M. (2015). New trend of biomimitics: Innovative material technology towards sustainability. *Engineering Materials, 63*, 18–22.
26. McMahon, M., & Hadfield, M. (2007). The butterfly effect: Creative sustainability design solutions through systems thinking. In *FAIM: Intelligent manufacturing now* (pp. 247–254). University of Limerick.
27. Blizzard, J. L., & Klotz, L. E. (2012). A framework for sustainability whole systems design. *Design Studies, 33*, 456–479.
28. Ramzy, N. (2015). Sustainable spaces with psychological values: Historical architecture as reference book for biomimitics models with biophilic qualities. *International Journal of Architectural Research ArchNet-IJAR, 9*, 248–267.
29. Vincent, J. F., Bogatyreva, O. A., Bogatyrev, N. R., Bowyer, A., & Pahl, A. K. (2006). Biomimitics: Its practice and theory. *Journal of the Royal Society Interface Interface, 3*, 471–482.
30. Royall, F. (2010). Defining biomimitics – Architectural applications in system and products. In *UTSoA-Seminar in Sustainable Architecture* (pp. 3–13). The University of Texas at Austin.
31. Zare, M., & Falahat, M. (2013). Characteristics of reptiles as a model for bionic architecture. *Advances in Civil Engineering, 1*, 124–135.
32. Fuiji, S., Sawada, S., Nakayama, S., Kappl, M., Ueno, K., Shitajima, K., Butt, H. J., & Nakamura, Y. (2016). Pressure-sensitive adhesive powder. *Materials Horizons, 3*, 47–52.

33. Edmondson, A. C. (1986). *A fuller explanation: The synergetic geometry of R.* Buckminster Fuller. ISBN 978-0817633387.
34. Alexander, C. (2002). *The nature of order: An essay on the art of building and nature of the universe* (Vol. 1). Center for Environmental Structure.
35. *Biophilic Design Overview | Living-Future.org*. International Living Future Institute. (2021) Retrieved September 28, 2021, from https://www.living-future/biophilic-design-overiew
36. Pederson Zari, M. (2018). *Regenerative urban design and ecosystem biomimicry* (1st ed.). Abingdon-on-Themes. ISBN 978113079489.
37. Kellert, S. R., Heerwagen, J., & Mador, M. (2008). *Biophilic design: The theory, science, and practice of bringing buildings to life*. Wiley.
38. Browning, W. D., Ryan, C. O., & Clancy, J. O. (2014). *14 Patterns of biophilic design*. Terrapin Bright Green, LLC.
39. Kellert, S. R. (2018). *Nature by design: The practice of Biophilic design*. University of Yale Press. https://yalebooks.yale.edu/book/9780300214536/nature-design
40. Environment, T. (2021). *The relationship between architecture and environment*. Ierek news. Retrieved September 29, 2012, from https://www.ierek.com/news/inde.php/2016/04/13/the-relationship-between-architecture-and-environment).
41. *14 Patterns of Biophilic design*. Terrapinbrightgreen.com. (2021). Retrieved October 6, 2021, from https://www.terrapinbrightgreen.com/reports/14-patterns
42. Newburger, E. (2021). *Disasters caused $210 billion in damage in 2020, showing growing cost of climate change*. Www.cnbc.com. Retrieved April 30, 2021, from https://www.cnbc.com/2021/01/07/climate-change-disasters-cause-210-billion-in-damage-in-2020.html.
43. *Adaptation – UN Climate Change Conference (COP26) at the SEC – Glasgow 2021*. UN Climate Change Conference (COP26) at the SEC – Glasgow 2021. (2021). Retrieved October 5, 2021, from https://ukcop26.org/cop26-goals/adaptation/
44. Giordani, A. (2014). *Cost of climate change in Europe could reach 4% of GDP*. Ec.europe.eu. HORIZON – The EU Research & Innovation Magazine. Retrieved October 5, 2021, from https://ec.europa.eu/research-and-innovation/en/horizon-magazine/cost-climate-change-europe-could-reach-4-gdp
45. Galey, P. (2019). *Climate impacts 'to cost world $7.9 trillion' by 2050*. Phys.org. Retrieved October 5, 2021, from https://phys.org/news/2019-11-climate-impacts-world-trillion.html.
46. *Climate Change* The World Bank. (2021). Retrieved October 5, 2021, from https://www.worldbank.org/en/topic/climatechange/overview#1
47. Pedersen Zari, M. (2010). Biomimitec design for climate change adaptation and mitigation. *Architectural Science Review, 53*(2), 172–183. https://doi.org/10.3763.asre.2008.0065.
48. Pedersen Zari, M. (2015). Can biomimicry be a useful tool for design for climate change adaptation and mitigation? In T. F. Pacheco, J. Labrincha, M. Diamanti, C. P. Yu, & H. Lee (Eds.), *Biotechnologies and biomimetics fir civil engineering*. Springer. https://doi.org/10.1007/978-3-319-09287-4_4
49. Suzelis, N. (2021). *Climate Change – Uneven Earth*. Unevenearth.org. Retrieved October 5, 2021, from https://unevenearth.org/tag/climate-change/
50. Cox, S., & Chomsky, N. (2020). *The green new deal and beyond – Ending the climate emergency while we still can.*
51. Miller, J. (2020). Stan Cox: The green new deal and beyond: Ending the climate emergency while we still can. *Agriculture and Human Values, 37*(4), 1321–1322. https://doi.org/10.1007/s10460-020-10122-9
52. Hojniak, D. & Hvid, C. (2021). *Healthy buildings, cities and you – How to design future living environments – Urban Insight*. Urban Insight. Retrieved September 18, 2021, from https://www.swecourbaninsight.com/healthy-buildings-cities-and-you-how-to-design-future-living-environments/
53. *Biodiversity and the economic response to COVID-19: Ensuring a green and resilient recovery*. OECD. (2020). Retrieved October 6, 2021, from https://www.oecd.org/coronavirus/policy-responses/

biodiversity-and-the-economic-response-to-covid-19-ensuring-a-green-and-resilient-recovery-d98b5a09
54. Leordean, L. (2016). *Biomimicry in Architecture: Mitigation and Adaptation to Climate Change (Master)*. Politecnico di Milano. Retrieved July 7, 2021, from https://issuu.com/ioanaleordean/docs/thesis_book_online-compressed
55. Climate change: EU to cut CO2 emissions by 55% by 2030. BBC News. (2021). Retrieved July 6, 2021, from https://www.bbc.com/news/world-europe-56828383
56. Abbas, G. M. & Banci, S. (2020). *ARCHI –TALKS 3: Branko Kolarevic – Building dynamics: Exploring architecture of change, University of Ankara, Turkey*. BOTT ETU Publications.
57. Smith, K. (2015). *Biomimicry Nature's Genius*. (BSSBE Capstone) University of Florida, USA. Retrieved October 10, 2021, from https://ufdcimages.uflib.ufl.edu/AA/00/03/74/59/00001/Smith_Katelynn_2015_Biomimicry%20Nature's%20Genius.pdf
58. Kolarevic, B., & Parlac, V. (n.d.). *Building dynamics: Exploring architecture of change* (pp. 2–43). University of Ankara, Turkey. BOTT ETU Publications.
59. Fairs, M. (2010). *Spanish Pavilion at Shangahi Expo 2010 by EMBT |Dezeen*. Dezeen. Retrieved October 10, 2021, from https://www.dezeen.com/2010/04/26/spanish-pavilion-at-shanghai-expo-2010-by-embt
60. www.fosterandpartners.com, F. (2021). *UAE Pavilion Shanghai Expo 2010 |Foster + Partners*. Fosterandpartners.com. Retrieved October 10, 2021, from https://www.fosterandpartners.com/projects/uae-pavilion-shanghai-expo-2010
61. Patterson, Z. (2021). *Everything you need to know about the UAE Pavilion at Expo 2020 Dubai*. TimeOut. Retrieved October 2, 2021, from https://www.timeoutdubai.com/news/473575-uae-pavilion-at-expo-2020-dubai
62. Calatrava, S. (2021). *Santiago Calatrava tops UAE Pavilion at Dubai Expo with 28 opening wings – Santiago Calatrava – Architects & Engineers*. Calatrava.com. Retrieved October 10, 2021, from https://www.calatrava.com/news/reader/Santiago-calatrava-tops-uae-pavilion-at-dubai-expo-with-28-opening-wings.html
63. *Architectural Design of the Slovenian pavilion at 2020 expo Dubai receives award in the USA |Gov.SI*. Portal Gov.SI. (2021). Retrieved October 11, 2021, from https://www.gov.si/en/news/2021-02-03-architectural-design-of-the-slovenian-pavilion-at-2020-expo-dubai-receives-award-in-the-usa
64. *Dubai Expo* 2020*: Slovenian Pavilion | Sempergreen® | Archello*. Archello. (2021). Retrieved October 10, 2021, from https://archello.com/it/project/dubai-expo-2020-slovenian-pavilion
65. *Oman unveils its Expo 2020 Dubai Pavilion Design. Arabian Business*. (2021). arabianbusiness.com. Retrieved October 11, 2021, from https://www.arabianbusiness.com/construction/407389-oman-unveils-its-expo-2020-dubai-pavilion
66. *IMPRESSIVE AZERBSIJIAN PAVILION – EXPO DUBAI 2020 | SCE PROJECT*. SCE Project (2021). Retrieved October 11, 2021, from https://www.sceproject.it/azerbaijian-pavilion-dubai-expo-2020
67. *Azerbaijan Pavilion – Expo 2020 Dubai – Special tensile structure and marquees*. Special tensile structure and marquees. (2020). Retrieved April 6, 2022, from https://www.macotechnology.com/portfolio/pavilion-azerbaijan-expo-dubai-2020/.
68. Sahar Ejaz, S. (2021). *50% of Expo 2020 Dubai's energy to come from renewable sources*. Gulfnews.com. Retrieved October 11, 2021, from https://gulfnews.com/expo-2020/news/50-of-expo-2020-dubais-energy-to-come-from-renewable-sources-1.1627207021737
69. Grimshaw Architects. (2021). *Terra –The Sustainability Pavilion Expo 2020 Dubai/GRIMSHAW*. Grimshaw.global. Retrieved October 11, 2021, from https://grimshaw.global/projects/dubai.expo-2020-sustainability-pavilion
70. Grimshaw Architects. (2022). *Terra — The Sustainability Pavilion Expo 2020 Dubai: Case Study/GRIMSHAW*. Grimshaw.global. Retrieved February 25, 2022, from https://grimshaw.global/sustainability/expo-2020-sustainability-pavilion-case-study/

71. www.fostersnd partners.com, F. (2021). *Alif – The Mobility Pavilion, Expo 2020 Dubai | Foster +Partners*. Fosterandpartners.com. Retrieved October 11, 2021, from https://www.fostersndpartners.com/projects/alif-the-mobility-pavilion-expo-2020-dubai
72. Yuko, N. (2021). *Japan Pavilion Architecture*. JAPAN PAVILION, EXPO 2020 DUBAI. Retrieved October 11, 2021, from https://expo2020-dubai.go.jp/en/architecture
73. Sullivan, B. (2021). *Second-Warmest Year on Record Cost the World $210 billion*. Bloomberg.com. Retrieved October 11, 2021, from https://bloomberg.com/news/articles/2021-01-07/second-warmest-year-on-record-cost-the-world-210-billion
74. *Rising to the Climate Change. World Resources Institute*. (2021). www.wri.org. Retrieved October 11, 2021, from https://www.wri.org/climate
75. *68% of the world Population projected to live in urban areas by 2050*. (*United Nations – Department of Economic and Social Affairs*. (2021). Www.un.org. Retrieved October 11, 2021, from https://www.un.org/developmet/desa/en/news/population/2018-revision-of-world-urbanization-prospects.html
76. Gurria, E. (2020). *Nature for life: The value of nature in achieving the Sustainable Development Goals | United Nations Development Programme*. United Nations Development Programme. Retrieved October 10, 2021, from https://www.undp.org/blog/nature-life-value-nature-achieving-sustainable-development-goals
77. Varshney, K., Pedersen Zari, M., & Bakshi, N. (2022). *Habitat provisioning: A regenerative building design approach*. Retrieved 16 February 2022, from https://doi.org/10.1007/978-3-030-51812-3_320-1
78. Blanco, E., Pedersen Zari, M., Raskin, K., & Clergeau, P. (2021) Urban ecosystem–Level biomimicry and regenerative design: Linking ecosystem functioning and urban built environments. *Sustainability*, *13*(404). https://doi.org/10.3390/su13010404
79. Blanco, E., Cruz, E., Lequette, C., Raskin, K., & Clergeau, P. (2021). Biomimicry in French urban projects: Trends and perspectives from the practice. *Biomimetics*, *6*(27). https://doi.org/10.3390/biomimetics6020027
80. Cruz, E., Hubert, T., Chancoco, G., Naim, O., Chayaamor–Heil, N., Cornette, F. A. R., Menezo, C., Badarnah, L., & Raskin, K. (2021). Design process and multi-regulation of biomimetic building skins: A comparative analysis. *Energy Build.* under review.
81. Pedersen Zari, M., & Hecht, K. (2020). Biomimicry for regenerative built environments: Mapping design strategies for producing ecosystem services. *Biomimetics*, *5*(2), 18. https://doi.org/10.3390/biomimetics5020018

Chapter 2
Biomimicry and Nature: Milieu, History, Approaches, and Design Methods and Process

2.1 Introduction

In nature, all individuals and creatures join into a group of life. However, culture is usually thought about as separated from nature, rather than a part of nature since in the industrial revolution, reductionist science has empowered us to outline an entire host of effectively manipulative advancements, which have changed the planet. Nature is an endless source of motivation and inspiration for engineers and scientists in different fields. Nature evolves and finds solutions that respond to organisms. Thus, every single organism has its characteristics and uniqueness to its environment. This exits through infinity generations while passing the test of survival to reach its next generation.

The field of biomimicry, where flora, fauna, or entire ecosystem are emulated as a basis for design, has attracted worldwide interest in the fields of architecture and engineering. In this context, an example of the beauty of biomimetic architecture blended with nature in Japan is illustrated in Fig. 2.1. Hence, it is imperative to define and understand biomimicry and what is meant by the term "biomimicry."

2.2 What Is Biomimicry?

In the twentieth century, the human populace has experienced a huge growth rate, making urban communities develop to huge extents. The assets of the earth have been exhausted by our utilization of energy consumption and building processes;

"Look deep into nature and then you will understand everything better"
Albert Einstein

M. Aboulnaga, S. E. Helmy, *Biomimetic Architecture and Its Role in Developing Sustainable, Regenerative, and Livable Cities*,
https://doi.org/10.1007/978-3-031-08292-4_2

Fig. 2.1 The beauty of biomimicry in architecture blended with nature. (Image credit and source: Mohsen Aboulnaga)

also regarding the problem of waste management, biomimicry may be a solution for those problems [1]. It is the examination of nature's models, frameworks, procedures, and components that can give answers for human issues. Biomimicry as a concept has several applications in science, but it is still not realized largely in the architecture field. This is due to the fact of the small number of case studies.

"Biomimicry" as a terminology appeared first time in the scientific literature in 1962. The term is utilized especially in the midst of materials researchers in the 1980s. There has been a gigantic motivation of enthusiasm amid the most recent decade; people like Janine Benyus, professor of Biology; Steven Vogel, American biomechanics researcher; and Julian Vincent, professor of Biomimetics and one of the most imaginative and thorough researchers, as well as an inspirational teacher and writer, in the area of bio-inspired materials, have all composed broadly in this branch of knowledge (Box 2.1) [2]. *"The abstraction of good design from nature"* is the definition of biomimicry from the vision of Julian Vincent, *"Conscious emulation of nature's genius"* is the definition of biomimicry from Janine Benyus point of view, and as a discipline, according to Steven Vogel, it could be *"emulation the mechanism of organism's forms or processes or systems to generate sustainable solutions"* [2]. Box 2.2 shows the definitions of biomimicry.

Box 2.1: Vision of Biomimicry

The abstraction of good design from nature. (Julian Vincent)

Box 2.2: Definitions of Biomimicry

Conscious emulation of nature's genius. (Steven Vogel)

Emulation the mechanism of organism's forms or processes or systems to generate sustainable solutions. (Janine Benyus)

2.3 Origins of Biomimicry

The origin of biomimicry is dated back to the 1960s [3]. Architects, engineers, planners, and designers have sought to biology for motivation since the beginnings of science in the mid-nineteenth century. They did not copy the types of plants and creatures, but they discover techniques in design analogous to the process of development and advancement in nature. Biological and natural thoughts are extraordinary in the works and writings of numerous innovative architects such as Le Corbusier and Frank Lloyd Wright, who are only the most celebrated [3]. One of the key examples of architecture – in modern history – that was blinded within nature is the Fallingwater house, built over a waterfall in nature by Frank Lloyd Wright in rural southwestern Pennsylvania, USA (Fig. 2.2). The house was designed in 1935 and the construction of the main house, guest house, and service wing was completed in 1939. The Förstes Weinterrassen, Bad Neuenahr-Ahrweiler in Germany is another example of architecture that is blended with nature as shown in Fig. 2.3.

Fig. 2.2 Fallingwater house, built over a waterfall blended within nature by Frank Lloyd Wright. (Image credit and source: Carol M. Highsmith, https://commons.m.wikimedia.org/wiki/File:Falingwater,_also_known_as_the_Edgar_J._Kaufmann,_Sr.,_residence,_Pennsylvania,_by_Carol_M._Highsmith.jpg#)

Fig. 2.3 Förstes Weinterrassen blended with nature, Bad Neuenahr-Ahrweiler. (Image credit and source: Udo Heimermann, https://commons.m.wikimedia.org/wiki/File:Förstehof-Bad-Neuenahr_Ahrweiler.jpg#)

However, the issue with the biological analogy of architecture in the past was the shallow imitating, but such analogy at a deeper level can be a most fundamental source of understanding as scientific discretion, as many writers on that subject have pointed out. Although there are many discoveries, new ideas, and applications in recent related to biological developments in the practice and theory of design, this shows echo thoughts in the prior history of biological analogy [3].

2.4 History of Biomimicry in Architecture and Its Outlook

The relationship between humans and nature has evolved and improved since the beginning of time, beginning with Adam's sons learning burial rituals by imitating the raven so that the fact that humans are from dust to dust is emphasized, returning to the closed loop of the ecosystem of human life. Since then, the relationship changed from fear to imitating nature, where cavemen were able to learn many things from observing natures' species and were able to adjust to fear of the unknown and lack of knowledge. In the Ancient Egyptian and Greek civilizations, early engineers and inventors have moreover turned and looked to nature for inspiration and ideas [4].

Several movements in the architectural field have taken their approaches to nature. The history of architecture depicted many exemplary models. In ancient

Egypt, building designers started to seek assistance from nature to specifically utilize designs and thoughts from nature, such as columns that were from palm trees and lotus plants as depicted in temples' walls. The proportions of natural forms are the perfect aesthetic achievements. Figures 2.4 and 2.5 present these examples of

Fig. 2.4 Ancient Egyptian examples inspired from nature. (**a**) Head of the column inspired by opened leaves of the lotus plant in Luxor, (**b**) Head of the column with closed leaves of the lotus plant, (**c**) A set of columns' heads with opened lotus plants, (**d**) A set of columns' heads with closed leaves of the lotus plant in Karnak temple, Luxor, (**e**) Columns' heads with opened lotus plant in Karnak temple, Luxor. (Images' credit and source: Mohsen Aboulnaga)

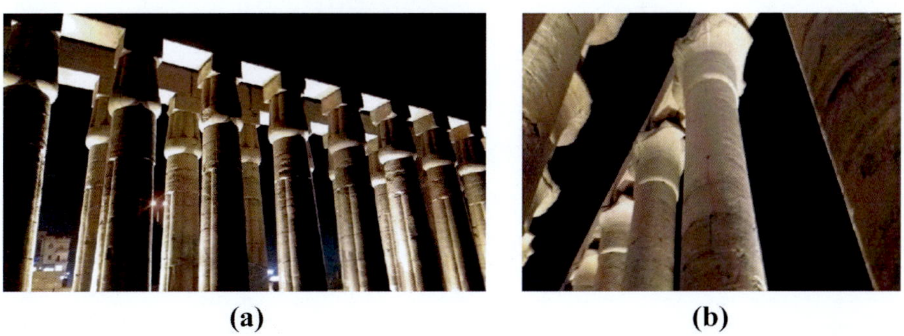

Fig. 2.5 Lotus set of columns in Ramesses II Temple in Luxor city, Ancient Egypt. (**a**) The lotus plants of the columns closed – night; (**b**) The opened lotus plants – night. (Images' credit and source: Mohsen Aboulnaga)

Ancient Egypt in the city of Luxor. Many of the most important advancements in the human history have been included from the observation of natural phenomena and shapes, but it was mainly inside the modern era that nature appeared in the scientific model [5]. Nothing is new in the idea of biomimicry itself, but nevertheless, human structures have been acquired from nature all through history.

Truly, it might be contended that biomimicry isn't something new, yet a *"Revival and Return"* to our most punctual motivations. Famous architects like Alvar Aalto, Frank Lloyd Wright, Le Corbusier, and others have laid accentuation on this association (Fig. 2.6), where architecture is being impacted by a few subjects of natural

(a) (b)

Fig. 2.6 Building with nature in Alborg – Finland and in Poissy – France. (**a**) Alvar Alto KUNSTEN Museum of Modern Art, (**b**) Le Corbusier example as a force of nature. (Images' credit and source: (**a**) Kunsten Museum of Modern Art, https://commons.m.wikimedia.org/wiki/File:KUNSTEN_Alborg_2006.jpg#, (**b**) Valueyou (talk), https://upload.wikimedia.org/wikipedia/en/3/3c/VillaSavoye.jpg)

(a) (b)

Fig. 2.7 Calatrava WTC building, inspired from bird's wings, New York City, USA. (**a**) The form inspired from the bird's wings, (**b**) The structure rips of the roof and façade. (Images' credit and source: (**a**) Shaun Dawson, https://commons.m.wikimedia.org/wiki/File:WTC_Hub_September_2016_vc.png and https://www.flickr.com/photos/iamsdawson/25656800868 (**b**) Kidfly182, https://commons.m.wikimedia.org/wiki/File:Oculus_Exterior_September_2021_002.jpg)

Fig. 2.8 The iconic skeleton of the WTC Hub, inspired from nature (bird's wings) in New York City during the day. (Image credit and source: Massmatt, https://commons.m.wikimedia.org/wiki/File:WTC_Hub_April_2016_vc.jpg)

Fig. 2.9 View of the iconic skeleton of the WTC Hub, inspired from nature (bird's wings) in New York City at night. (Image credit and source: Steve from Austin, TX, USA, https://commons.m.wikimedia.org/wiki/File:WTC_Hub_January_2017_2_vc.jpg)

Fig. 2.10 Interior of the WTC hub depicting the beauty of the structural rips – New York, USA. (Image credit and source: Kidfly182, https://commons.m.wikimedia.org/wiki/Category:Interior_of_the_WTC_Transportation_Hub#/media/File:Oculus_Interior_017.jpg)

Fig. 2.11 Sunlight impinging through the WTC Hub façades' rips. (Image credit and source: Kidfly182, https://commons.m.wikimedia.org/wiki/Category:WTC_Transportation_Hub#/media/File:Oculus_Interior_011.jpg)

Fig. 2.12 Top-up view of the interior structure of the WTC Hub façades' rips. (Image credit and source: Kidfly182, https://commons.m.wikimedia.org/wiki/Category:WTC_Transportation_Hub#/media/File:Oculus_Interior_011.jpg)

Fig. 2.13 The iconic structure from inside at night highlighting the façades' rips. (Image credit and source: Spurekar, https://commons.m.wikimedia.org/wiki/Oculus_(2566004800).jpg)

and sociologies, while numerous subjects hold an undeniable impact on engineering. Natural sciences and inspirations from nature become the dominant era, it is the source of inspiration for architects that encourage them to observe, learn and emulate so that they could create better architectural designs for better living; for instance, vivid examples of Calatrava's designs are the WTC Hub and Milwaukee Art Museum (Figs. 2.7, 2.8, 2.9, 2.10, 2.11, 2.12, 2.13 and 2.14). This biological structure developed and moved toward various methodologies and fields [6].

(a)

(b)

Fig. 2.14 The Milwaukee Art Museum inspired from nature. (**a**) The exterior of Milwaukee Art Museum by Calatrava mimicking a sailing ship, (**b**) Daylight through the rips of the wings from inside Milwaukee Art Museum, Wisconsin, USA. (Images' credit and source: (**a**) Andre Engels, https://commons.m.wikimedia.org/wiki/File:Milwaukee_Art_Museum_exterior.jpg# (**b**) T. Canaan, https://commons.m.wikimedia.org/wiki/File:Interior_Milwaukee_Art_Museum.jpg)

A defined starting point of biomimicry in the field of architecture cannot be declared. However, nature is the source of motivation and extraction since humans existed, yet natural shelters were the first homes for humans such as caves have been utilized and so do the trees, as models for architectural design throughout history.

(a) **(b)**

Fig. 2.15 Natural materials used to fabricate shelters and houses (huts) in Africa. (**a**) Natural material used in a house, Kisii village (**b**) A Japadhola hut depicting natural material. (Images' credit and source: (**a**) Alexander Leisser, http://commons.m.wikimedia.org/File:Kisii_village_01.jpg; (**b**) Erinamukuta, http://commons.m.wikimedia.org/File:Japadhola_house.JPG)

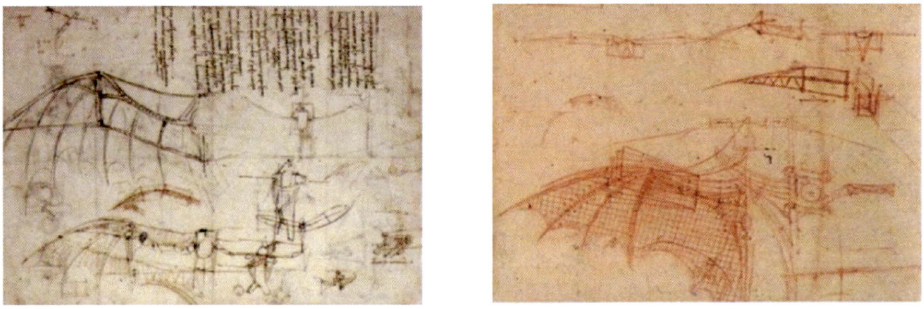

Fig. 2.16 Drawings trials for flying machine by Leonardo Da Vinci. (Images' source: http://commons.m.wikimedia.org/Leonardo_Design_for_a_Flying_Machine,_c_1488.jpg; http://commons.m.wikimedia.org/Leonardo_Design_for_a_Flying_Machine,_c_1505.jpg)

Natural materials were exploited to fabricate shelters and houses (African huts) and innovations to collect and process. Materials were created and constantly grew and promoted until today as shown in Fig. 2.15. In Egyptian sanctuaries, numerous different illustrations refer to emblematic and sensible utilization of models [6].

The first pioneer of biomimicry in architecture was Leonardo da Vinci, who took inspiration from nature's models [6]. Looking back to history, there are some historical examples of problem-solving inspired by nature for creating a flying machine. Da Vinci found that it is critical to study the birds' anatomy and flying techniques as presented in Fig. 2.16. Although his machine was never completed, the abstract principle of capturing the inspiration from nature let da Vinci be a pioneer in biomimicry, who got the idea to construct the first airplane from a flying pigeon.

Biomimicry was not existing at that time yet, but history was already a source of inspiration and motivation. For example, Gaudi in his iconic buildings – Casa Batlló building, Casa Milà, and Sagrada Familia church in Barcelona – articulated such a

(a) (b)

Fig. 2.17 The Casa Batlló building as a biomimetic example in Barcelona exhibiting the inspiration of the human body pelvic. (**a**) Exterior of the Casa Balto at night, (**b**) Exterior structure of Casa Batlo. (Images' credit and source: (**a**) and (**b**) Mohsen Aboulnaga)

(a) (b)

Fig. 2.18 The Casa Milà building as a biomimetic example in Barcelona inspired from the human body's pelvic. (**a**) The exterior of Casa Milà at night, (**b**) Façade's structure of Casa Milà. (Images' credit and source: (**a**) and (**b**) Mohsen Aboulnaga)

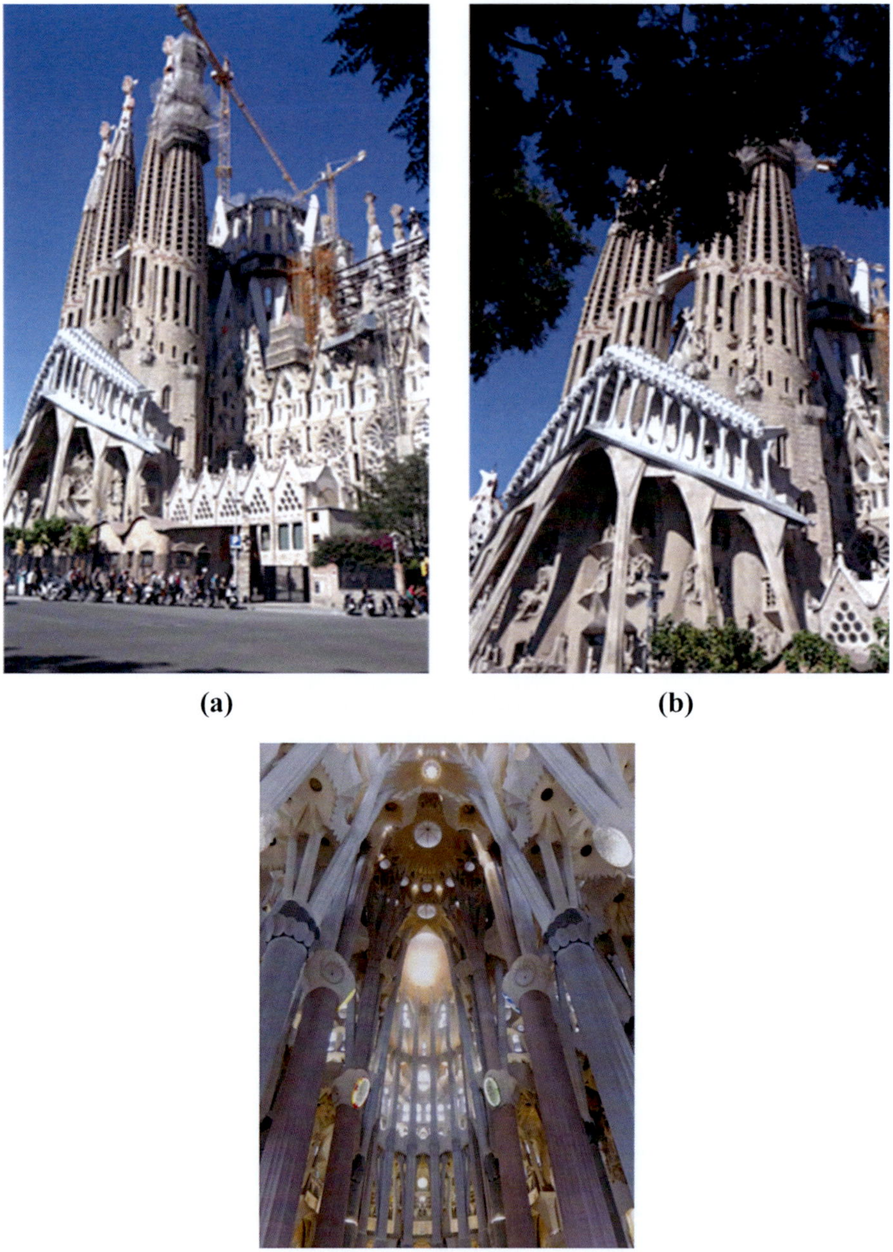

(a) (b)

(c)

Fig. 2.19 The Sagrada Familia church inspired from sequoia trees, Barcelona. (**a**) View of Sagrada Familia, (**b**) The exterior with details, (**c**) Interior of the church. (Images' credit and source: (**a**) and (**b**) Mohsen Aboulnaga; (**c**) Tim Bary, http://commons.m.wikimedia.org/File:Sagrada_Familia_intrior_over_altar.jpg)

Fig. 2.20 Crystal palace inspiration from nature, London – United Kingdom. (**a**) Doum leave inspiration, (**b**) Water Lily leaves, (**c**) Licuala orbicularis leaves, (**d**) Crystal Palace with the Lilly-shaped vaulted roof, (**e**) Crystal Palace with the Lilly-shaped vaulted roof. (Images' credit and source: (**a**) Boumlik Messaïli, https://commons.m.wikimedia.org/wiki/File:Doum.jpg; (**b**) Nguyễn Tấn Phát, https://commons.m.wikimedia.org/wiki/File:South_Vitnam%27s_Water_Lily.JPG; (**c**) Eric in SF, https://commons.m.wikimedia.org/wiki/File:Licuala_orbicularis.jpg; (**d**) https://commons.m.wikimedia.org/wiki/File:Crystal_Palace_1910.jpg; (**e**) https://commons.m.wikimedia.org/wiki/File:Crystal_Palace_Sydenham.png)

concept in his buildings' design (Figs. 2.17, 2.18, 2.19, and 2.20). The inspiration for exterior structure of the Casa Batlló building in Barcelona was from the human pelvic bones as shown in Fig. 2.17. Gaudi has also exhibited the same concept in Casa Milà in Barcelona, where he created one of his legendary buildings, inspired by the analogy of human structure (Fig. 2.18); the Casa Milà is now a museum. Another iconic example is the legendary Sagrada Familia church in Barcelona, where Gaudi looked to nature in his design by emulating the tree structure that was inspired by sequoia trees in the main façade and the interior, presented in Fig. 2.19.

Additionally, James Paxton in 1851 designed the structural system of the Crystal Palace in London, UK, from his observation of giant water lilies' leaves of the Amazonian water lily plant, as depicted in Fig. 2.20. The leaves of this water lily look brittle but can support a weight. The Crystal Palace's structure is founded on the ribs and trunk of the lily leaves [1, 7].

2.5 The Evolution of Mankind and Relationship with Nature

Biomimicry moves our present-day ideology closer to nature by using it as a design inspiration to sustainably solve human problems. Biomimicry also links the built environment to the natural world by striving to use Mother Nature as a model of

Fig. 2.21 Biomimicry strives to use Mother Nature – Upper Kachura Lake near Skardu City, Western Himalayas, Pakistan. (Image credit and source: Sufyan785, https://commons.m.wikimedia.org/wiki/File:Mother_nature.jpg)

Fig. 2.22 Mother Nature as a rich source of inspiration for biomimicry – Pordenone, Italy. (Image credit and source: Mohsen Aboulnaga)

inspiration and a standard of measure and a mentor [8]. Figures 2.21, 2.22, and 2.23 depict Mother Nature examples.

According to the increase of technology, the relationship with nature diverted from inspiration to "Domain over Nature" as sought by designers like Francis Bacon, when he began to isolate inside spaces in architecture from outside context

Fig. 2.23 Mother Nature, a rich source of inspiration for biomimicry – Finland. (Image credit and source: Mohsen Aboulnaga)

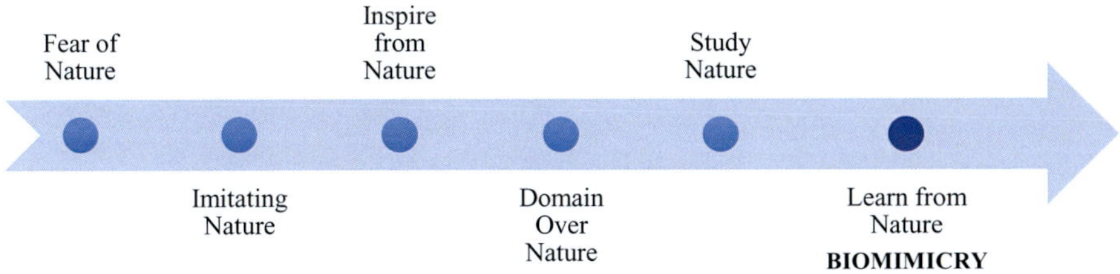

Fig. 2.24 Evolution of mankind and nature relationship. (Image source: Developed by authors after http://csfs.bue.edu.eg/files/Library/Papers/Sustainability%20and%20the%20Future/12.pdf)

and surroundings, as being warm in the space by using technologies no matter how cold outside is and how heaters and air conditioners could affect the surrounded nature, or being cool in the space no matter how hot outside is with the freedom in designing and using technologies [8].

The evolution of humankind and nature's relationship is presented in Fig. 2.24. Nonetheless, architecture moved quickly away from considering nature, as a model of inspiration, toward the domain over nature phase. It was the machines that made these changes possible to the extent that Le Corbusier went as far as saying "houses are machines for living in" [9].

The relationship with nature transformed to a more adaptive form through

Box 2.3: Slogans of Notions By Famous Legendary Architects About Exploiting Nature as the Best Example of Integration

Form follows function. (Louis Sullivan)

Form and function are one. (Franck Lloyd Wright)

organic architecture in an attempt to integrate the spaces into a coherent whole and the adaptation between the site and the built environment. In modern architecture, Louis Sullivan's slogan *"form follows function"* was later altered by Franck Lloyd Wright to *"form and function are one,"* using nature as the best example of this integration (Box 2.3) [9]. Due to the severe effects of the domain over nature phase nowadays, the whole perspective of human need is being shifted toward a different slogan *"function follows needs."*

Our built environment should be adapted to nature, by learning from it instead of learning about it, in order to find a method that contributes in solving problems such as the use of non-renewable energy sources and climatic change impacts. This is mainly an attempt to improve the quality of life for upcoming generations and live sustainably on the earth, hence moving mankind from the industrial age into the biological age (Box 2.4) [9].

Thus, it is time to paradigmatically shift our way of thinking about solving design problems toward fulfilling human needs, diverting the slogan to "function follows needs" and identifying a strategy for teaching biomimicry as a tool for sustainability, along with a recommendation to include it in design curriculums [9].

Box 2.4: Enhance Livability for the New Generations
Shifting mankind from the industrial age into the **biological age**

2.6 Integrating Biomimicry Approach as a Way of Thinking

Teaching biomimicry needs a technique and a long-lasting relationship between people and nature. Through the correlation of ethics and virtue, this relationship could be achieved. Observing living beings and creatures, witness the greatness and evidence of God's creation found anywhere. Hence, linking nature to urban areas and considering it as a role model in creating a sustainable built environment is the main goal to protect and sustain the planet. Without this connection, any sustainable approach including biomimicry will simply supplant ordinary practices and accelerate the expanding degradation of the natural and built environment. In order to achieve this goal, students, professors, and researchers should be inspired and motivated to look deeply into nature in search of insights, emulating and learning from

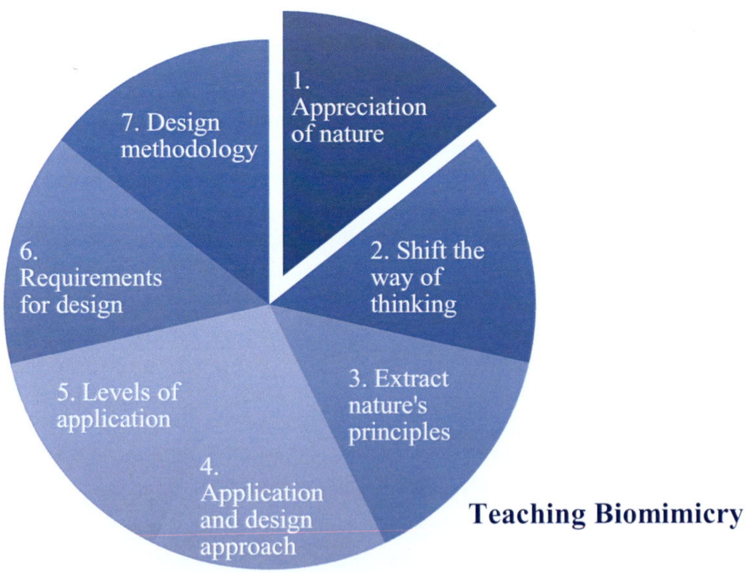

Fig. 2.25 The seven stages of teaching biomimicry. (Image source: Developed by authors after https://www.academia.edu/432839/BIOMIMICRY_A_21ST_CENTURY_DESIGN_STRATEGY_INTEGRATING_WITH_NATURE_IN_A_SUSTAINABLE_WA)

its form, function, dynamic equilibrium, and characteristics of the ecosystem as much as possible.

Learning from nature is a learning process that includes plants and animals' species, supports the whole nature, and holds absolute power and shrewdness. Teaching biomimicry comes through its seven stages [9] as illustrated in Fig. 2.25. The portrayal of natural frameworks includes models of organic frameworks that guide in shifting principles from natural design to synthetic design. According to Wilson et al., such a shift can be expressed as follows [10, 11]:

> Analogical interpretation includes distinguishing proof and systematizing transfer of organic standards to the designing area

In this context, several researchers have developed systems pertaining to different schools of thought and suitability varying from various fields that help the systematic transfer of biology to technology. Most of these approaches are being used mainly in the field of product design and in their own area of specialization, such as mechanical engineering. The difficulty found in accurately identifying relevant biological solutions is the enormous number of answers found in nature. These have been identified as the large analogical distance, lack of cross-domain knowledge, and identification of relevant strategies [10].

Quite a few researchers developed systematic approaches for leveraging biological examples into architectural design. The identified user-friendly analogical translation systems for architectural design have been recognized through reviewing of literature. It is examined and identified; the most appropriate analogical translation systems currently applicable to architecture are Bio-TRIZ and Design Spiral. However, typological analysis explores more architectural design approaches, and nature study analyses investigate more ecologically sustainable design approaches to architecture [11].

Box 2.5: Values of Nature

We still do not know one thousandth of one percent of what nature has revealed to us. (Albert Einstein)

2.7 Biomimicry and Early Conceptual Design

As put by Albert Einstein, nature has great values to inspire from. Box 2.5 highlights Einstein's quote about the enormous values of nature: *"We still do not know one thousandth of one percent of what nature has revealed to us"* [12, 13]. Architecture is a creative activity; there is less consideration of inspiration from nature in most projects in the early stage of design. For this reason, designers should be familiar with the relationship between the conceptual design of any project and nature.

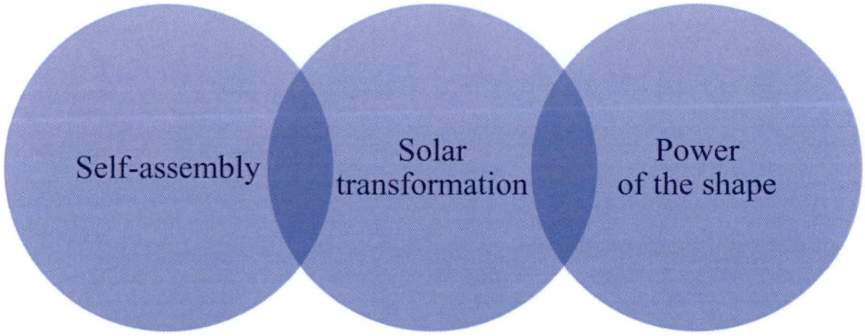

Fig. 2.26 Some of the ideas of biomimicry. (Image source: Developed by authors after J. Benyus)

Based on conceptual design decisions and principles of inspiration from nature, the best project can be considered the one that will have inspiration from different aspects of nature.

In 1998, Janine Benyus established the new science under the name of biomimicry, which means "innovation and inspiration from nature looking to nature as a teacher" [12, 13]. She says:

> *in the 3.8 billion years since, life has learned to do some amazing things to fly circumnavigate the globe, live at the top of mountains and the bottom of the ocean, lasso solar energy, light up the night, and make miracle materials like skin, horns, hair, and brains. In fact, organisms have done everything we humans do or want to do, but without guzzling fossil fuels, polluting the planet, or mortgaging their future* [12, 13].

Janine Benyus also acquired 12 main ideas of biomimicry from nature notably self-assembly, solar transformation, the power of the shape, and so on. Fig. 2.26 summarized some of the main ideas.

An important question! What if we – as architects forming a small part of this society – could learn from nature, to optimize our design from the economical, functional, structural, and material points of view within this context as nature does? Nature solves the problems, somehow, in a context which is similar to the milieu we have – the earth. Many architects believe now that nature can be a powerful and trustful source as Frank Lloyd Wright, who said:

> *Study nature, love nature, stays close to nature. It will never fail you.*

Louis Kahn describes the power of nature by saying:

> *It is my feeling that living things and non-living things are dichotomous, but I feel that if all living plants and creatures were to disappear, the sun would still shine and the rain still falls. We need Nature, but Nature does not need us.*

In the discussion of architectural theory and the curricula of the school of architecture, one finds little concern for inspiration from nature, especially in the early stage of design, and the central focus of biomimicry concept in architecture is the link between conceptual designs and nature because there are many interconnections between conceptual design and nature [12, 13].

Fig. 2.27 Nature inspiration in Ancient Egypt, Greek civilizations. (**a**) Ancient Habu temple in Luxor City, (**b**) Ancient Egyptian columns based on stylized plants, (**c**) Lotus columns inside Habu temple, (**d**) Greek columns. Ancient Egyptian columns based on stylized plants . (Images credit and source: (**a–d**) Part 1, 2, and 4 Mohsen Aboulnaga and (**d**) Part 3: Brent Rostad, https://commoms.m.wikimedia.org/wiki/File:Ionic_columns_(9651323383).jpg)

2.8 Revival of Nature in Architectural Field

Building development, technology and machines are in parallel pace together. Machines are considered the icon of progress and mankind's overriding of nature for the last hundred years. Machines are the extraordinary metaphor for the structures of today, but in contrast, technologies are advancing rapidly in the digitized world in the time of COVID-19. Le Corbusier – one of the twentieth century's pioneer architects – revealed the fact that *"houses were machines for living in."* The built environments are increasingly acting as machines, but are neglecting the integration of nature as well as cultural impacts [14]. Tragically, buildings utilize energy and materials like machines do. Hence, reducing natural resources and minimizing energy consumption is a prime target when blending buildings with nature. The starting point of the issue is the shortsighted conviction that technology joined with a lot of energy was the response to any answers to the design problem. Nature becomes distant in solving the design problems compared with its condition in the past few decades – in the prehistoric era. There was a strong connection between man and the environment, but nevertheless, it was lessened with time, became very weak, and even lost sometimes [14].

As indicated, in the early civilizations of Ancient Egypt, Greek, and Rome, nature was the source of inspiration (Fig. 2.27). Architecture is getting motivation from nature, and it frequently goes further than emulating only the shape, surface, or highlights of nature. In the field of architecture, designers seek to inspire their thoughts and ideas from the surroundings, and with passing time, the feeling of returning to nature in their design ideas increases. A few examples in real life obtained their inspiration from different points of view like formal, structural, and functional [14]. It is imperative to identify the overlying layers between nature and architecture.

2.9 Overlap Between Architecture and Nature

The importance of connecting nature and architecture was investigated through a few trials. The focus is on the morphological aspect where Paolo Portoghesi worked on the combination of both nature and architecture in the encyclopedic reference "Nature and Architecture." This book is a comprehensive aggregate of identification between nature and architecture and examines both structure and function [15].

A book entitled "Bau-Bioni" and published by Werner Nachtigall concentrates on analogies in construction, especially imitating structures that exist in the construction of nature such as reinforced structures, membranes, folded structures, and shells. In the 1970s, a Russian Juri Lebedew published the first book "Architekturbionik," in which he presents a comprehensive collection incorporating the latest developments and trends in architectural developments worldwide and showed a basic extracted from nature [15]. The overlap between architecture and nature can be summarized in three main layers: model layer, measure layer, and mentor layers (Fig. 2.28). More explanations of the 3-Ms' connection can be discussed in the next sections.

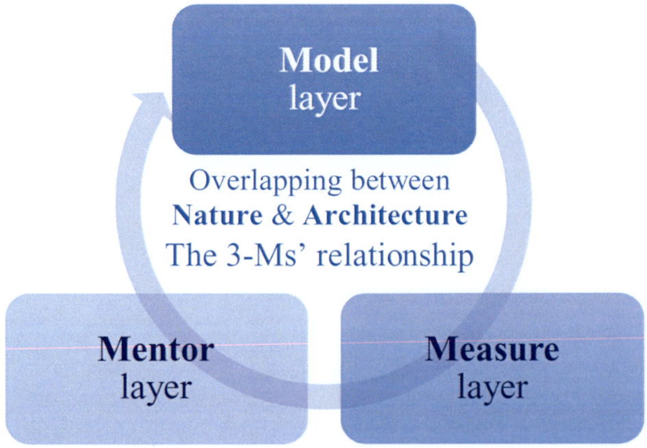

Fig. 2.28 The three layers connecting architecture to nature (Image credit and source: Developed by authors)

2.9.1 *Nature as a Model*

Biomimicry, as an approach, studies the perfect models of nature, getting inspiration from their design and processes' mechanism to solve human problems in a sustainable way. This has been noticed in the way plants and animals behave in the structures they utilize, using the sun and simple compounds to produce totally biodegradable materials used in interiors and architecture such as fibers, ceramics, plastics, solar cells emulating leaves, steel fibers, and forms of natural aesthetic motifs and non-tinctured colors from butterflies – all perfect models for humanity to adopt and exploit [9].

2.9.2 *Nature as a Measure*

In the biomimicry approach, nature's life principles are the ecological standards and the reference when following this approach in developing buildings, urban areas or inspiring new innovations, but the fact that the biological adaptation system is cultivated to provide optimum functionality – it is a prodigy of God's existence to all who have wisdom – which lasts due to 3.8 billion years of evolution [8, 9].

2.9.3 *Nature as a Mentor*

Biomimicry will change the relationship with nature, transforming the vision of considering the environment as a source of raw materials, ideas, and inspiration for issue fixing. The mentor has the foresight and awareness in a sustainable way for the shift in thinking about nature, yet stimulates designers to look beyond the surface and navigate deeply to the unseen, play with the depth of field, bounce with nature, and hence find new perceptions [8, 9, 16].

After understanding the layers that connect architecture with nature, it is vital to learn about the methods and processes of biomimicry design. The next section highlights this important part for architects and designers.

2.10 Biomimicry Design Methods and Processes

The process of transforming nature's methodologies into creative and sustainable design solutions is a step-by-step process. Such a process is called "design spiral" as shown in Fig. 2.29. Biomimicry design spiral drives outward; however, the time needed for achieving revolutionary sustainable innovation is relatively minimized

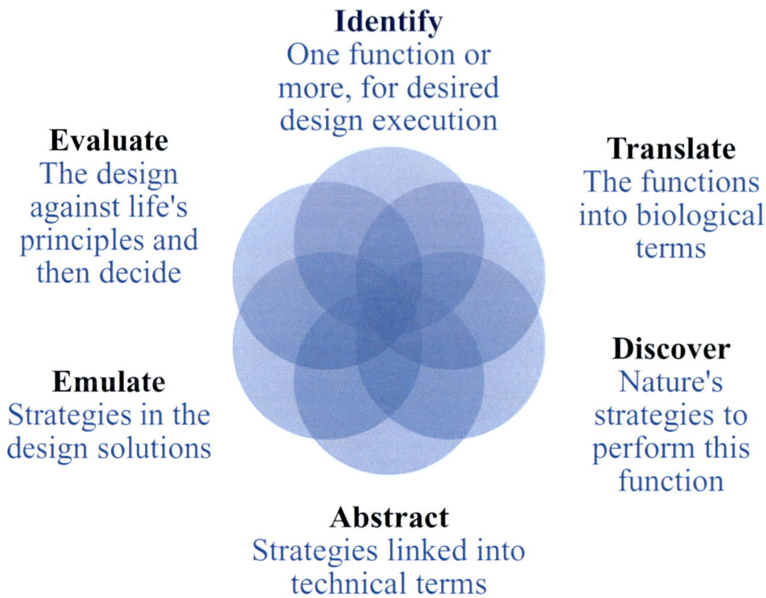

Fig. 2.29 Biomimicry design processes. (Image credit and source: Developed by authors)

by using this approach [17]. Biomimicry design spiral is used when the problem you are trying to solve is known. This encompasses the following six main steps:

(a) **Identify/define** the tasks that the design needs to operate or perform; hence, a list of functions should be created.
(b) **Translate/biologize:** This function into keywords or terms that make sense in the natural world. This is followed by explaining the design brief, asking about the performance of nature towards that: "How does nature behave in such functions or solve this issue?"
(c) **Discover:** The strategies that are used by nature to accomplish these functions and discover natural models.
(d) **Abstract:** It is "reverse engineer" the strategies that have been discovered and describe how they work in terms of the design profession and abstract design principles.
(e) **Emulate:** Utilizing proficient abilities to make a design solution in light of copying at least one of the techniques that have been found and abstracted and imitating nature's systems as profoundly as conceivable in the design, emulating form, function, and ecosystem.
(f) **Evaluate:** Incorporates three measures: assess your design solution against your first original one; examine your design against nature's unifying patterns or "Life's Principles," nature's guidelines for sustainability; and reflect on numerous thoughts and lessons that developed in the above steps and strategize the use of the laps around the spiral [17].

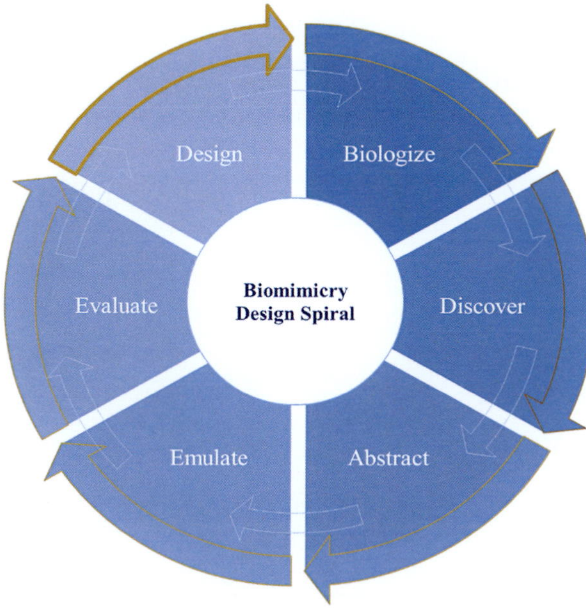

Fig. 2.30 The biomimicry design spiral developed by the Biomimicry Institute. (Image credit and source: Developed by authors after https://toolbox.biomimicry.org/methods/process)

This spiral flow eases the idea of the design method and assists in understanding the processes of biomimicry. Fig. 2.30 presents the steps for biomimicry design spiral. Most of the challenges humans are facing today, such as mechanical and structural problems, have been solved by nature but without generating waste or residuals. Such biological processes are continuously evolving [17]. Biomimetics tries to treat such mistakes by designing effective systems and items. However, this concept has boldly emerged as the circular economy by Pearce and Turner in 1990 [19] and received adoption by Cooper in 1994 and 1999 [18].

Circular economy (EC) is an important concept, especially in developing countries. This concept includes several principles of process, and these processes include making, using, reusing, remaking, and recycling, as shown in Fig. 2.31. A gracious number of studies have focused on endeavoring to minify the utilization of structures' materials and focus on discovering substitutes for these non-sustainable sources from normal or reused content. CE principles are based on the recirculation of building materials that can support the disposal of depleted energy resources [19]. In using materials, circular economy (CE) can be defined as a closed loop system, where products can be recuperated from wastes to be raw materials again. Manufacturing recognizable value-added new green or natural materials and products is the ultimate target of CE [20]. Humankind waste resources in a way are unsustainable over time, which makes it necessary and desirable to implement waste recycling policies that benefit all of humanity. The arrival of the CE is

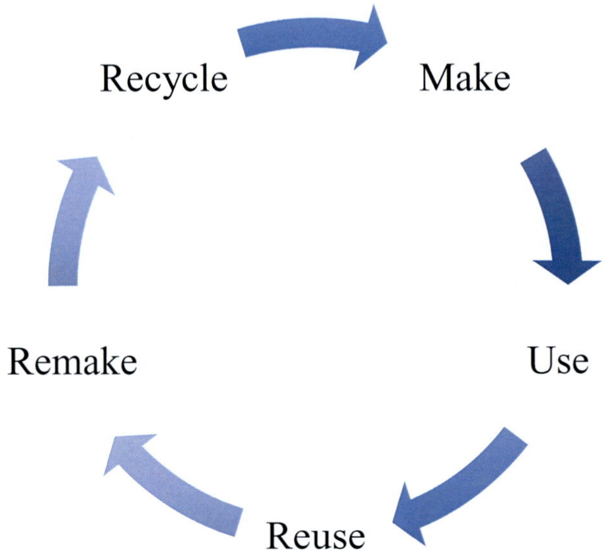

Fig. 2.31 The main principles of circular economy. (Image credit and source: Developed by authors after https://better-business-alliance.org/green/do-consumers-understand-what-circular-economy-means/)

strengthening the environmental awareness of the population as well as more efficient use of scarce resources on the planet [20].

The CE is understood as "an economic system that represents a change of paradigm in the way that human society is interrelated with nature and aims to prevent the depletion of resources, close energy and materials loops, and facilitate sustainable development through its implementation at the micro (enterprises and consumers), meso (economic agents integrated in symbiosis) and macro (city, regions, and governments) levels. Attaining this circular model requires both the cyclical and regenerative environmental innovations, but in the way society legislates, produces, and consumes" [20].

Biomimicry is a continual motivation process, for creating "organs," which are singular items, or "organisms," which are systems and procedures with the end goal of reconciliation into a sustainable system. Moreover, biomimicry can produce precedent photovoltaic, "the organ," inspired by photosynthesis, or a smart house system "the organism." Additionally, solutions will be found to both common and complex problems. Thus, every organism can teach us something that will improve the quality of life or be a solution to the difficulties that exist nowadays [17].

The circular economy concept is partially related to biomimicry aspects. For billion years, nature and ecosystem are in equilibrium because they are following the rules of life by having no waste, harvesting natural energy, adapted to constant change, the flow of the resources is a closed loop, optimized as a full system, regenerative, local resources use, and densely interconnected and symbiotic. In the

Fig. 2.32 The HELIX model of biomimicry. (Image credit and source: Emily Binet Royall, https://soa.utexas.edu/sites/default/disk/analytic_tools/analytic_tools/09_02_su_royall_emily_paper_ml.pdf)

context of CE, the idea of having a closed loop in using materials or going through a process is applied. Furthermore, it is already inspired from the natural system that is created around us, and the key point of biomimicry concept is being sustainable, and this is applied in both biomimicry and circular economy. In this regard, design with nature, as Buckminster Fuller put, the best way to predict the future is to design it (Box 2.6).

Box 2.6: Design with Nature
The best way to predict the future is to design it. (Buckminister Fuller)

2.11 The HELIX Model of Biomimicry

The HELIX model of biomimicry, which is developed by Emily Royall, reflects a number of small differences as illustrated in Fig. 2.32 [21, 22]. The spiral shape of the model indicates the meaning of biomimicry as a constantly developing procedure, boundlessly looking for a closer fit to the evolving environment. Similarly, the continual feedback and chronic fine fitting required to adjust "organs" and "living beings" to the earth [21, 22]. It is noticed, however, that the spiral idea is a critical structural building composition in nature, and it is encoded into the environment. The "organs" and "organisms" make up the two strands of the HELIX, which reverberate their queue uniformity, where organs incorporate particular products [21, 22]. The branch of biomimicry combines the organ and organism strands. These are "sustainability" branches, confirming the cross-reliance of organs and organisms. "Organs" include singular products such as photovoltaic cells or fiber optics, and organisms are systems such as smart grids/cities [21, 22]. By integrating the organs and organisms produced by biomimicry into a continuously evolving system, this relationship with sustainability is also related to a basic biological rule that nature seeks to minimize the amount of energy used in a given period of time. Biomimicry and nature connection could be clarified by taking nature as mentor, measure, and model as shown in Fig. 2.32 [18, 21, 22].

2.12 Biomimicry Relationship with Environment and Sustainability

A specific time and known starting point for biomimetics in architectural design cannot be declared. Nevertheless, getting inspiration from nature in design has existed consistently. Discoveries in natural sciences hardly impact architecture and arts, but this has appeared through numerous cases involving the work of the scientist of biology Ernst Haeckel, who has been cited and clarified from the turn of the twentieth century until today [15].

Mimicking nature to attain sustainability is a vital approach to meet today's challenges, especially in the time of COVID-19 pandemic and its related crises. Architecture that emulates it tends to imitate how the organisms behave rather than just imitating its form. This is achievable with a better and super understanding of the organisms, biology, ecology, and environment, and then the architectural form will change [12, 13]. If the built environment becomes truly regenerative, it will increase human and community well-being as well as that of ecosystems. In recognizing the links between the two, there is growing research identifying people's need and preference to be close to nature or images of nature. This is described by advocates of the biophilia theory and has been translated into built form in various ways [12, 13, 15].

2.13 Biomimitic Design Strategies and Goals

The pervasiveness and convenience of fossil fuels, as the main sources of energy, have enabled extreme inefficiency to develop cities and buildings [23]. Nonetheless, nature educated us numerous vernacular ways of designing and manufacturing buildings which were consequently ignored and largely lost from our aggregate memory. Now that the thoughtlessness of releasing many millennia of stored carbon is becoming progressively obvious, there is a chance to investigate the extraordinary effectiveness of the responses that natural organisms have evolved to better inform cities and building development and design. For all intents and purposes, each issue that we presently confront, whether it is producing energy, finding fresh water or manufacturing materials could have inspired solutions through biomimicry approach [23].

Fig. 2.33 presents the three major challenges addressed through biomimicry thinking. However, there will be various cases in nature that could benefit from studying, while fascination with nature without a doubt backpedals as long as the human presence itself, now we have a chance to return, gain, and learn more [19, 23].

There are advantages of scientific knowledge, better tools, and tasteful sensibilities unconstrained by chronicled historical tenets. Numerous current ways to deal with ecologically sustainable design depend on mitigation and adaptation. In some cases, buildings could be cut out of being static consumers and can become net producers of useful resources [19, 23].

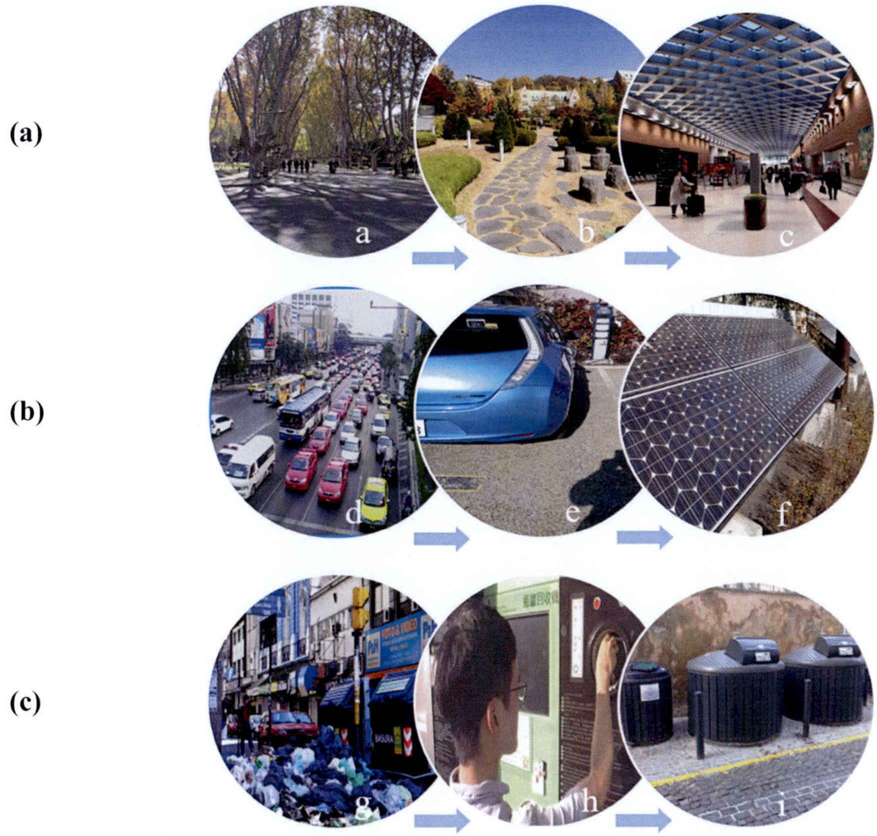

Fig. 2.33 Major challenges addressed through biomimicry thinking. (**a**) Achieve resources efficiency, (**b**) Shift from fossil fuel to solar energy, (**c**) Transform the waste. (Images' credit and source: (**a–f**) Mohsen Aboulnaga (**g**) Ursis, https://commons.m.wikimedia.org/wiki/File:Garbage_in_the_Wind.JPG#, (**h**) Courtesy DW, (**i**) Mohsen Aboulnaga)

2.14 Importance of Biomimicry Architecture in the Time of COVID-19

During COVID-19 pandemic, it's clear that building for virus resilience and flexibility will be a fundamental element of architecture design going forward. As the world is slowly reopening and easing lockdown measures, everyone is adapting to new realities. Imposing drastic adjustments to our lives, the coronavirus has introduced a new "normal," changing our perceptions and altering our priorities. Driven toward questioning and evaluating our environment, we are constantly reacting and anticipating a relatively unknown future [24]. While we cannot foresee the future, we can create logical analogies that are based on the current situation and our first responses. Architects around the world have put their knowledge to use in the fight against the coronavirus and have created innovative solutions that might stick around for a while. Few are designing facilities, while others are rethinking the city.

Some of the key points that are elaborated on how the profession can evolve are highlighted as follows [24]:

- **Interdisciplinary approaches:** One thing that the pandemic has taught us is that architecture cannot save the world, at least not on its own. As the industry resumes, our take on the built environment will become different. In fact, interdisciplinary approaches are already building up, bringing new expertise and views to the practice, to address more efficiently these urgent global issues. The collaborative take on "shared" worldwide challenges will integrate notions of public and personal health, mobility and transportation, environmental psychology, biophilia, and even agriculture, to name a few [24].
- **Future projects:** Emergency architecture and crisis architecture are topics that will start taking center stage as the world changes [24].
- **Rethinking the concept of home:** As we move forward in the different phases of the pandemic, we will focus again on our intimate spaces.
- **New parameters:** Focusing on health-oriented approaches, new standards are going to be set. Design and materials are rethought according to the current situation, generating new forms of living;
- **The adaptable city:** Cities, the epicenter of transmission of diseases, have been hit the hardest by the coronavirus. Re-questioning their systems, they are adapting to new realities, innovating, creating, and experimenting along the way.
- **Public spaces:** While popular public spaces have always been the most engaging and the most crowded, the pandemic has taught us that sharing can also be possible under strict social distancing measures.
- **Density:** For a while now, urban designers and policymakers have been advocating for densification as a sustainable process to expand the city, rather than relying on devastating urban sprawls.
- **Transportation/mobility:** One of the biggest challenges during this pandemic has been transportation. With a high concentration of people, these public networks do not conform to the new social distancing norms. In fact, many cities around the world are already planning for an alternative future, replacing traffic lanes with pedestrian paths.
- **Economy:** Opening the streets to people can also mean opening the realm for their businesses. While the economy is hit hard by the COVID-19 pandemic, cities are mobilizing their creativity in order to prevent further damages and help business owners bounce back.
- **The firms:** As businesses begin opening back up, architecture firms are looking at how to manage their offices while students and recent graduates take their first steps into a new job market. Offices are trying to anticipate workload, especially as specific building types like commercial offices and sectors of the market shrink.
- **Work models:** Across the world, firms implemented remote and digital working methods practically overnight. For those that didn't already have these in place, the move has tested the normative model of where and how work is completed.

Most of the AIA-recommended measures are tactical and programmatic, focusing on subtle alterations of space and different ways of using buildings. Based on CDC guidance with input from architects, public-health officials, and facility managers, the measures are a reminder that architecture is as much about how spaces are used as what is built [25].

The question is, does the biomimicry approach in architecture have an impact on adapting to COVID-19 pandemic? The expected answer is YES, as any new adaptation could be inspired by looking around nature for inspiring us to solve the challenges that face the humankind. This will be presented profoundly in chapter seven.

2.15 Conclusion

A sustainable building is characterized as a building that seeks integration of qualities (economic, social, cultural, and environmental as well as naturally and executed) in an exceptionally expansive manner. Along these lines, the sound utilization of common assets and proper administration of buildings' stock is required, with a specific end goal to achieve saving the resources from scarcity, diminishing energy consumption, and improving environmental quality and attain sustainability and the sustainable development goals (SDGs).

As sustainable design becomes widely accepted and essentially needed, a new compose of sustainability such as biomimicry was advanced to achieve a sustainable future city. In this context, environmental designers and architects are major players in this approach while they should ensure the identification of biomimicry as one of the most important principles of sustainability. Architects acquired inspiration from biology in the early nineteenth century, and biomimicry has been applied to building designs in several countries around the world. Hence, this should be the path to achieving sustainable building, especially in the age of COVID-19. Having clean and fresh air with good and natural cross-ventilation in buildings' spaces is essential and required new measures.

A great understanding of biomimicry including its origins and background has been illustrated to foster better buildings and cities. Also, the overlap between architecture and nature has been discussed. The biomimicry design spiral for knowing in depth the meaning of biomimicry will be the new era of design and it is time to revive nature amidst COVID-19. Moreover, the main strategies and goals of biomimicry have been addressed by achieving better use of resources in an efficient way and with less carbon emissions through solar energy, less waste and recycling.

In the next chapter, this information will be linked with the influence of biomimicry on architectural design, highlighting the approaches and levels of biomimicry, and linking it with sustainability plus learning about the future of biomimicry.

References

1. Snijders, A., & Pronk, L. (2016). *Urban biology.* www2.projects.science.uu.nl. Retrieved November 15, 2020, from https://www2.projects.science.uu.nl/urbanbiology/articlepage-biomim.html
2. Pawlyn, M. (2016). *Biomimicry in architecture* [ebook] (2nd ed., pp. 1–5). RIBA. Retrieved May 10, 2020, from https://www.routledge.com/Biomimicry-in-Architecture/Pawlyn/p/book/9781859466285
3. El Ahmar, S. (2011). *biomimicry as a tool for sustainable architectural design towards morphogenetic architecture (master of science).* University of Alexandria.
4. Rust, W. (2011). *Blueprints from Nature Investigating a biomimitic approach to architecture. (Master of Architecture).* University of Cape Town.
5. Ancient Egyptian architecture | Types, History, & Facts. (n.d.). Encyclopedia Britannica. https://www.britannica.com/art/Egyptian-architecture
6. Yedekei, G. (2014). *Biomimitic architecture anew interdisciplinary approach to architecture* (pp. 29–35). University of Putra Malaysia. Retrieved October 2, 2020, from https://frsb.upm.edu.my/upload/dokumen/FKRSE1_157-542-1-PB.pdf
7. ElDin, N., Abdou, A., & ElGawad, I. (2016). Biomimetic potentials for building envelope adaptation in Egypt. *Procedia Environmental Sciences, 34*, 375–386. https://doi.org/10.1016/j.proenv.2016.04.033
8. Shahda, M., Elmokadem, A., & Abd Elhafeez, M. (2014). Biomimicry levels as an approach to the architectural sustainability. *Port-Said Engineering Research Journal, 18*(2), 117–125. https://doi.org/10.21608/pserj.2014.45298
9. Mansour, H. (2010). *Biomimicry at 21st century design strategy integrating with nature as a sustainable way*. British University in Egypt (BUE). Retrieved December 6, 2020, from https://www.academia.edu/432839/BIOMIMICRY_A_21ST_CENTURY_DESIGN_STRATEGY_INTEGRATING_WITH_NATURE_IN_A_SUSTAINABLE_WAY
10. Wilson, J. (2008). *A systematic approach to bio-inspired conceptual design.* (Ph.D.) Georgia Institute of Technology.
11. Gamage, A., & Hyde, R. (2011). Can Biomimicry, as an approach, enhance Ecologically Sustainable Design (ESD)? In *45th Annual Conference of the Architectural Science Association.* ANZAScA. University of Sydney. Retrieved October 29, 2021, from https://www.anzacan.net/paper/can-biomimicry-as-an-approach-enhance-ecologically-sustainable-design-esd
12. Oguntona, O. A., & Aigbavboa, C. O. (2017). Biomimicry principles as evaluation criteria of sustainability in the construction industry. In *Energy Procedia* (Vol. 142, pp. 2491–2497). Elsevier. https://doi.org/10.1016/j.egypro.2017.12.188
13. Oguntona, O., & Aigbavboa, C. (2017). Biomimicry principles as evaluation criteria of sustainability in the construction industry. In *9th International Conference on Applied Energy*, ICAE 2017. Retrieved October 27, 2021, from https://www.sciencedirect.com/science/article/pii/S1876610217359222
14. Vahedi, A. (2009). *Nature as a source of inspiration of architectural conceptual design.* (Master of Science), University of Eastern Mediterranean University.
15. Gruber, P. (2011). *Biomimetics in architecture.* Springer. Retrieved December 8, 2020, from https://link.springer.com/book/10.1007%2F978-3-7091-0332-6
16. De Pauw, I., Kandachar, P., Karana, E., Peck, D., & Wever, R. (2010). Nature inspired design: strategies towards sustainability. In *Knowledge collaboration & learning for sustainable innovation ERSCP-EMSU conference.* Delft University of Technology. Retrieved December 8, 2020, from https://www.researchgate.net/publication/254906399_Nature_inspired_design_Strategies_towards_sustainability
17. The Biomimicry Institute. (2016, June 14). *The power of the biomimicry design spiral.* Biomimicry Institute. Retrieved December 8, 2020, from https://biomimicry.org/biomimicry-design-spiral/#.V-Khra1S1-4

18. Pearce, W. Turner, R. (1990). Economics of natural resources and the environment. Retrieved December 8, 2020, from https://www.researchgate.net/publication/31662420_Economics_of_natural_resources_and_the_environment_DW_Pearce_RK_Turner
19. Eberhardt, C., Birgisdottir, H., & Birkved, M. (2019). Potential of circular economy in sustainable buildings. In *IOP Conference Series: Materials Science and Engineering 471* (pp. 1–8). IOP Publishing. Retrieved December 8, 2020, from https://doi.org/10.1088/1757-899X/471/9/092051
20. European Materials Research Society. (2018). Materials innovation for the global circular economy and sustainable society. VI World Materials Summit, Strasbourg, 20th & 21st November, 2017. Retrieved December 8, 2020, from https://www.esf.org/fileadmin/user_upload/esf/2018_WMS_online.pdf
21. Royall, E. (2010). *Defining biomimicry: Architecturals applications in systems and products*. The University Of Texas. Retrieved December 8, 2020, from https://soa.utexas.edu/sites/default/disk/analytic_tools/analytic_tools/09_02_su_royall_emily_paper_ml.pdf
22. Pedersen-Zari, M. (2007). *Biomimetic approaches to architectural design for increased sustainability*. School of Architecture, Victoria University. Retrieved December 8, 2020, from https://www.academia.edu/9509268/Biomimetic_approaches_to_architectural_design_for_increased_sustainability
23. Leordean, L. (2016). *Biomimicry in architecture – Mitigation and adaptation to climate change*. Politecnico di Milano. Retrieved January20, 2020, from https://issuu.com/ioanaleordean/docs/thesis_book_online-compressed
24. Architecture post COVID-19: the Profession, the Firms, and the Individuals | ArchDaily. Retrieved January 28, 2021., from https://www.archdaily.com/939534/architecture-post-covid-19-the-profession-the-firms-and-the-individuals
25. What Will Architecture Design Look Like After COVID-19? (2020). Redshift EN. https://redshift.autodesk.com/architecture-design-covid-19

Chapter 3
Biomimicry Influences Architecture and Design: Thinking, Approaches, Levels, Application Types, and Inspiration

3.1 Introduction

In a world that suffers from scarce natural resources, there is a need to manage these resources effectively and efficiently. Cities including buildings are responsible for 60–80% of all the global energy use and accountable for 75% of the carbon emissions [1–3]. Buildings account for a significant part (42%) of the world's energy consumption [4–6]. According to International Energy Agency (IEA), the buildings and building construction sectors combined are responsible for over one-third of the global final energy consumption and nearly 40% of the total direct and indirect CO_2 emissions. These energy consumption rates tend to rise rapidly due to vast urban population growth [7, 8]. In developing nations, energy use in buildings accounts for more than 50% [8]. With the increasing environmental awareness and the need to reduce energy demands and mitigate carbon emissions, developing more sustainable and resilient solutions for buildings and cities is vital to attain the sustainable development goals (SDGS), mainly SDG 11 – Sustainable Cities and Communities. Resilience is typically associated with the capacity of an element to recover from a change or to withstand a shock and manage it effectively and/or respond appropriately to variant conditions [10].

In nature, this problem is tackled by applying strategies of adaptation – the process in which an organism becomes better suited to its environment that is fundamental for efficiency and survival over the short and long terms [11]. Designing building envelopes that have the capacity to adapt to their environments will enhance not only its resilience but also its sustainability, by requiring less energy to operate and employing resources more efficiently as illustrated in the distinctive curved shape of the building "Can of Ham" at 70 St. Axe, in Central London, England, UK (Figs. 3.1, 3.2, 3.3) [12, 13]. Figure 3.4 also shows the iconic and legendary envelope of the "Gherkin" building, at 30 St. Axe in Central London, UK – designed by Sir Norman Foster, where it is inspired from nature particularly – the stem shape of

M. Aboulnaga, S. E. Helmy, *Biomimetic Architecture and Its Role in Developing Sustainable, Regenerative, and Livable Cities*,
https://doi.org/10.1007/978-3-031-08292-4_3

Fig. 3.1 The distinctive curved shape of the building "Can of Ham" at 70 St. Axe, in Central London, England, UK. (Image credit and source: The Frog001, https://commons.m.wikimedia.org/wiki/File:70_St_Mary_Axe.jpg)

cylindrical shape of columnar cacti or cactus, generating clean energy and operating resources efficiently [14].

To this end, environmental adaptation strategies from nature are studied, in terms of air, water, heat, and light. These environmental aspects were chosen among others due to their considerable impacts on occupants' needs and comfort and their immediate impacts on global energy use.

3.2 The Cause of Biomimicry Thinking

Learning from nature has a strong basis for new solutions, so it contributes to attaining sustainable development. Biomimicry is a developing worldview that can assist in driving designers into their new path to sustainability. Biomimicry is also a new set of inquiries that can be applied to the design processes and in addition outcome (Box 3.1). Biological designs are described as strong, adaptable, multifunctional, regenerative, durable, and for the most part zero-waste [15] (Fig. 3.5). At the point

Fig. 3.2 The unique flowing curve engulfs the 70 St. Axe building's edge in London, England, The UK. (Image credit and source: Bobulous, https://commons.m.wikimedia.org/wiki/File:70_St_Mary_Axe_seen_from_Bevis_Marks.jpg#)

Fig. 3.3 View of the biomimetic architecture of 70 St. Axe building in London, England, UK. (Image credit and source: Bobulous, https://commons.m.wikimedia.org/wiki/File:70_St_Mary_Axe_seen_from_Bevis_Marks_and_Heneage_Lane.jpg#)

Fig. 3.4 The iconic "Gherkin" building at 30 St. Axe in London, England, UK. (**a**) View behind St. Andrew church, (**b**) Main view of the "Gherkin", (**c**) Main entrance of the Gherkin. (Images' credit and source: (**a**) TxIIxt TxIIxt, https://commons.m.wikimedia.org/wiki/File:London_-_St._Andrew_Undershaft _Church_-_The_Gherkin.jpg#mw-jump-to-license; (**b**) Jey Han, https://commons.m.wikimedia.org/wiki/File:Gherkin_(30_St_Mary_Axe).jpg; (**c**) FASTILY, https://commons.m.wikimedia.org/wiki/Category:30_St_Mary_Axe#/media/File:30_St_Mary_Axe_142012_07_03.jpg)

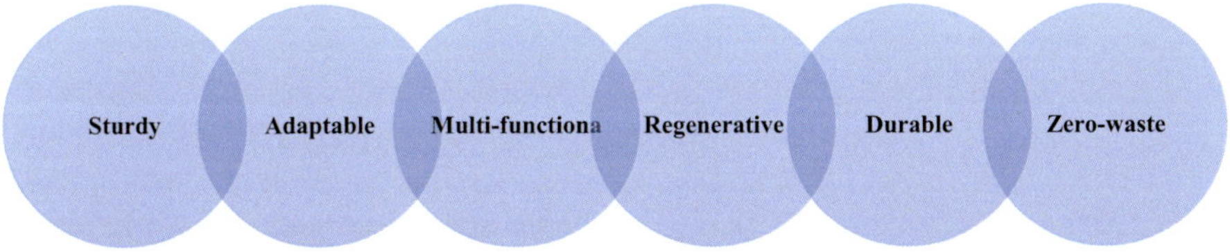

Fig. 3.5 Description of biological designs' features. (Image source: Developed by authors)

when profoundly informed by biology, design thinking shifts from a human-centric model and considers product life cycle and earth system confinements.

Box 3.1: Biomimicry

Biomimicry is a developing worldview that can assist in driving designers into their new path to sustainability interventionists.

Biomimicry is also a new set of inquiries that can be applied to the design processes and in addition outcome. (Blok, V., and Gremmen, B. – Ecological Innovation, 2016)

Table 3.1 compares human-made systems and biological systems to find out how nature responds to environmental conditions and changes. The biological systems respond by default in a sustainable way, based on transformative survival and natural optimization processes [15].

Table 3.1 Human-made systems and biological systems

Human systems	Biological systems
Simple	Complex
Flow of the resources is linear	Flow of the resources is closed loop
Disconnected and monofunctional	Densely interconnected and symbiotic
Resistant to change	Adapted to constant change
There is waste	There is no waste
Long-term toxins frequently used	No long-term toxins frequently used
Often centralized and monocultural	Distributed and diverse
Depend on fossil fuel	Depend on solar income
Engineered for one goal maximization	Optimized as a full system
Extractive	Regenerative
Global resources use	Local resources use

Source: Developed by authors

Looking for alternatives for the ecosystem destructing technologies of the industrial age is progressively recognized, due to the increment of the pressure and consumption on resources and the environment. Thus, the interest in recyclable and biodegradable materials is expanding. With regard to the European Union (EU), for example, it is contended that "In a world with growing pressures on resources and the environment, the EU has no choice but to go for the transition to a resource efficient and ultimately regenerative circular economy." Throughout the years, biomimetic ways to deal with innovation and technology have progressively gotten consideration as an alternative for the ecosystem demolition. Biomimicry or biomimetics is "science that reviews nature's models and after that emulates or takes inspiration from these designs and procedures to solve human problems" [15]. It is the time to think how the design decisions can be environmentally friendly and make sense in the overall context of the natural world, rather than ask how to meet today the environmental standards following the laws of nature and life's principles. Understanding these regenerative characteristics enables us to perceive that all the materials we use as creators and designers even high technological and manufactured materials can likewise be viewed as supplements or nutrients.

The laws of nature are the bedrock of good design; they also illuminate a cohesive and durable set of science-based design practice, which is called "Cradle to Cradle design" [16, 17]. As per the opposes of biomimicry, nonetheless, the idea is wide that one may admire and ask whether it is more than a motto fashioned by scientific experts so as to play a role in the field of sustainability. They suspect that biomimicry is real revolution and admire whether it can accomplish its desire under which conditions. If the current discussions and researches about this approach and the potentials of this field as a new way of thinking and acting ecologically indicate a certain inspiration [18], thus the design in nature is different! It is the synthesis of a whole, not a collection of parts. People should care about nature and respect it too, by obeying not only its functions but also its patterns and forms. This way of thinking may lead to sustainable and environmentally friendly buildings, where sustainability goes hand in hand with a respect for nature. In clarifying this concept, Salingaros says [19]:

> Part of humans' conscious system looks for information, whereas another part looks for meaning. By imposing an artificial meaning on the built environment, contemporary architects contradict physical and natural processes, and thus create buildings and cities that are inhuman.

Due to the anxiety about climate change (CC) impacts, the world is currently on an urgent roadmap for sustainable practices and hence has attracted the designers and architects to bestow more attention to biomimicry for the efficient solution to the CC problems. The practice of biomimicry has resulted in numerous penetrations in fields of transportation, materials science, and medicine. However, the extension of biomimicry principles into the field of architecture remains in its infancy. Currently, the integration of biomimicry principles in the field of architecture appears to often be just aesthetic or the result of a random, unrepeatable process [19].

3.3 Biomimicry in Architecture

In the field of architecture, numerous cases learned from nature and impacted it. Constructions like trees' branches, blossoms' analogies, network configurations, etc. inspired architectural design thinking since the antiquated times. Inspiration is done in two ways: (a) replicating the form and (b) transferring the procedure of development of living structure as material, shape, and structure to design thinking (Fig. 3.6). In form finding, it does not refer to a functional or environmental approach, and it is just imitating a shape from the surrounding environment and nature. However, the second way is different; it is observing and comprehending the functionality in nature. Thus, it is important to notice that each organism's functionality is to make a home, survive in its environment, and bear its condition while performing this behavior in harmony and with no damage to its surrounding environments [20].

The reduction of the impact of humans on the earth turns out to be more critical, in this manner, the need for sustainable building. A sustainable building is characterized as building practices, which strive for integral quality (economic, social, and environmental execution) in an extremely expansive manner. Using natural resources and suitably managing a building stock will save scarce resources, diminishing energy consumption and enhancing quality of the environment [20]. The designers and clients now realize the terrific impact buildings have on the natural environment. As sustainable design becomes widely accepted, a new form of sustainability such as "biomimicry" was advanced to achieve a sustainable future. In this context, biomimicry – as a sustainable design tool – caught the prime attention of environmental designers, where the architect is a sole player, after it was identified as one of the most important principles of sustainability. Architects acquired inspiration from biology in the early nineteenth century [20]. Architects not only explored

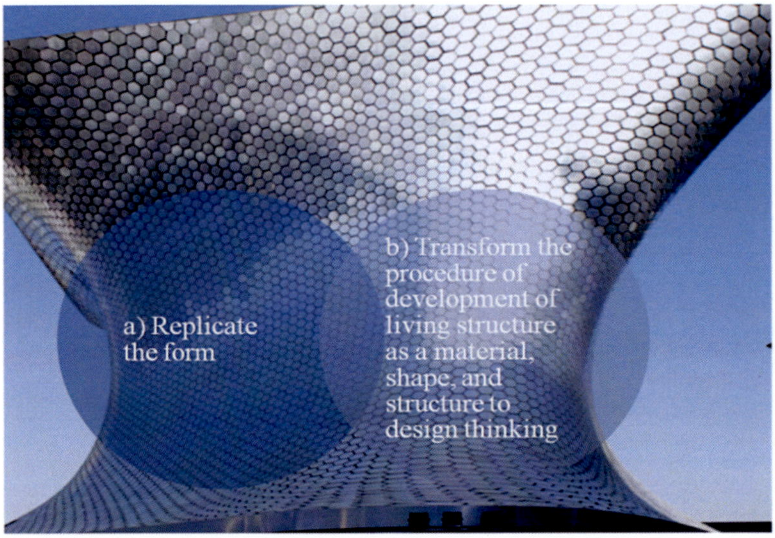

Fig. 3.6 Ways of biomimicry inspirations. (Source: Developed by authors after https://www.re-thinkingthefuture.com/architects-lounge/a351-20-incredible-building-facades-that-will-make-you-rethink-building-envelope/; Image credit and source: Raystormxc, https://commons.m.wikimedia.org/wiki/File:Museo_Soumaya_Plaza.jpg)

imitating the forms of nature but also investigated to find out methods in design similar to the process of growth and evolution in nature.

Biomimicry in architecture has been applied to many building designs in different parts of the world. These biomimicry examples are spread all over the world, including the United States, Europe (Germany, United Kingdom, and Poland), Asia (China, Singapore, and Qatar), and Australasia (Australia and New Zealand) as well as Africa (Zimbabwe and Namibia); among these buildings, which will be elaborately presented and discussed in Chap. 4, are:

1. Building in Harare, Zimbabwe
2. CH2 building in Melbourne, Australia
3. Hydrological Centre for University of Namibia
4. Munich Olympic stadium in Munich – Bavaria, Germany
5. The Ultima Tower in San Francisco – CA, United States
6. Dog building in Tirau, New Zealand
7. The Gherkins Tower in London, United Kingdom
8. Cell-shaped in Chengdu – Sichuan province, China
9. Rain Collector Skyscraper in Nowak, Poland
10. Tardigrada Botanica in Budapest, Poland
11. Beijing Olympic Stadium in Beijing, China
12. The Eden Project in Cornwall – England, UK
13. Habitat 2020 in Beijing, China
14. Qatar's giant Cactus building, Doha, Qatar

15. Esplanade Theatre in Singapore, Singapore
16. Sagrada Família church in Barcelona, Spain
17. The Lotus Temple in Delhi, India
18. Recent innovative and biomimetic buildings (2018–2020)

3.4 Approaches of Biomimicry

Approaches to biomimicry are typically characterized by two classifications defining people's needs or the design problem and how it is solved in the ecosystem. There are two approaches that govern biomimicry: Approach 1, design looking to biology (problem-based approach), and Approach 2, biology to design, (design solution-based approach) [21].

3.4.1 Design Looking to Biology (Problem-Based Approach)

Throughout the literature review, the design looking to biology – the problem-based approach that is known by several titles – all referring to the same meaning includes "design looking to biology," "direct," "top-down," and "problem-driven biologically inspired design" as shown in Fig. 3.7. This approach takes the problem of technology as a starting point; it requires designers and biologists to identify exactly their design problems and then link and relate these to organisms that solved same as this problem. This approach also starts by tackling the initial goals by the designers and identifying the parameters for the design [22].

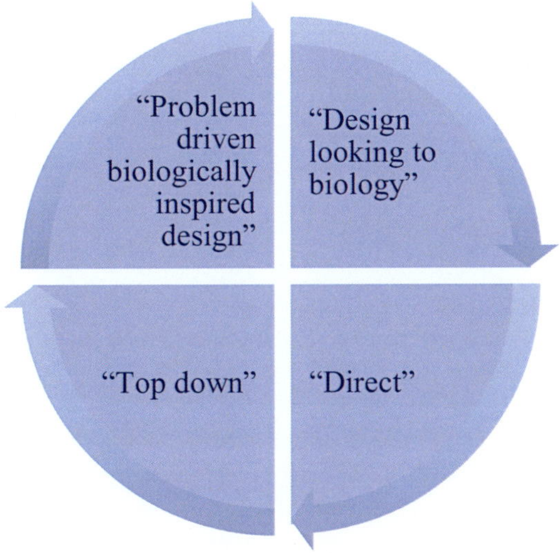

Fig. 3.7 Different titles of biomimicry's problem-based approach *"design looking to biology"*. (Image source: Developed by authors)

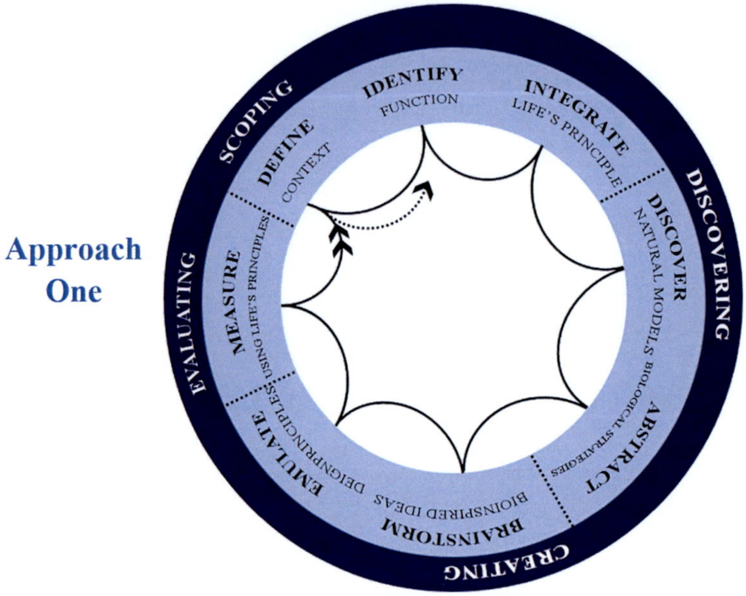

Fig. 3.8 Steps of design looking to biology design – a biomimicry approach. (Image source: Developed by authors after https://www.aast.edu/pheed/staffadminview/pdf_retrieve.php?url=63415_46_1_Biomimicry%20as%20a%20Design%20Approach%20for%20Aadaptation%20Paper.pdf&stafftype=staffpdf)

Designers in this case will be able to reach and apply biomimicry when they understand nature profoundly and rigorously. However, the translation of biological knowledge to a design setting will remain at a shallow level with limited scientific understanding. It is said that the problem-based approach starts by transforming the built environment from an unsustainable configuration to a different effective paradigm. The steps of design looking to the biology approach are presented in Fig. 3.8. These encompass four main steps, which are (a) scoping, (b) discovering, (c) creating, and (d) evaluating. For each one of these steps, there are measurements [23–25].

3.4.2 Biology to Design (Design Solution-Based Approach)

Through the literature review, this approach is known with different naming such as "biology influencing design," "bottom-up approach," "solution-driven biologically inspired design," "biology to design," and "indirect approach" as shown in Fig. 3.9 [23–25]. They all refer to the same meaning. Consider the phenomena found in nature as a starting point, this approach depends on previous knowledge of a biological research and discovered solutions and not on a search for solutions in nature and then apply this knowledge on the existent design problem. Distinguishing the specific attributes or behaviors in an organism or environment and afterward

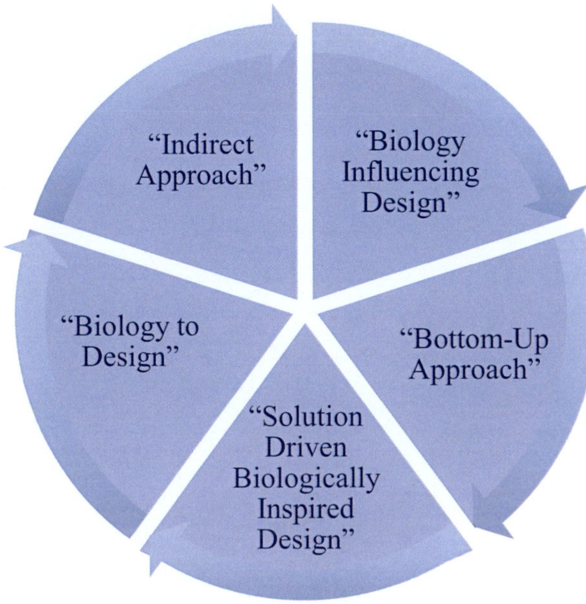

Fig. 3.9 Different titles of biomimicry's problem-based approach "*design looking to biology*". (Image source: Developed by authors)

utilizing them as guidelines for developing architectural or industrial designs could be an inspiring process for biomimicry architecture.

At the point when biological knowledge influences human design and architecture, the cooperative design process is subject to individuals who know about ecological and biological research instead of human design problems. There is an advantage of this approach that biology may impact people in a way that is outside a predetermined design problem. The potential for a genuine move is in the methodology of how the human design focuses to find a solution to an existing problem from a design point of view; nevertheless, the disadvantage of this approach is in fact that biological research should be managed as relevant to a design context [23–26].

The steps of biology to design approach, illustrating both of the two approaches together, are shown in Figs. 3.10 and 3.11.

3.5 Levels of Biomimicry

It is imperative to look for the levels of biomimicry based on several books, papers, theses, and articles. The levels of biomimicry can be identified into three levels: (a) organism level, (b) behavior level, and (c) ecosystem level (Fig. 3.12). Inside each level, a further five conceivable measurements to the mimicry exist as shown in

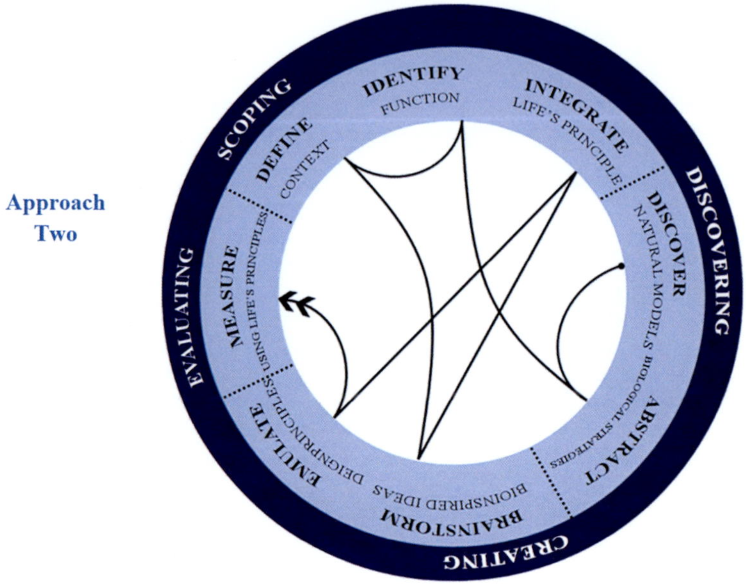

Fig. 3.10 Steps of biology to design approach. (Image source: Developed by authors after http://www.sciencedirect.com/science/article/pii/S187802961630055X)

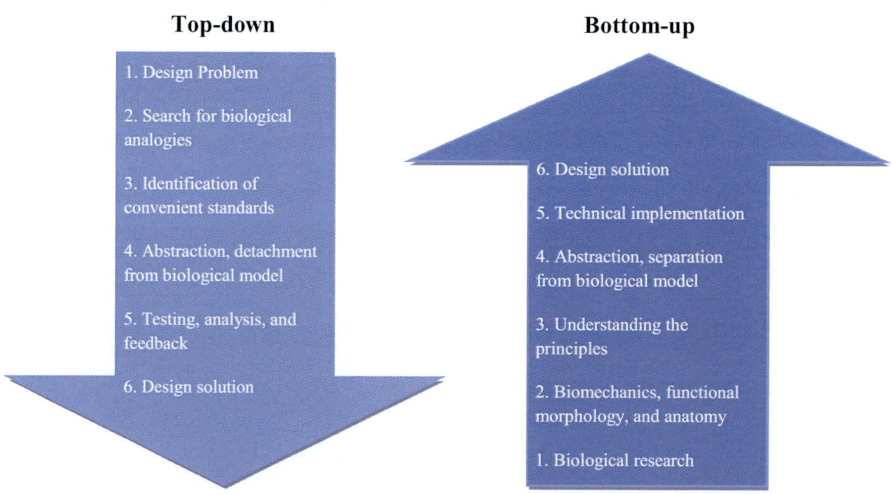

Fig. 3.11 Biomimicry top-down and bottom-up approaches. (Image source: Developed by authors after http://www.sciencedirect.com/science/article/pii/S1110016815001702; and https://doi.org/10.1016/j.aej.10.015)

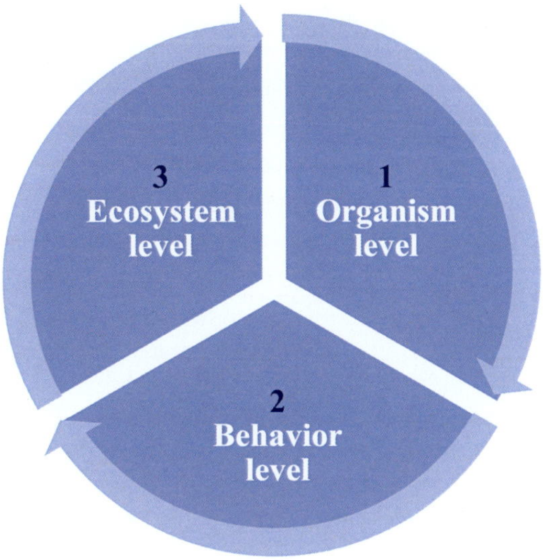

Fig. 3.12 Three levels of biomimicry. (Image source: Developed by authors)

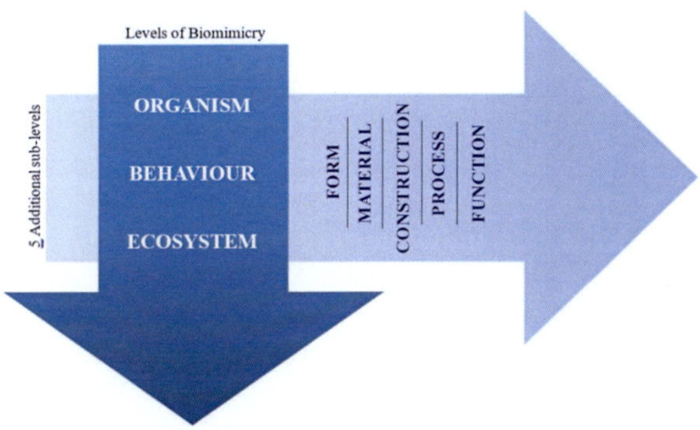

Fig. 3.13 Biomimicry levels and five additional sub-levels. (Source: Developed by authors after http://www.academia.edu/1739669/BIOMIMICRY_AS_A_TOOL_FOR_SUSTAINABLE_ARCHITECTURAL_DESIGN_towards_morphogenetic_architecture)

Fig. 3.13. The design might be biomimetic, for instance, regarding what it looks like (form or material or construction) and/or how it works (process) or what it can do (function) [27].

3.5.1 Organism Level

This level means mimicking a particular organism, as a plant or creature, which may include imitating some portions of or the entire organism's form, material, function, and process. In this level, it inspires the innovation that exists in the organism and then emulates it in order to create an imaginative, simple, man-made sustainable solution for what was considered before a problem. The design may come with a new, fancy-looking building, yet the environmental impact is not important at this level [27].

3.5.2 Behavior Level

The behavior level refers to mimicking conduct or act (behavior) that the organism does to survive or reproduce regularly and observe how it functions. It might incorporate translating a part of how an organism behaves or identifies with a larger context. It is imaginable to comprehend behavior level with watching the way an organism operates in the surrounding conditions and the natural limits of energy and material availability. Behavior level mimicry cannot be reasonable for all situations. Time and context ought to be taken into consideration [27].

3.5.3 Ecosystem Level

The ecosystem level is creating and mimicking an ecosystem (bio-network) and mutual principles that allow them to successfully function. This level integrates the other two levels (organism and behavior) to accomplish a sustainable environment. It is, however, a profound appreciation of the environment and nature's regenerative procedures and processes. This level may start from a little scale and tends to prompt a greater size of thinking such as green urban communities or eco-cities [27].

The imitating of biological communities is a fundamental part of biomimicry as depicted by Benyus (Box 2.2, Chap. 2). The term eco-mimicry has likewise been utilized to describe the mimicking of ecosystem in design. Nonetheless, it is based on the deconstructing of the design problem into the performance of characteristics, functions, processes, and past solutions and then mimicking the common principles of the genius of nature that are relevant to the design problem. This would allow it to function in a successful sustainable way as a part of a complex system and utilize the relationships between different processes in nature, thereby reaching a total emulation of the whole ecosystem.

A structure for understanding the use of biomimicry is suggested to redefine these different levels, as shown in Table 3.2. Furthermore, illuminating the potential of biomimicry as an incremental tool for regenerative capacity of the built environment. Within the three levels of biomimicry, there are five sub-levels which are form, material, construction, process, and function as illustrated in Fig. 3.14 [28].

Additionally, the structure depicted here is applicable to both biomimicry approaches: design looking to biology and biology impacting design. The advantages of this level are that it can be used in integration with the organism and

Table 3.2 Application of levels and sub-levels of biomimicry

Level of biomimicry	Sub-levels	Example – a building that mimics termites
Organism level (mimicry of a specific organism)	Form	The building resembles a termite.
	Material	The building is produced using an indistinguishable material from a termite; a material that mimics termite exoskeleton.
	Construction	The building is made similarly as a termite; it experiences different development cycles for instance.
	Process	The building works similarly as an individual termite; it produces hydrogen proficiently through meta-genomics.
	Function	The building capacities are like a termite in a bigger setting; it reuses cellulose squander and makes soil.
Behavior level (mimicry of how an organism behaves or relates to its larger context)	Form	The building seems as if was made by a termite; a reproduction of a termite hill.
	Material	The building is produced using similar materials that a termite is built with; utilizing processed fine soil as the primary material.
	Construction	The building is made similarly that a termite would build and work in; heaping earth in specific spots in specific circumstances and times.
	Process	The building works similarly as a termite mound would; by careful orientation, shape, materials choice, and regular ventilation, or by mimicking how termites cooperate.
	Function	The building capacities would be similar if made by termites; interior conditions are controlled to be ideal and thermally steady.
Ecosystem level (mimicry of an ecosystem)	Form	The building resembles an ecosystem (a termite would live in).
	Material	The building is produced using a similar kind of materials that (a termite) biological community is made of; it utilizes normally occurring common compounds, and water as the essential chemical medium.
	Construction	The building is amassed similarly as a (termite) ecosystem; standards of succession and expanding complexity over time are utilized.
	Process	The building works similarly as a termite ecosystem (it catches and changes energy from the sun, and stores water).
	Function	The building can work similarly that a (termite) ecosystem forms a part of a complex system, by using the connections between processes; it can take an interest in the hydrological, carbon, nitrogen cycles, and so in a similar way to an ecosystem.

Source: Pedersen Zari (2007)

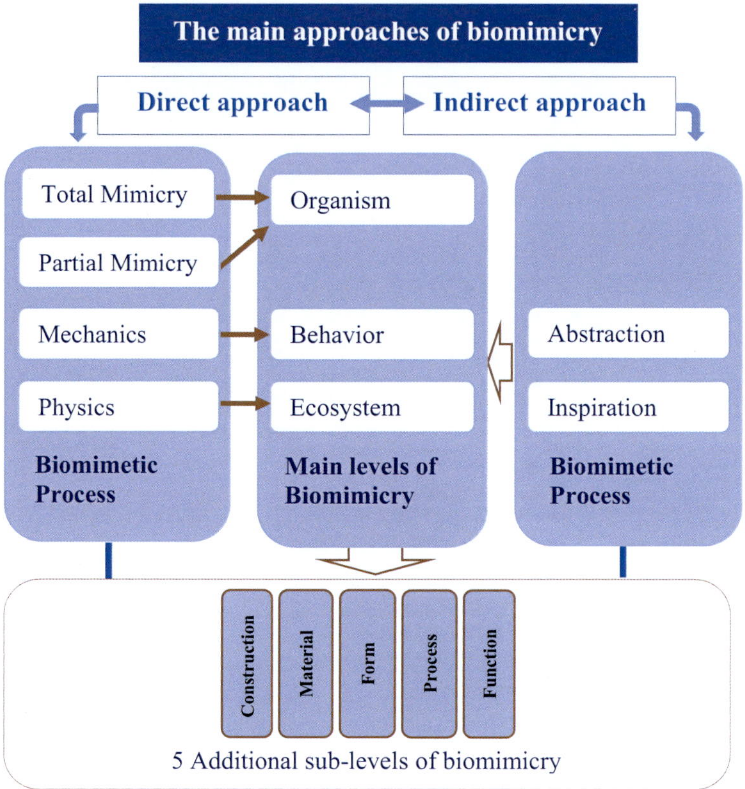

Fig. 3.14 Design approaches of biomimicry with different levels and dimensions. (Source: Developed by authors after http://www.researchgate.net/publications/30603521_Biomimicry_for_ecologically_sustainable_design_in_architecture_a_proposed_methodological_study)

behavior levels; hence, this level is applicable to a domain of temporal scales and serves as an initial benchmark or goal for what constitutes truly a sustainable or even regenerative design for a specific place [28].

3.5.3.1 Ecosystem and Sustainability

The ecosystem and sustainability are the highest level of biomimicry that imitate the ecosystem in a sustainable way. This level is the most difficult one based on the fact that it requires talented systems thinking to ensure the designs fit consistently inside the biosphere. The interface of the world has been influenced by design and architecture, so that balancing the profound innovation and inspiration through biomimicry – with a focal point of ecological and social investigation – is necessary. It needs exertion with respect to specifically transferring eligible aspects of the models in nature to the final design to utilizing positive ends [29].

3.5.3.2 Mimicking the Process Strategies of Ecosystem

Designers and planners have a fruitful model to follow in formulating how the systems in buildings or urban environment ought to be assembled, yet how they should function, by mimicking process strategies in the living organisms and creatures. Formulating a collection of ecosystem strategies deduces from a comparative analysis of ecology, biology, biological plan, industrial ecology, and biomimicry [30].

For understanding the cross-disciplinary of ecosystem work mechanism, many aspects of ecosystems are convenient for the designers, not only for investigation and creation of regenerative built environments but also contribute through ecosystem under climate change. These are achieved by the dependence on local materials, the diversity in organism's relationships, and the complex information feedback loops as well as self-correction.

Several researches in the field of aquaculture systems demonstrate – in several cases – the creation of resilient and effective systems. Such as mimicking the ecosystems in the aspect of waste that turns into a resource for another segment of the system, or where energy is shared guaranteeing the system itself winds up cyclical and with no duplication of effort [30]. The happy shrimp agriculture in the Netherlands, for instance, is deliberately situated next to Benelux, where it is a pulverized coal-let power generating station in Rotterdam's Maasvlakte (a part of the city's harbor and industrial region). Greenhouse-enclosed basins for raising shrimp are kept at a relentless at 308 °C because of a 2.5 km pipe, which exchanges the power plant's waste coolant water at 608 °C through an exchange system to the homestead. Already this coolant was released straightforwardly into the encompassing ocean. Table 3.3 lists the ecosystem functions and process strategies for mimicry of the built environment. As shown in Table 3.3, it records the biological system processes and recommends existing design techniques or strategies that might be helpful as starting points in determining how such processes could be coordinated into the built environment [30]. It also assists in further understanding the potential of ecosystem biomimicry. In the process level of ecosystem biomimicry, a logical framework is provided to apply the existing technology for achieving sustainability.

3.5.3.3 Mimicking the Function of Ecosystem

Several products that people use could be categorized as an ecosystem functions by many ecologists. The benefit of mimicking the ecosystem's elements is that it furnishes designers with the chance to detect the quantifiable goals in the ecology, specifically to suit a specific climate and area for an improvement and to integrate with the ecosystem that exists instead of draining it. The ecosystem services include regulations, supplying, supporting, and life-fulfilling [31].

Examples of the regulation services are pollination and climate regulations; supplying services' examples are food and medicines, and supporting services include soil formation and fixation of solar energy as well as life-fulfilling services. It is

Table 3.3 Ecosystem functions and process strategies for mimicry of the built environment

Mimicry of ecosystem function strategies, existing related design techniques, and climate change				
Ecosystem services that the built environment may be most able to mimic	Mimicking the ecosystem functions by the built environment		Existing related design techniques or technologies	Climate change implications
	1. Services supplying; food, materials, fuel/ energy, and water	Providing food, raw materials, and fuel for future Freshwater collection and distribution is a must.	Many techniques such as green rooftops, vertical farming, plan for deconstruction, renewable energy generation, solar heat, rain water collecting, and gray water reuse	More independent and in this way vigorous urban situations even with change Decreased transport and vitality age related GHG emissions
	2. Services organization; it is a shorter scale due to the time, climate regulations, prevention of disturbance, and modesty of extreme conditions	Developing and regulating climate by sequestering carbon; providing protection from decreased ozone and remediating the heat island effect Preventing disturbance and extremes such as wind, wave forces, erosion and floods, droughts for surrounding ecosystems Water/air/soil should be purified before return to nonhuman ecosystems	Capacity of carbon building structure; green rooftops; urban forests; permeable clearing surfaces; engineering; and bio-remediation; filtration method.	Decreased rate of a worldwide temperature alteration. More secure and more beneficial constructed condition Sound groups and biological communities can acclimate to change all more effectively Solid biological systems might be a better mean to keep on providing administrations to people
	3. Services supporting; it is a longer scale due to the time, soil, nutrient recycling and habitat's provisioning	Materials ought to be biodegraded or reused and held in the system Common surroundings for nonhuman species ought to be arranged	Dark/gray water recycling, composting techniques, cradle to support design, recycling and reuse methods; natural life corridors, green belts, and urban permaculture	Sanitary soil means greater potential for healthy biomass and production of food, better human health, and resilience Mitigation of GHG emissions and waste protect ecosystem Disturbance

Source: Pedersen Zari and Storey [22]

commonly known that the built environment has a huge negative impact on the ecosystem services, if not well managed. Making assembled conditions that duplicate, mimic, or give these biological community administrations will assist in diminishing the pressure on the ecosystems since the advancement of both urban conditions and atmosphere is persistently evolving. This is not quite the same as the mimicking process strategies of the ecosystem. The fact is that it moves behind mimicking the environment as a metaphor, which can undoubtedly be misjudged. Biomimicry, at the function level, transforms into the general hypothetical idea and targets generator for the change and enhancement, while specific procedures to achieve these goals can be drawn from a broad assortment of outline methods and techniques. Ecosystem biomimicry, at the function level, likewise characterizes criteria that ensure that advancements or frameworks – to be composed in an improvement – are fitting as far as overall environmental impact [31].

All ecosystem services could not be imitated by the built environment, treatment, and control of species conventional assortment, for instance, it cannot be emulated. In any case, commitment or positive joining with numerous vital ecosystem services could be an objective for new advancements [32]. Table 3.4 lists the levels of biomimicry (biological system) benefits that the built environment may be most ready to imitate. The built environment sometimes has similar functions as the ecosystems. A building could be intended to form a system in the same way as the ecosystem functions; this could be in the form of producing food, sustainable power source, and raw materials for the future's built environment where it is conceivable. In addition, the system is gathering and decontaminating water, clean air, and soil; regulating the atmosphere through relieving GHG emissions; mitigating the heat island effect; and controlling disintegration and conservatives' extreme climate [32]. Accordingly, the reasons for CC risk mainly attribute to the increasing GHG emissions due to the urban sprawl and the built environment. The distinction here is that quantifiable targets, for example, along with the emissions levels, carbon stockpiling, water harvesting, resource, and energy pre-production, are resolved through a comprehension of appropriate ecosystems or the pre-development ecosystem of the site. In this way, it depends on natural reality instead of human political needs.

The design team, consisting of Myth Architects and Green Works Landscape Architecture Consultants, examined how the site's unique biological ecosystem worked before improvement, keeping in mind the target to decide for attaining the objectives for the environmental execution of the advancement over 50 years. The expressed objectives of the project include (a) saving urban density, (b) accomplishing carbon balance, (c) accomplishing water advancement, and (d) living inside the site's sunlight budget related to energy production and utilization [33].

The investigation of the ecosystem demonstrated that it was conceivable to meet such objectives in that area. Furthermore, it gave a fruitful case to emulate in accomplishing them. Anticipated advantages of nature emulation incorporate expanded local biodiversity, diminish the heat island effect (HIE), and decrease interest for

Table 3.4 Levels of biomimicry information and biological system benefits

Levels of biomimicry information	Aspects of the level
Organism features (features of the organism itself)	Formal traits incorporate shape, shading, volumetric treatment, transparency, rhythm
	Association and hierarchy of the importance of parts and systems
	Structure, stability, and gravity resistance
	Development materials and procedures and lifecycle
	Function and behavior
	Motion and aerodynamics
	Morphology, anatomy, modularity, and patterns
	Portability and mobility
	Self-assembly
	Healing, recovery, survival, and maintenance
	Homeostasis that balance internal systems while external forces change
	Systems which include organ, digestive, circulatory, respiratory, skeletal, muscular, nervous, excretory, sensory, and locomotive systems
Organism-community relationship (the organism's relationships to its community of similar organisms as well as other creatures that it may deal with)	Survival techniques
	Interaction with other creatures
	Trans-generation knowledge transfer and training
	Hierarchy of community members
	Group management and coordination
	Communication
	Collaboration and teamwork.
	Self-protection
	Sensing responding and interaction.
	Risk management
Organism-environment relationship (how an organism may deal with and fit in its environment?)	The contextual fit
	Adjustment to change
	Response to climate by cooling, heating, and ventilation solutions
	Response to context, for example, by camouflage, self-protection, and self-cleaning
	Adaptation to ecosystems include adjustment to various light or sound levels, shading, and self-illumination
	Shelter building
	Limited resource management (adaptations to lack of water, light, or food)
	Waste management
	Input/output/process cycling

Source: El-Zeiny [21]

new consumable and waste water systems as well as the infrastructure foundation improvement, in addition to expanding the sources of renewable energy [33]. The latter includes 10 MW of photovoltaics and 1.4 MW of wind turbines, making it a net-energy maker [33].

It is imperative, however, that the built environment has some weaknesses in its own system adaptation, but it is a suitable response for dealing with environmental changes and impacts, since it watches out for a noteworthy number of the crucial issues based on the current urban situations that need re-evaluation and adaptation [34].

The refinement can be compared with a long-term treatment of the fundamental reasons for an ailment in an individual instead of a short-term treatment of side effects, which may in certainty disturb the basic condition. For this circumstance, the condition is the way that the dominant part of human urban settlements is liable to nonrenewable energy sources to feed, heat, sustain, and transport people in a linear system, which makes sullying driving to a limited extent to CC.

Since humans attempted to comprehend their surrounding environment, observed nature, learned from it, and utilized it as well, this is most effortlessly found in more primitive communities, where everything is made by hand from the available materials that exist and designed with particular themes from the natural surroundings that they know so well. It is comprehended that nature already has the ideas of design, which are for the most part agreed to be needed in every design. The three factors, which can be found in nature, are (a) structure, (b) function, and (c) system as presented in Fig. 3.15.

The characteristics are wholeness, integrity, and unity in structure, where all parts contribute to the impact or reason for the whole and no part might be removed without some harm to the entirety. These are focal ideas and concepts in the aesthetics and in the natural history of Aristotle and are characteristics in the Aristotelian view both of living beings and of the best works of art [35].

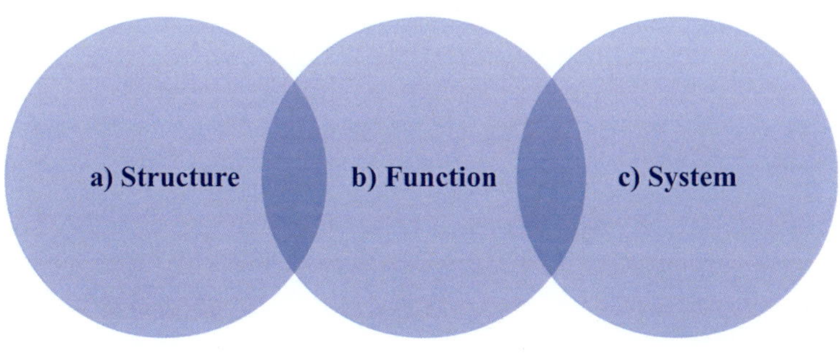

Fig. 3.15 Factors found in nature to inspire biomimetic architecture. (Source: Developed by authors)

3.6 The Applications and Types of Biomimicry in Architecture

It is important to know the applications and types of biomimicry; therefore, some architectural designs and projects, which acquired inspiration, ideas, and solutions from nature, will be discussed under three levels of biomimicry's inspiration: (a) organism level, (b) behavior level, and (c) ecosystem level. Moreover, pointing out the architects who are inspired from nature usually developed ideas through forms, functions, structures, or materials and could also be visual, conceptual, or computational inspiration.

3.6.1 Inspiration Sources

Inspiration source is core in biomimicry, and nature has three types of inspiration sources; these are visual, conceptual, and computational resources as highlighted in Fig. 3.16. Each type plays a different role, and understanding those levels is important in the design progress [36].

3.6.1.1 Visual Inspiration in Nature

Visual inspiration is moderately well comprehended and broadly utilized. For this situation, the perspective of different living organism, or their system, is utilized to create a similar form. The buildings and the design acquire the inspiration from nature through visual principles like forms, shapes, and surfaces, which can be added to visual inspiration. Examples of visual inspiration in nature are London City Hall in London, UK, and Denver International Airport, Denver–Colorado, USA.

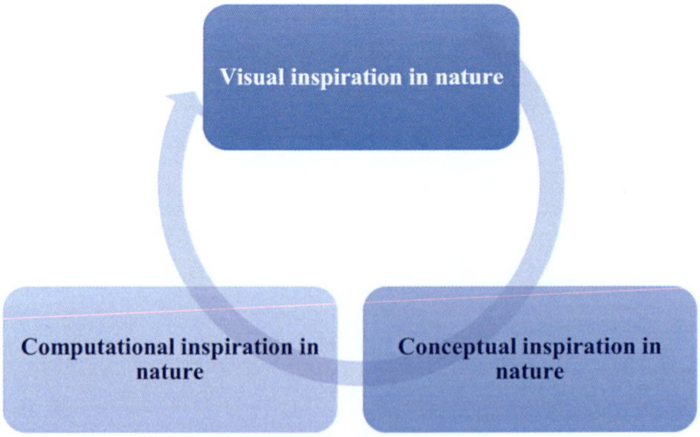

Fig. 3.16 The three inspiration sources in nature. (Source: Developed by authors)

The iconic City Hall building at 110 the Queen's Walk, London, is on the south bank of the Thames between the Tower Bridge and London Bridge in Central London – England, UK. It is home to the greater London authority and London's local government. The unique building has some formal ideas inspired from nature as shown in Figs. 3.17, 3.18, 3.19, 3.20, 3.21, 3.22, and 3.23. It is one of the models

(a) (b) (c)

Fig. 3.17 London City Hall in London, England, United Kingdom. (**a**) General view of London City Hall and Shard Tower, London, (**b**) Front view of the City Hall façade, (**c**) A side view of the building. (Images' credit and source: (**a**) Sevku, https://commons.m.wikimedia.org/wiki/File:2013_London_Thames_Shard_City_Hall_2.jpg#; (**b**) Gyula Péter, https://commons.m.wikimedia.org/wiki/File:St_Khtharine%27s_%5E_Wapping,_London,_UK_-_panoramio_(7).jpg#; (**c**) CGP Grey, https://commons.m.wikimedia.org/wiki/File:2005-04-09_-_Untited_Kingdom_-_England_-_London_-_City_Hall_-_Miscellenaeous_4887188437.jpg#)

Fig. 3.18 Birds' eye view of London City Hall from the Tower Bridge, Central London, England, UK. (Image credit and source: Nigel Chadwick, https://commons.m.wikimedia.org/wiki/File:City_Hall_from_The_Tower_Birdge_Experience_-_geograph.org.uk_-_1231330.jpg#)

Fig. 3.19 London City Hall near London Bridge in London – England, United Kingdom. (Images' credit and source: Steve F–E Cameron, https://commons.m.wikimedia.org/wiki/Category:City_Hall,_London#/media/File:003SFEC_LONDON-200705.JPG)

Fig. 3.20 The biomimetic architecture and unique location of London City Hall become a viable, social, and livable place in the city. (Images' credit and source: Steve Sea, https://commons.m.wikimedia.org/wiki/File:Politics_in_a_wraped_building_-_panoramio.jpg#)

developed to generate energy, mitigate CO_2 emissions, and adapt to climate change. The iconic building and its form – a distorted glass shepherd which is seen as head-shaped – is designed by Sir Norman Foster + Partners. The building is justified in terms of two forms of function:

1. Environmental – through reducing the total glass surface area in addition to the fact that it is a nonpolluting building as illustrated in Figs. 3.17, 3.18, 3.19, and 3.20).
2. Democratic as the building – depicting the democratic process in London – is formed around a splendid interior ramp under which the citizens can symbolically walk above the debating chamber of their elected representatives. It also reflects trusts that changing the spots where people live in could change the world. Even the form of spiral-shaped ramp is inspired from nation the storm shape as presented in Figs. 3.21, 3.22, and 3.23 [37, 38].

- *Denver International Airport, Denver Colorado, USA*

The iconic Denver International Airport in Denver, Colorado, is the largest airport terminal (1.5 million square feet) near the magnificence of the Rocky Mountains in the United States as shown in Figs. 3.24 and 3.25. The building design is inspired from the naturally occurring magnificence of the Rocky Mountains highest as presented in Fig. 3.26. The airport building illustrates a combination of both historic and modern architecture. The wonderful splendid peaked roof of the airport

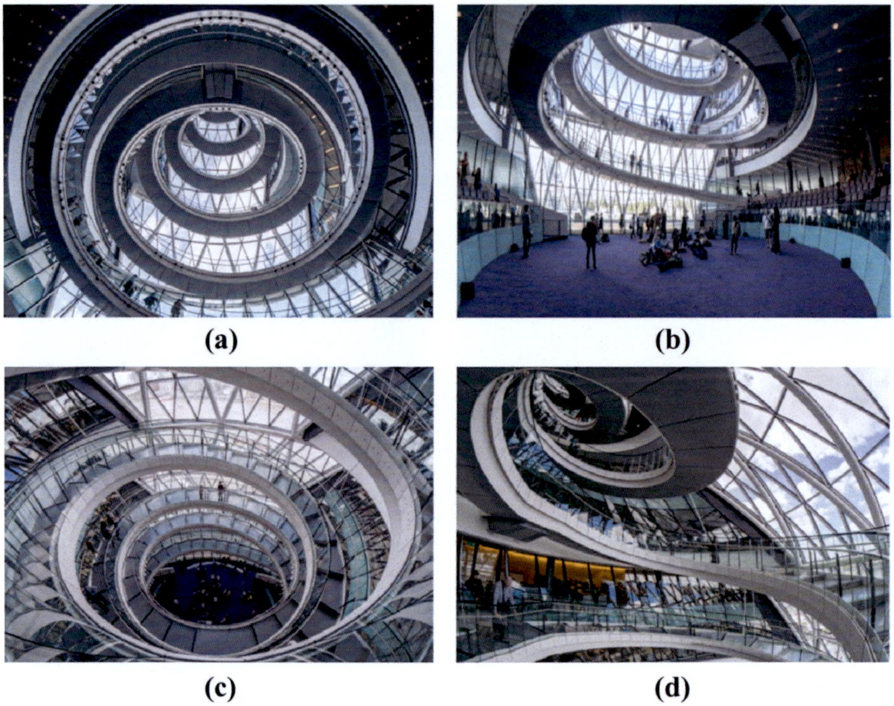

Fig. 3.21 Interior of London City Hall in London spiral-shaped ramp. (a) The ramp inspired from a cyclone shape, (b) The Council Chamber and spiral staircase, (c) The spiral ramp from the top floor, (d) People walking the stairway built on ramp. (Images' credit: Colin, source: (a) https://commons.m.wikimdia.org/wiki/File:City_Hall,_London,_Spiral_Staircase_-_0.jpg; (b) https://commons.m.wikimdia.org/wiki/File:City_Hall,_London_Chamber.jpg; (c) https://commons.m.wikimdia.org/wiki/File:City_Hall,_London,_Spiral_Staircase_-_7.jpg; (d) https://commons.m.wikimdia.org/wiki/File:City_Hall,_London,_Spiral_Staircase_-_4.jpg)

Fig. 3.22 Nature (hurricane) inspiration for the spiral staircase of London City Hall. (a) Cyclone formation, (b) Hurricane inspiring the staircase design. (Images' credit and source: (a) ESA, https://commons.m.wikimedia.org/wiki/File:Cyclone_Idai_west_of_Madagascar.jpg; (b) NOAA, https://commons.m.wikimedia.org/wiki/File:John_1994-08-23_0200Z.png#mw-jump-to-license)

Fig. 3.23 The spiral ramp seen from the glazed façade looking over the Thames and London City Hall, London, England, UK. (Image credit and source: Uwe Aranas, https://commons.m.wikimedia.org/wiki/File:London_UK_City_Hall-and-More-London-01.jpg#)

(a) **(b)**

Fig. 3.24 Denver International Qirport and nature inspiration. (**a**) Arial view of the airport with the roof's caps, (**b**) The conical tent of white Teflon structure. (Images' credit and source: (**a**) PéterQuinn925, https://commons.wikimedia.org/wiki/File:Denver_International_Airport_Main_Terminal_at_dusk.jpg#; (**b**) https://commons.wikimedia.org/wiki/File:Denver_international_airport.jpg)

terminal's building at night emulates the white snow caps of the Rocky Mountains as illustrated in Fig. 3.27 [36, 39].

In addition, a second example is Selfridges Department Store at Upper Mall East in Birmingham – England, UK [40]. It is designed and completed in 2003 by the architects Jan Kaplicky and Amanda Levete, who are the authors of a book entitled Future Systems. The iconic building includes a stunning space-age design that briefs the aesthetic objective of bionics and architecture. The idea was inspired from

Fig. 3.25 The Teflon tensile texture covering Denver International Airport. (Image source: https://commons.m.wikimedia.org/wiki/File:Denver_International_Airport.jpg)

(a) (b)

Fig. 3.26 The inspiration idea of Denver International Airport. (a) Cathedral Spires of Rocky Mountains, (b) Rocky Mountains snow caps. (Images' source: (a) https://en.m.wikipedia.org/wiki/Garden_of_the_Gods; (b) https://commons.m.wikimedia .org/wiki/File:Rocky _Mountains_Snow_Cap.jpg)

nature – wasps' nest, where the organic natural form and structure are combined with high technology to accomplish optimization as illustrated in Figs. 3.28 and 3.29 [36]. The Selfridges Building facades are covered by 1500 shiny aluminum discs, giving the building an almost splendid alien-like sheen [41]. Details and views of Birmingham's iconic building as well as the design inspiration idea from the larva stage of the wasp with the egg are shown in Figs. 3.30 and 3.31, while the buildings linkage to Park Street car park are shown in Fig. 3.32.

A third example of the visual inspiration in nature is the marvelous Museo Soumaya magnificent museum which is located in the new cultural and commercial heart district in Mexico City, Mexico (Fig. 3.33). The design idea is inspired by the honeycombs resulting in a form of a rotated rhomboid supported by 28 curved steel columns of varying sizes and shapes. The building is clad in a membrane of 16,000

Fig. 3.27 The roof of the west terminal of Denver International Airport at night. (Image credit and source: PéterQuinn925, https://commons.wikimedia.org/wiki/File:Denver_International_Airport_Main_Terminal_at_dusk_1.jpg)

Fig. 3.28 Selfridges' iconic building inspired from nature in Birmingham (Image credit and source: Bs0u10eu0, https://commons.m.wikimedia.org/wiki/File:Selfridges_Building,_Birmingham_(2012).jpg)

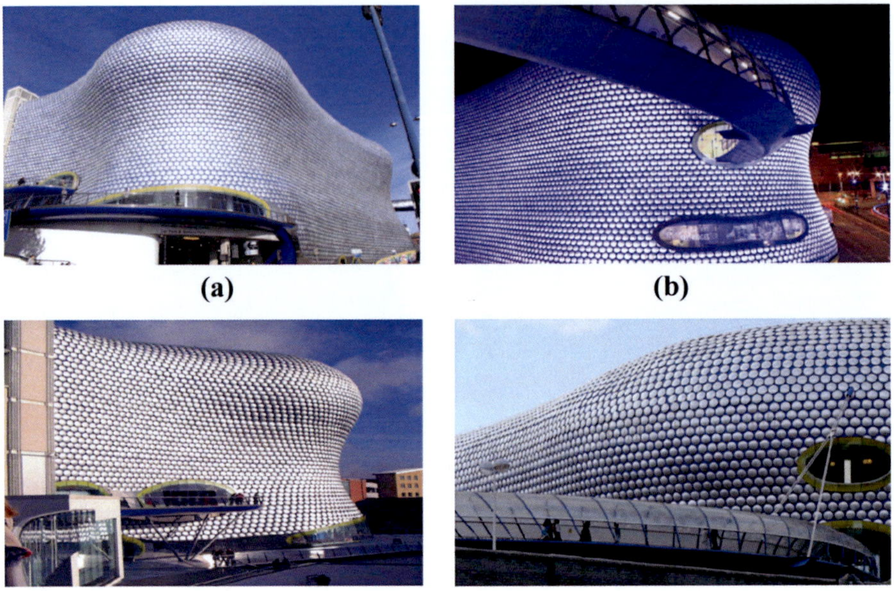

(a) (b) (c) (d)

Fig. 3.29 Façades of Selfridges' Building in Birmingham. (**a**) Selfridges' buildings entrance and façade, (**b**) Main façade with aluminum discs, (**c**) Side façade showing the blobitecture style, (**d**) Walkway from Selfridges to Train station. (Images' credit and source: (**a**) Christophe Finot, https://commons.m.wikimedia.org/wiki/File:Birmingham_-_Ringbull_-_Selfridges_5.JPG; (**b**) Some Guy, https://commons.m.wikimedia.org/wiki/File:Selfridges_Birmingham_at_night.jpg; (**c**) G-Man, https://commons.m.wikimedia.org/wiki/File:Birmingham_Selfridges_Building.jpg; (**d**) Mattbuck, https://commons.m.wikimedia.org/wiki/File:Birmingham_Moor_railway_station_MMB_08_Birmingham_Bullring.jpg#)

hexagonal mirrored-steel elements in reference to the traditional colonial ceramic-tiled building facades, which creates a dazzling appearance to the entire building facades during daylight and at night as presented in Fig. 3.34. The iconic 150-feet high museum – with an area of 16,000 m^2, designed by FR EE, Fernando Romero Enterprise – has windowless façades, which are inspired from the bees' honeycomb nest, giving the whole building a unique mystical air as shown in Fig. 3.35. The façade is made of a dazzling "trapezoid in motion," a shiny silver cloud-like structure reminiscent of a Rodin sculpture and known as the world's flashiest museum, as shown in Fig. 3.36 [42–44]. Figure 3.37 shows the innovative interior of the museum and its roof's truss.

A fourth example of visual inspiration in nature is Barcelona's iconic tower the Agbar Torre Glories, which is located in the City of Barcelona, Spain (Fig. 3.38). It has a dynamic movable glass that controls heat and daylight as shown in Fig. 3.39 [45]. The envelope, which is visually inspired from the cactus plants (Fig. 3.40), encompasses tinted glass and windows to regulate the daylight inside the building space [46]. The Agbar Torre Glories Tower has 472-foot unique façade features,

(a) (b)

Fig. 3.30 Part of the 1500 aluminum shinny discs on the Selfridges' façades in Birmingham, England, UK. (**a**) The 1500 aluminum discs' facade, (**b**) Close-up view of the metal disc. (Images' credit and source: (**a**) Tony Hisgett, https://commons.m.wikimedia.org/ wiki/File:Selfridges_2_(3424720462)jpg#; (**b**) Kkrumina, https://commons.m.wikimedia.org/wiki/File:Birmingham_.JPG#)

(a) (b) (c)

Fig. 3.31 The wasp nest idea inspiring the design of Selfridges' Department Store building. (**a**) Larva stage of the wasp with the egg, (**b**) Close-up view of the eggs, (**c**) The grub-like organisms. (Image credit and source: (**a**) Richerman, https://commons.wikimedia.org/wiki/File:Wasp_Nest_01.jpg; (**b**) Richerman, https://commons.wikimedia.org/wiki/File:Wasp_nest_3.jpg; (**c**) Richerman, https://commons.wikimedia.org/wiki/File:Wasp_Nest_03.jpg)

4500 windows with temperature-controlled levers, and an equal number of LED lights that illuminate at night [47, 48] as illustrated in Fig. 3.38c. The detail of the innovative façade is presented in Fig. 3.39.

The Aquatics Center in Stanford, London, UK, is the fifth example. It is part of the London Olympic 2012 complex (Fig. 3.41). The iconic and sustainable Aquatics Center is designed by the legendary and late architect Zaha Hadid [49]. The idea of this magnificent building was inspired from nature – the fluid geometry of water in

Fig. 3.32 Selfridges' façades from Park Street Cap Park showing the pedestrian bridge in Birmingham, England, UK. (Image credit and source: Jimmy Guano, https://commons.m.wikimedia.org/wiki/File:Selfridges_Birmingham_from_Park_Street_car_park.jpg#)

motion to create spaces and a surrounding environment in sympathy with the river landscape of the Olympic Park. Also, an undulating roof sweeps up from the ground like a wave, enclosing the pools of the Aquatics Center with a unify gesture, as illustrated in Figs. 3.42 and 3.43. The form of the Aquatics Center has also been described by Rowan Moore as a vast turtle waving oversized flippers [50]. In 2014,

Fig. 3.33 The amorphous shape Soumaya Museum Plaza in Mexico City, Mexico. (Image credit and source: Raystormxc, https://commons.m.wikimedia.org/wiki/File:Museo_Soumaya_Plaza.jpg)

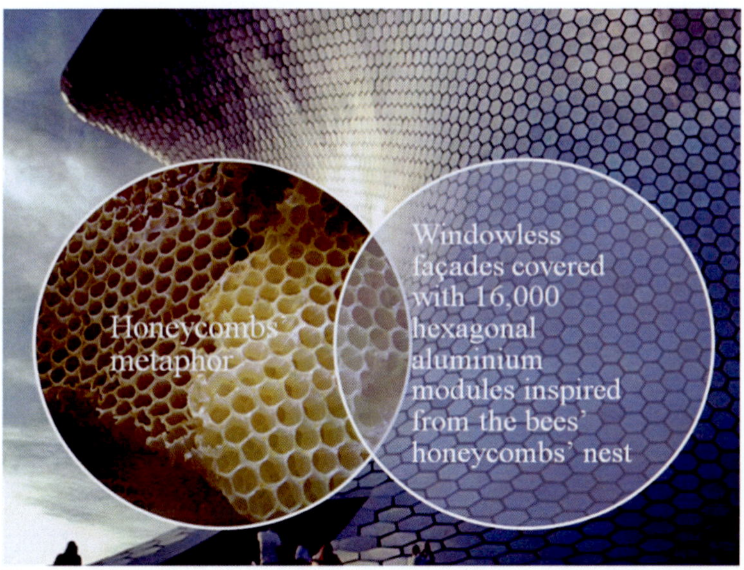

Fig. 3.34 Visual inspiration of the honeycombs nest in design of the Museo Soumaya, Mexico City. (Image credit and source: Merdal, https://commons.m.wikimedia.org/wiki/FileHoney_comb.jpg#; dbasulto, https://commons.wikimedia.org/wiki/File:Museo_Soumaya_Mexico.jpg)

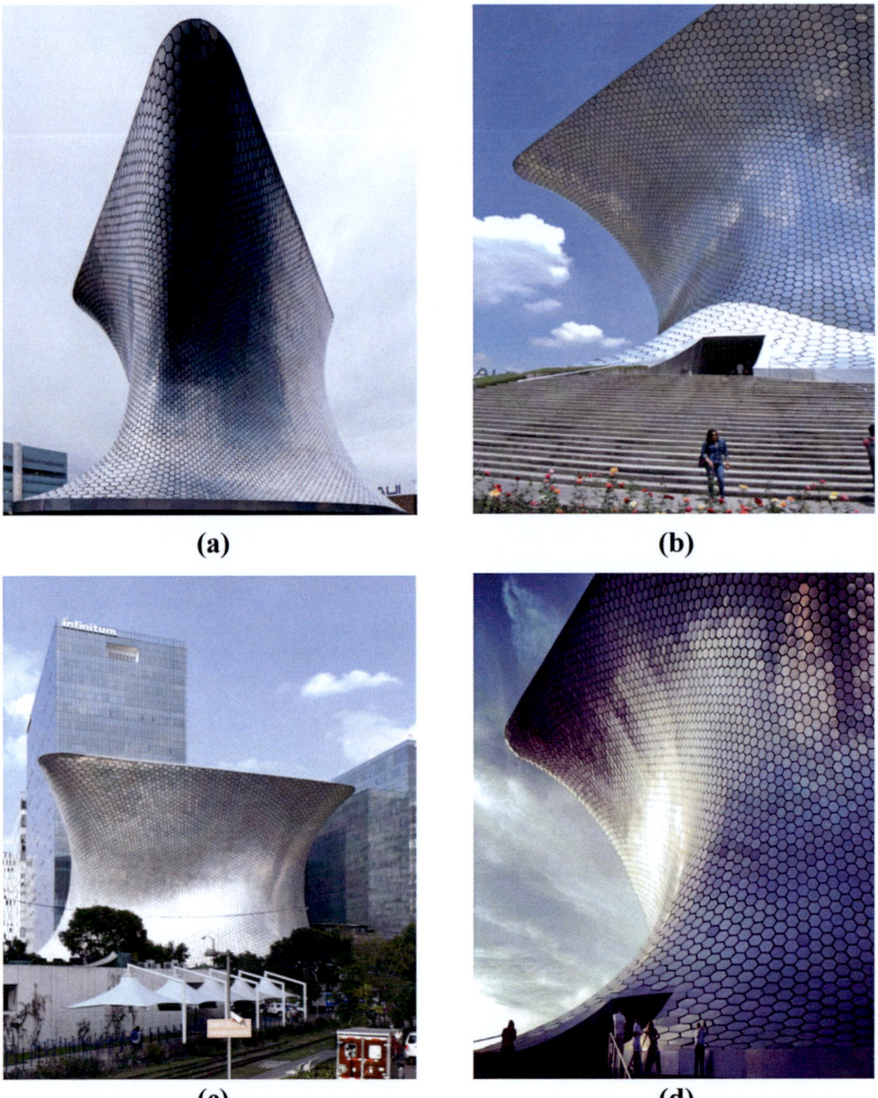

Fig. 3.35 Various façade views of Museo Soumaya museum in Mexico City. (**a**) The windowless hexagonal mirrored-steel, (**b**) The entrance of the museum, (**c**) Façade's hexagonal aluminum membrane, (**d**) Side entrance with mirrored-steel. (Images' credit and sources: (**a**) Diego Delso, https://en.m.wikimedia.org/wiki/Museo_Soumaya.jpg; (**b**) ProtoplasmaKid https://commons.wikimedia.org/wiki/File:Exterior_del_Museo_Soumaya.jpg; (**c**) Gzzz, https://commons.wikimedia.org/wiki/File:Plaza_Carso_Mexico.jpg; (**d**) dbasulto, https://commons.wikimedia.org/wiki/File:Museo_Soumaya_Mexico.jpg)

Fig. 3.36 Close-up view of the façade of Soumaya museum in Mexico City. (Image credit and source: Jespa, https://commons.m.wikimedia.org/wiki/File:Museo_Soumaya_(acceso).JPG)

(a) (b)

Fig. 3.37 The interior of Museo Soumaya museum in Mexico City. (**a**) Interior and space tress ceiling, (**b**) Daylighting impinging through the interior. (Images' credit and source: (**a**) Francisco Anzola, https://commons.wikimedia.org/wiki/File:Rodins_at_Museo_Soumaya_(5605439388).jpg#; (**b**) Adriàn Cerón, https://commons.m.wikimedia.org/wiki/File:Galer%C3ADa_6_del_Museo_Soumaya.jpg#)

Fig. 3.38 The iconic Agbar Torre Glories Tower in Barcelona, Spain. (**a**) General view of the tower in heart of the city, (**b**) A 4500-window façade with temperature-controlled levers, (**c**) Close-up view of the movable glass controlling heat and daylight. (Images' credit and source: (**a**) Axelv, https://commons.m.wikimedia.org/wiki/File:Torre_Agbar_and_Glories.jpg#; (**b**) Ardfern, https://commons.m.wikimedia.org/wiki/File:Torre_Agbar,_July_2009.JPG#; (**c**) Diliff, https://commons.m.wikimedia.org/wiki/File:Torre_Agbar_-_Barcelona,_Spain_-_Jan_2007.jpg#)

London's Aquatics Center was opened to the public, after removing the wings that house the extra seating following the Olympic Games. The iconic Aquatics Center depicts the stunning sinuous building in all its glory, where over 2.5 million visited the Center [50] and the number of visitors increased each year by 47% since March 2016 with over 14,000 people every week [51]. Authors perceive part of the Aquatics Center's edge roof also inspired from nature – the shark's head as presented in Figs. 3.41 and 3.42a, c.

An additional example is Lisbon's Gare do Oriente Train station in Portugal which is the hub for high-speed trains and regional trains and buses as shown in Figs. 3.44 and 3.45. The magnificent structure is designed by Santiago Calatrava and is located in the station in Parque das Naçöes and completed in 1998 for the World Expo [52]. The munificent concert structure and intersecting paths are creatively delineated and inspired from nature where architect Calatrava mimics a seashell from nature in the main entrance shed of Lisbon Railway Station as presented in Fig. 3.44a. The palm trees in a mushroom by recalling the gothic spires of Brookfield palace in Toronto are shown in Figs. 3.44b and 3.44c [52]. Moreover, Figs. 3.45 and 3.46 illustrate the details of inspiration from nature in the iconic railway station.

Fig. 3.39 Details of the façade and its movable window levers – Torre Agbar Tower in Barcelona. (Image credit and source: Lars20070, https://commons.m.wikimedia.org/wiki/File:DetailsTorreAgbar.jpg#)

Fig. 3.40 The cactus plants inspire the design of Torre Glories Tower. (a) African cactus, (b) Genera *Mammillaria* with white flowers. (Images' credit and source: (a) Mohsen Aboulnaga; (b) Thibdx, https://commons.m.wikimedia.org/wiki/File:Close_view_of_a_white_cactus_flower_in_Marrakech_at_Jardins_de_Majorelle.jpg)

Fig. 3.41 Queen Elizabeth Olympic Park Aquatics Centre in Stanford. (Image credit and source: Fred Romero, https://commons.m.wikimedia.org/wiki/File:London_-_QE_Olympic_Park_London_Aquatics_Centre_%282%29.jpg)

Legendary architects, such as the Finnish architect Alvar Aalto and the British architect Zaha Hadid, have inked many exemplary buildings that mimicked nature as shown in Figs. 3.46, 3.47, and 3.48. These iconic buildings are the splendid design of the Dongdaemun Cultural Exhibition and Design Plaza (DPP) in Seoul, South Korea – designed by Zaha Hadid Architects – was inspired from nature by ensemble fluid that reflects a Korean tradition [53, 54] (Fig. 3.46a, b);

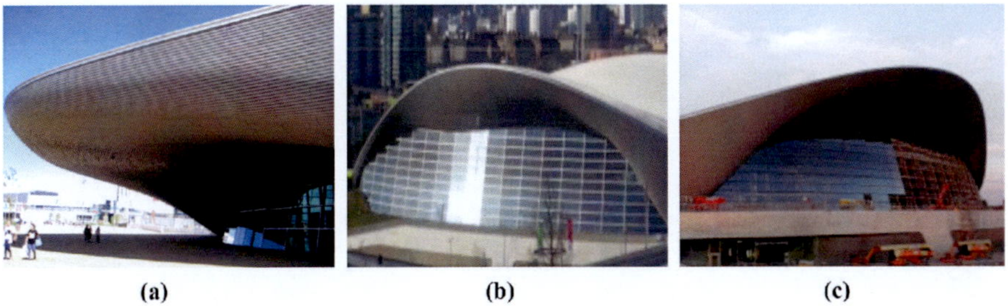

Fig. 3.42 London's Aquatics Centre in Stanford. (a) Main swimming pool entrance, (b) Flowing roof edge agility, (c) Roof edge mimics a shark head. (Images' credit and source: (a) Olaf Janssen, https://commons.m.wikimedia.org/wiki/File:London_Aquatics_Centre_Exterior_Aug_2014_05.JPG; (b) Bert Seghers, https://commons.m.wikimedia.org/wiki/File:LondonOlympicPark-ArcelorMittalOrbitAndAquaticcsCentre.JPG; (c) James Mitchell, https://commons.m.wikimedia.org/wiki/File:London_Aquatics_Centre.jpg)

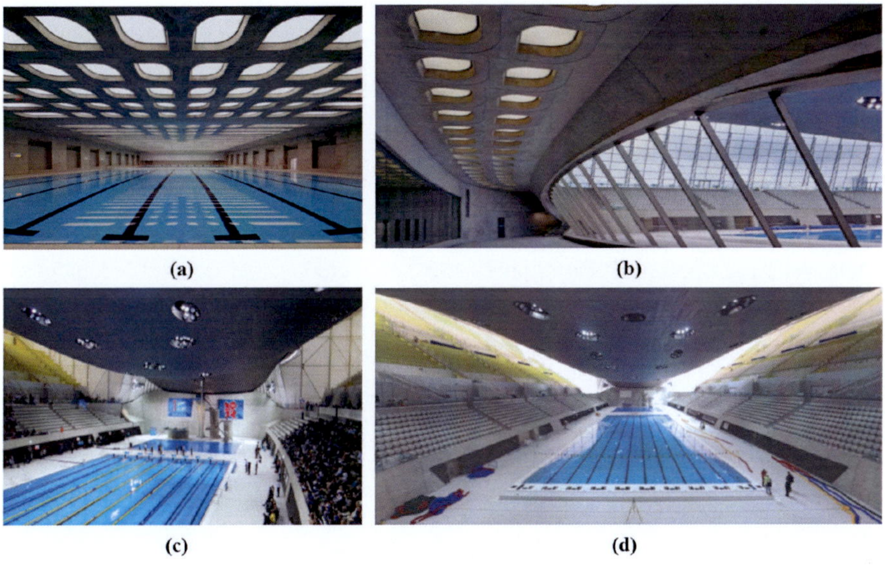

Fig. 3.43 London's Aquatics Centre in Stanford, London. (a) Ceiling inspired from nature, (b) Spectators ally to recycled seating areas, (c) The pool beneath the wave roof, (d) View of the main swimming pool. (Images' credit and source: (a) and (b) Zaha Hadid Architects – London, https://zaha-hadid.com; (c) Ben Sutherland, https://commons.m.wikimedia.org/wiki/file:London_2012_Olympics_Aquatics_Centre_01.jpg; (d) Alexander Kachkaev, https://commons.m.wikimedia.org/wiki/file:London_Aquatics_Centre_panorama.jpg#)

the famous Pavilion of Expo 2012 in Yeosu, South Korea inspired from nature, where the kinetic façade hinged louvers mimic the water wave's movement and shark's gills (Fig. 3.46c, d); and Alvar Alto's famous Kamppi Chapel in Simonkenttä 7 – downtown Helsinki is mimicking nature, where the iconic envelope emulating the Calla Lilies, funnel-shaped spathe white flower, but in a vivid orange color (Fig. 3.46e, f) [55].

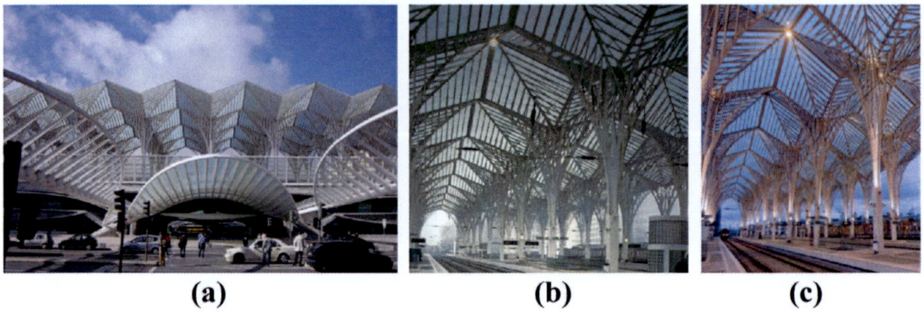

Fig. 3.44 Gare do Oriente Railway Station inspired from nature in Lisbon, Portugal. (a) Main entrance shed of the train station mimicking a seashell, (b) A structure mimicking palm trees in a mushroom form, (c) The platforms' structure at night. (Images' credit and source: (a) Hannesrb Commonswiki, https://commons.m.wikimedia.org/wiki/File:Estação_do_Oriente_or_(Gare_do_Oriente)_by_Santiago_Calatrava.jpg; (b) Feliciano Guimarães, https://commons.m.wikimedia.org/wiki/File:Vistas_do_cais_de_embarque_da_Estação_do_Oriente_08.jpg#; (c) Kent Wang, https://commons.m.wikimedia.org/wiki/File:Lisbon_Oriente_Station_(41172867580).jpg)

3.6.1.2 Conceptual Inspiration in Nature

Conceptual inspiration in design occurs by using a principle found in nature, such as the biological principle of balancing. This principle states that any living organism reacts correspondingly to recover its vital functions when attacked by an external object. Therefore, designers and architects can apply this principle, for example, to determine the optimal shape of shell roofs subjected to thermal and mechanical loads. Using conceptual inspiration requires a solid understanding of both nature and structural engineering; however, it cannot be used in a mechanistic way by an automated designing system only [36].

The Urban Cactus, a 19-storey private sky garden round tower designed by AEMSEN, is located in the city of Rotterdam's harbor in the Netherlands. The 16,000 m^2. innovative building at the head of the spine belongs as an object to reflect more green layers than to the surrounding urban structure; it is inspired by the irregular shape of outdoor plants in open-air spaces, as shown in Fig. 3.49. By rotating this division on a grid in an all-around building, the Urban Cactus tower provides a private garden view to all the 98 apartments' garden of the city as the ultimate in urban living as illustrated in Fig. 3.49b, c [56].

The Jasper Jägers et al. design of the building to provide natural daylight and a remarkable design on the harbor, giving it the sign of biomimetic architecture, nevertheless, it is not 100% green or sustainable; thus, it just gets a good mention on most biomimetic architecture records [36]. Figure 3.50 presents a general view of the Urban Cactus building in Rotterdam.

Jasper Jägers et al., AEMSEN Rotterdam also design another project 'Barbizon' which is a proposal for redevelopment of the plots at Barbizonlaan 94–104 in Capelle aan den IJssel into a future-proof and affordable housing project [56]. The Barbizon is an innovative and sustainable concept of new urban living that depicts

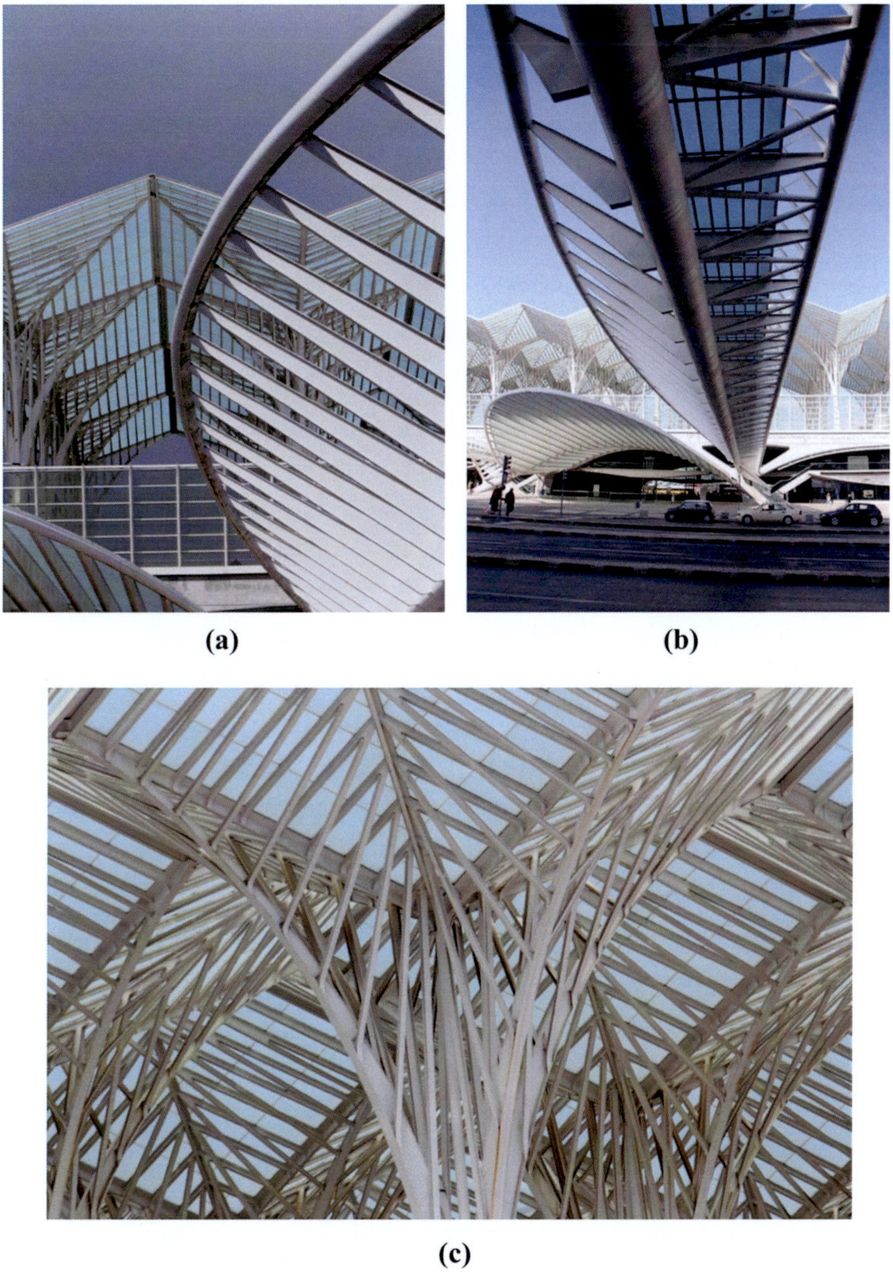

(a) (b) (c)

Fig. 3.45 The Gare do Oriente Railway Station inspired from nature, Lisbon, Portugal. (**a**) Close-up view of the front façade of the station, (**b**) A detailed view of Lisbon Railway Station and its bridge by Santiago Calatrava (**c**) Close-up view of the mimicked skeleton of Gare do Oriente train station in Lisbon, Portugal. (Images' credit and source: (**a**) FrDr, https://commons.m.wikimedia.org/wiki/File:Lisboa_142.jpg#; (**b**) David Prieto, https://commons.m.wikimedia.org/wiki/File:Detalle _ de_la_estción_de_Oriente_(105975705).jpg#; (**c**) FrDr, https://commons.m.wikimedia.org/wiki/File:Lisboa_143.jpg#)

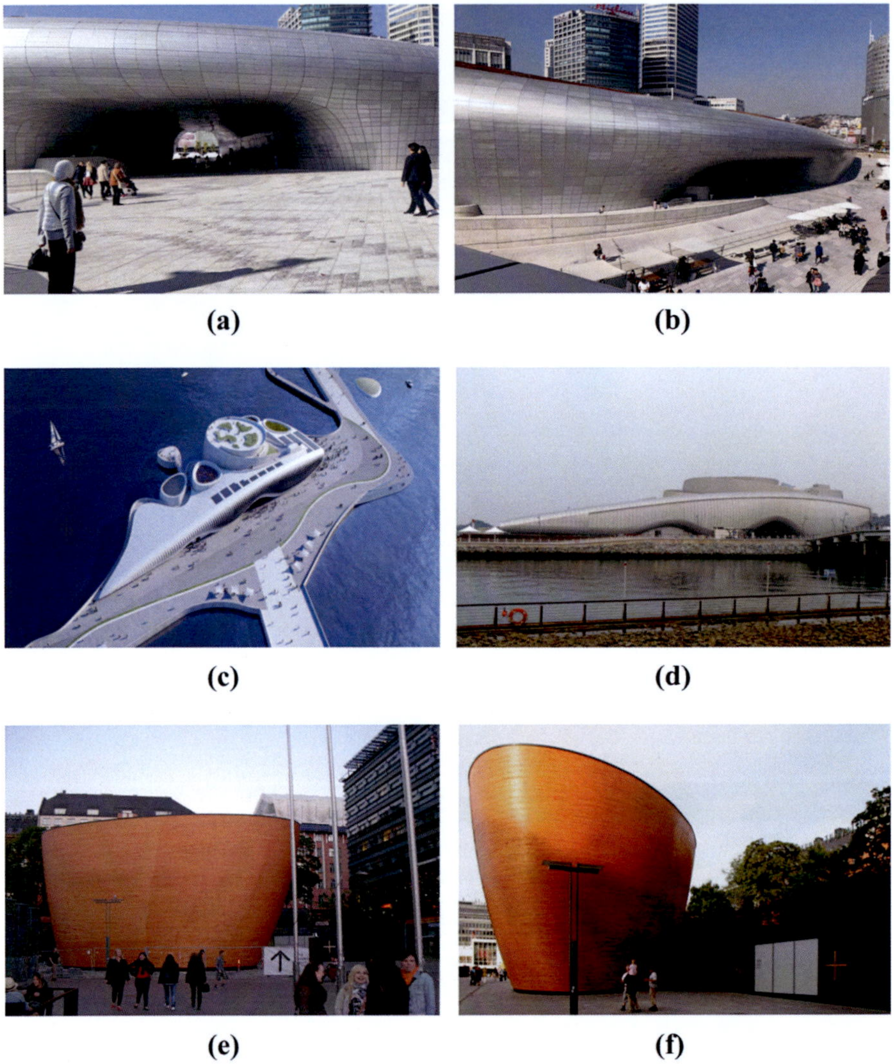

Fig. 3.46 Iconic architectural examples in South Korea and Finland mimicking nature. (**a**) Dongdaemun Design Plaza building in Seoul, South Korea, (**b**) Zaha Hadid's inspired and impressive design from nature, (**c**) The Theme Pavilion, Expo 2012 in Yeosu, South Korea inspired from nature as a large fish shape, (**d**) The façade kinetic hinged louvers mimicking the water waves and a shark's gills, (**e**) Alvar Aalto's famous Kamppi Chapel mimicking nature in Helsinki, Finland, (**f**) The iconic envelope emulating the calla lilies, funnel-shaped spathe flower. (Images' credit and source: (**a**) and (**b**) Nada Nidal and Mohsen Aboulnaga; (**c**) Hyolee2, https://commons.m.wikimedia.org/wiki/File:Expo2012_Theme_pavilion.JPG#; (**d**) Cesco 82, https://it.m.wikipedia.org/wiki/File:Expo2012-Thematic-Pavilion-Day-Top.gif#; (**e**) Mohsen Aboulnaga (**f**) Tappinen (shifted by Rabanus Flavus), https://commons.m.wikimedia.org/wiki/File:Kamppi_Chapel.jpg)

Fig. 3.47 Aerial view of the mimicked Dongdaemun Design Plaza at night, Seoul, South Korea. (Image credit and source: Eugene Lim, https://en.m.wikipedia.org/wiki/File:Dongdaemun_Design_Plaza_at_night,_Seoul,_Korea.jpg#)

AEMSEN's vision of healthy buildings in cities (Fig. 3.50). The project is a green and modular residential building that blends nature into buildings and spaces. In addition, the project concept covers many environmental and social aspects as seen in Fig. 3.50a. Moreover, the shared green outdoor spaces, developed in the form of a green valley, are accessible to all residents of the complex by many corridors (Fig. 3.50d, e). Social sustainability has been addressed through the social interaction between residents, and that plays a major role in this concept in addition to greenery and biodiversity (Fig. 3.50b, e) [56].

An additional model of inspiration from nature has been manifested in the city of Rotterdam, the Netherlands (Fig. 3.51). The Cube Houses are designed and built by architect Piet Blom. The concept was centered on the living as an urban roof represented high-density housing. Each house looks like a tree with the top cubes that mimic a head of a tree, and it is tilted 45 degrees and rested upon a hexagon-shaped pylon [57]. The whole design symbolizes a village within a city where each house looks like a tree by itself as well as all the houses together like a forest (Fig. 3.51a) inclined cube [57]. Figure 3.51b–e shows the conceptual design and the Cube Houses in the City of Rotterdam as an inspiration from nature.

Other examples of conceptual inspiration are manifested in Milan, Italy (Fig.3.52). The Bosco Twin Vertical Forest towers designed and built in 2014, Milan, North of Italy, by the Italian architect Stefano Boeri, who inspired and

Fig. 3.48 General view of the Urban Cactus Building in the city of Rotterdam, the Netherlands. (Images' credit and source: AEMSEN, https://aemsen.com/urban-cactus)

created the concept of towering forests in Italy and the world (Fig.3.52a). The vertical forest high-rise towers were developed as part of a massive redevelopment of the new Porta Nuova district in Milan. The towers, 116 m and 85 m high, are featuring 700 trees and over 20,000 different shrubs and plants, as shown in Figs. 3.52a, b and 3.53 [58].

In addition, Stefano Boeri designed more creative conceptual inspiration projects in both Nanjing Pukou District, China, and Lausanne, Switzerland, where vertical

(a) (b) (c)

Fig. 3.49 Conceptual inspiration of the Urban Cactus building in the city of Rotterdam, the Netherlands. (**a**) Urban Cactus building and irregular pattern plant, (**b**) Sky gardens for each apartment to provide livability, (**c**) Arial view of the rotated on grid private garden. (Images' credit and source: AEMSEN, https://aemsen.com/urban-cactus)

forest towers are built. The first tower will stand 200 m, while the second one is 108 m high as illustrated in Fig. 3.54 [59]. Moreover, the Nanjing's Green Towers are Asia's first vertical forest towers which house over 1000 trees and about 2500 shrubs of 27 various local species (over 3000 plants) covering 4500 m^2 in area (Fig. 3.54a). These trees are developed to generate 60 kg of oxygen every day (about 22 tons yearly); this would offset 25 tons of CO_2 every year, hence mitigating the city's air pollution (smog) and regulating air temperature [59]. Figure 3.54b shows the vertical forest tower in Lausanne, Switzerland.

3.6.1.3 Computational Inspiration in Nature

The computational inspiration in nature is encouraging automated conceptual design viewpoint. It is considered the most interesting but still inadequately comprehended and difficult. One of these computational inspirations in architecture is anti-smog building in Paris France, which is completed by French architect Vincent Callebaut. The project is making some genuine waves in the world of biomimetic architecture. It's a mixed-use building, raised over neglected railroad tracks in Paris. Vincent Callebaut transforms the building design into useful reused energy resources and an eco-friendly building [36]. In addition, the project design incorporates green technologies that actually suck the smog from its boulevards reflecting the architect's objectives in relation to the biomimetic architecture and tree's root inspiration [36].

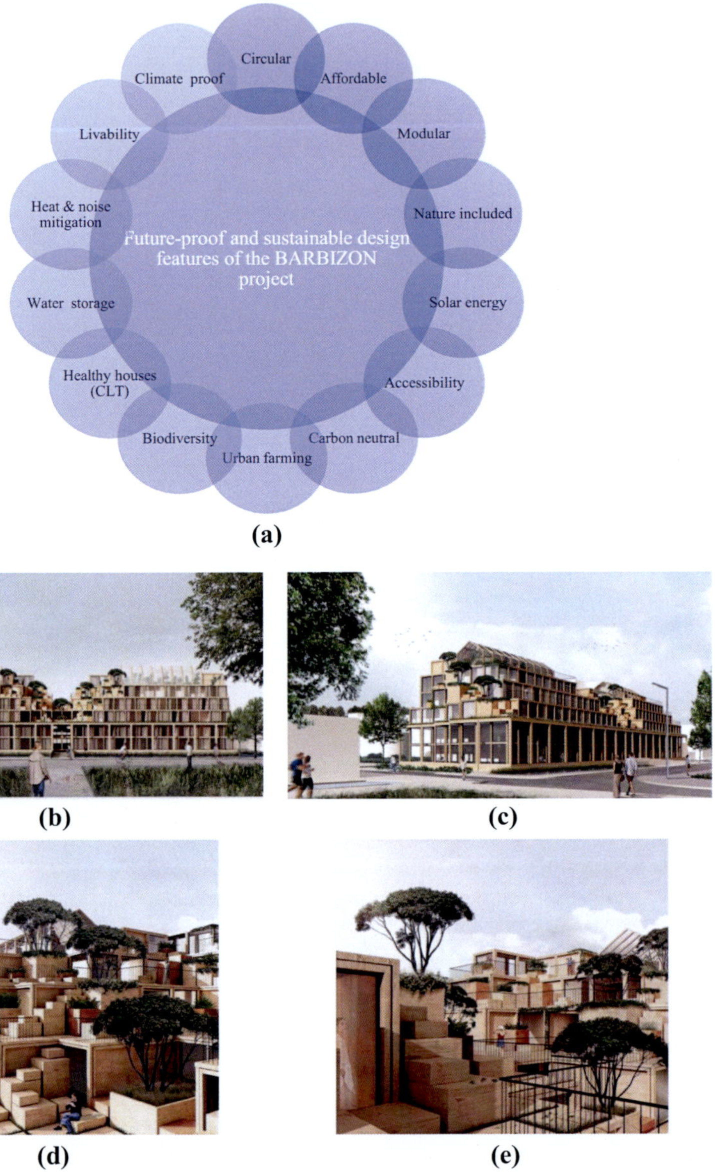

Fig. 3.50 The Barbizon project illustrating the new urban living. (**a**) Features of future-proof and sustainable design of the Barbizon project (Image source: developed by authors after AEMSEN, https://www.aemsen.com/barbizon, (**b**) Front view of the project integrating nature, (**c**) Modular housing with urban farming, (**d**) Pre-manufactured timber for CO_2 neutrality (**e**) Nature inclusive and social sustainability. (Images' credit and source: AEMSEN, https://www.aemsen.com/barbizon)

(a)

(d)

Fig. 3.51 The Cube Houses' inspiration from nature creating an urban living roof, the city of Rotterdam, the Netherlands. (**a**) Aerial view of the urban living roof, the city of Rotterdam, the Netherlands, (**b**) Conceptual inspiration of the urban living roof, the city of Rotterdam, (**c**) The Cube Houses' caps inspired from nature a tree, (**d**) The titled cube emulating a tree, (**e**) The inspired roof of the urban living, the Cube houses. (Images' credit and source: (**a**) Hanselpedia, https://commons.m.wikimedia.org/wiki/File:Kubuswoningen_in_rotterdam.jpg#; (**b**) Raul Ayres, https://commons.m.wikimedia.org/wiki/File:Rotterdam_Cube_House_street _view. jpg#; (**c**) Alexander Migl, https://commons.m.wikimedia.org/wiki/File:Cube_houses_(Rotterdam)_ IMG_ 5090. jpg#; (**d**) Greet C. Smulders, https://commons.m.wikimedia.org/wiki/File:Kubuswoningen_ Speelhuisplein _Helmond.jpg#; (**e**) Heush, https://commons.m.wikimedia.org/wiki/File:Cube-houses-Rotterdam.jpg#)

(e)

(f)

(g)

Fig. 3.51 (continued)

Fig. 3.52 Conceptual inspiration of Bosco Vertical Forest Towers in the city of Milan, Italy. (a) The innovative twin vertical forest towers, (b) View of the high-rise green building achieving livability and adapting to climate change. (Images' credit and source: (a) and (b) Stefano Boeri Architetti, www.stefanoboeriarchitetti.net)

Fig. 3.53 Close-up view of one of the Bosco Vertical Forest Towers in the city of Milan, Italy. (Image credit and source: Stefano Boeri Architetti, www.stefanoboeriarchitetti.net)

(a) (b)

Fig. 3.54 Vertical forest towers' projects in China and Switzerland. (**a**) Nanjing's vertical forest towers in the city of Nanjing, China, (**b**) Vertical Forest tower in Lausanne, Switzerland. (Images' credit and source: (**a**) and (**b**) Stefano Boeri Architetti, www.stefanoboeriarchitetti.net)

In 2017, Vincent Callebaut created an innovative design of the (NAUTILUS Eco-Resort), Biomimetic Learning Center in Palawan, Philippines, which is inspired from nature. This project is created to be zero-emission, zero-waste, and zero-poverty. The schematic design phase of this innovative project is to cover 27,000 m^2 and the buildings' height about 68 m [60]. Figures 3.55 and 3.56 present the iconic Biomimetic Learning Center.

Another example of computational inspiration in nature is the Italian Expo Pavilion 2015, Milan, Lombardy region, Italy. The pavilion claimed the first concrete building in the world to effectively improve the quality of the air around it (Fig. 3.57). The building's façade has 900 biodynamic panels made with photocatalytic concrete and titanium dioxide; the technology was invented by Luigi Cassar, a chemist at cement manufacturer Italcementi as shown in Fig. 3.57. The Palazzo Italia "eats" smog and helps reduce pollution with the biodynamic material that envelopes each panel on each façade; thus polluted air is captured when it comes into contact with light and is then transformed into inert salts that can reduce smog levels in the environment adjacent to the facades [61] as illustrated in Fig. 3.57. These panels are also incorporated in the interior to provide better air quality as shown in Fig. 3.58.

Fig. 3.55 The spiral innovative design of the NAUTILUS Eco-Resort in water – Biomimetic Learning Center in Palawan, Philippines. (Image credit and source: https://www.bing.com/images/search?view=detailV2&ccid=SK%2bu8BgJ&id=E9749B81B2A0BAF0EA3FD1FC7BBD311DDC9A4794&thid=OIP.SK-u8BgJYdthBpmxEhTBqwHaFj&mediaurl =https%3a%2f%2fwww.innaturale.com%2fwp-content%2fuploads%2f2017%2f09%2fNautilus-il-visionario-resort-green-delle-Filippine.jpg&cdnurl=https%3a%2f%2fth.bing.com%2 fth%2fid%2fR.48afaef0180961db610699b11214c1ab%3frik%3dlEea3B0xvXv80Q%26pid%3dImgRaw%26r%3d0&exph=1200&expw=1600&q=NAUTILUS+Eco-Resort&simid=607999165 5918 13833&FORM=IRPRST&ck=56C088E1C EEB8168D3E44AB17240BB60&selectedIndex= 0&qft=+filterui%3alicenseType-Any

3.7 Biomimicry Design Lens and Nature's Design Principles

Biomimicry design lens encompasses nature's principles. There are basic principles of life, according to those principles; biomimicry design lens has been designed as presented in Fig. 3.59 [62]. It is important to fully understand that nature's attributes and fundamentals are connected and assisted in developing architecture and design before formulating solutions to comprehend the relation of this concept with the life's principles, as shown in Fig. 3.60. Nonetheless, the 13 principles of biomimicry are listed as follows [63]:

(a) Nature runs on sunlight; it responds actively to the sun to maximize energy absorption. It relies on energy from geothermal vents on the ocean floor.
(b) Nature only gathers and uses the energy it needs efficiently.
(c) Nature fits form to function.

Fig. 3.56 Buildings with green façades mimicking nature in the form of seashells – The Biomimetic Learning Center by Vincent Callebaut. (Image courtesy: Vincent Callebaut Architectures, https://vincent.callebaut.org/category/projects)

Fig. 3.57 The biodynamic panels made with photocatalytic concrete and titanium dioxide facades of the Italian Expo Pavilion 2015, Milan, Italy. (Image credit and source: Wiki rkit, https://commons.m.wikimedia.org/wiki/File:Italia_Expo_2015.JPG#)

Fig. 3.58 The creative panels in interior to clean indoor air, Italian Expo Pavilion 2015, Milan, Italy. (Image credit and source: Luca Nebuloni, https://commons.m.wikimedia.org/wiki/File:Expo_2015_Milan_(19773341851).jpg#)

(d) Nature is resilient to changes and self-healing.
(e) Nature recycles everything and uses waste as a resource, using nontoxic materials and fully recycled at the end of their life in creating structures.
(f) Nature rewards cooperation.
(g) Nature demands local expertise.
(h) Nature self-assembles and generates structural organization on all scales, from molecules to galaxies.

Fig. 3.59 Biomimicry design lens. (Image source: Developed by authors after https://bio-sis.net/wp-content/uploads/2016/04/Biomimicry38_DesignLens_Lifes_Principles_WEB.pdf/)

(i) Nature banks on diversity, constantly mutating and adapting in a flexible and dynamic flow of change, which increases its complexity, diversity, and efficiency over time.
(j) Nature taps the power of limits.
(k) Nature is beautiful, a model and indisputable evidence of creation.
(l) Nature optimizes rather than maximize (Benyus 1997), using the least materials for optimal structure and function. Material in the honeybee hexagonal colony shape.
(m) Nature remains in balance with the biosphere.

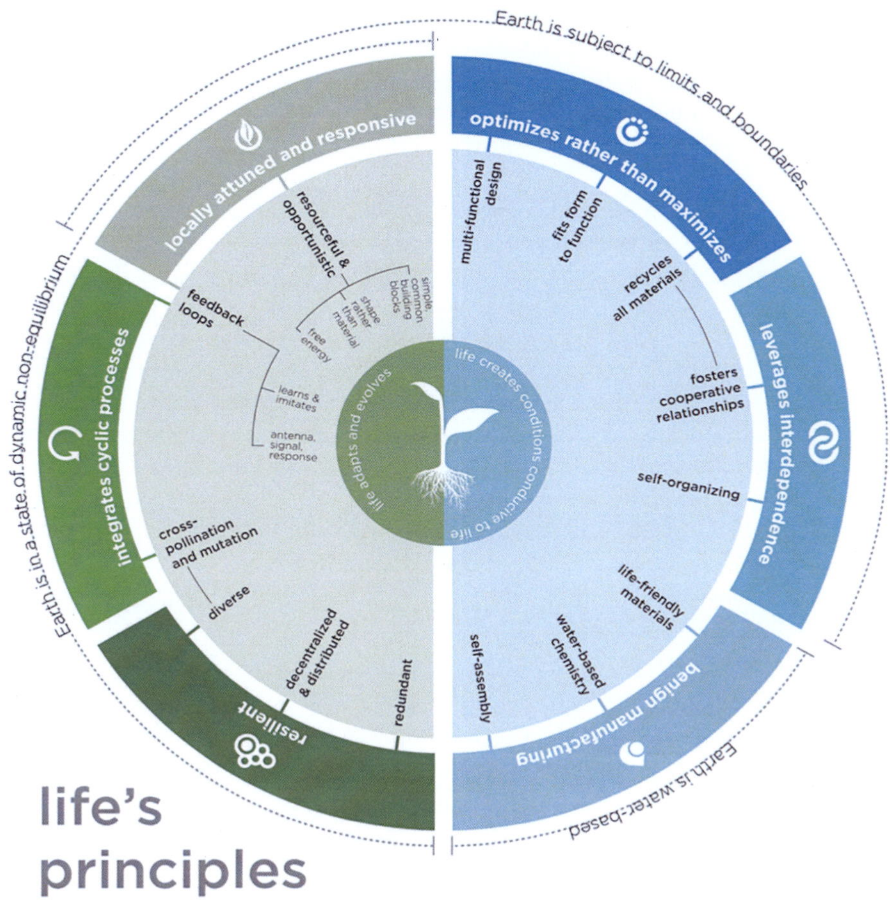

Fig. 3.60 Life's principles. (Source: Biomimicry guide 2009)

3.8 Biomimicry as a Tool for Achieving Sustainability

When looking for a qualified tool for sustainability targets of designed products or materials and architecture, the solution in our view could be biomimicry. Most cases of biomimicry are organism mimicry, while biomimicry at the living level might be persuasive for its capability to deliver novel architecture and design. The likelihood exists where a building acts as a part of a larger system that can copy and imitate natural processes and can function like an ecosystem in its creation, can operate and end of life (demolition), and can possibly add to the built environment that goes beyond sustainability and begins to end up regenerative [64, 65]. This does not prevent creature biomimicry at a detail or material level.

Ecology and architecture are comparable and similar in one principle; they are both expressions of the arts of space shaping and making relations among humans and their surroundings. This type is called eco-architecture. The re-evaluation of the relationship between nature and architecture is a must. In this context, architects use the principles of nature for the efficient functioning of the building and take inspiration from the deeper laws of nature [64, 65].

At the core of ecological architecture, it lays the fact that buildings and natural systems remain in symbiotic connection and positively influence each other. Both participants start to help each other, work together, and adapt. Environmental and social problems such as waste production, energy consumption, material use, and greenhouse gas (GHG) emissions attribute to the built environment since the human has been created.

Biomimicry will respond to the quick urban development in an efficient way and environmentally friendly approach. Biomimicry helps in creating innovative and eco-friendly approaches that can provide convenient and flexible solutions for human beings' problems. Organisms in nature are efficient and adapt to their context. There are no disposals in nature; it reuses and recycles everything, adjusts to nearby conditions, exploits sun and daylight and other natural sources of energy, and only utilizes the energy and resources that it needs. In addition, the biomimicry approach provides solutions for structural and water efficiency, zero-waste systems, thermal environment, and energy supply, which are fundamental for any sustainable building design [64, 65].

3.9 Biomimicry in Egypt

To the authors' knowledge, there are no architectural mimicked examples of biomimicry here in Egypt yet, but there is a group of students who use biomimicry as a concept and approach of sustainable and environmentally friendly designs. This was depicted in an irrigation system in Fayoum in an international competition. Lotfi, N., is a design student at the German University in Cairo, Egypt, as part of the "biomimicry student design challenge 2012." They created a team to enter this competition as the term biomimicry was new, but it inspired several ideas into their mind which helped her link design with sustainability. It was the start of a journey to the stagnant canals and over 5000 miles east to Boston to accept a first-place award from biomimicry pioneer Janine Benyus. The challenge's theme of that year was related to water; Lotfi's team was attracted to the potential benefits of biomimicry in addressing water-related problems in Egypt [66]. The team name was Egy-Osmo (a combination of Egypt and osmosis), which decided to think profoundly to create a solution to benefit their country and contribute in solving a problem. It was carried out by looking locally, especially in the city of Fayoum, a well-known city for its agricultural activities. In Egypt, long canals created by humans and known as "Teraa" are commonly seen and are used for crop irrigation; however, they run through Fayoum and are most probably used in different ways.

Creating sustainable solution to the irrigation problem was the team's main mission. The team inspired the idea and inspiration from the giraffe and dromedary camel after a long time of searching and thinking, designing a bio-irrigation system called “dromedarily sustainable” [65]. Same as the dromedary camel’s technique in circulating water in its group, the team produced a system that circles water by utilizing the energy caught from a waterfall to keep microbial growth down. The group likewise looked at the giraffe and how it directs blood through its neck, for motivation on the best way to push water through a tight compartment to increase speed and pressure. Essentially and sustainability water is circled around and continuously available until an agriculturist needs it. At the point when water is required, the agriculturist can open an entryway along the border of the perimeter to enable it to veer off into the homestead. Egy-Osmo presented their plan to the biomimicry understudy configuration challenge and had made it to the second round and was welcome to Boston to contribute with their ideas in front of judges [66].

3.10 The Future of Biomimicry

Biomimicry has a very bright future especially in the time of the COVID-19 pandemic, if we decide to be connected to nature and get inspiration from it. Nature evolves and adapts to the surroundings and thus forces natural systems to become highly optimized and efficient. In the same way, our buildings and their structure will be forced to become living organisms in the future and yet have to adapt, not only to their surrounding environments, reducing and reusing the waste they produce, but also to climate change and COVID-19 pandemic as well as post-COVID. This will require all to learn from natural systems, which serve as a guideline for rational design of target functionalities. Having said that, we are not constrained by feedstock availability since nature does not produce waste in a stable ecosystem as compared to modern manufacturing economy, which produces vast amounts of waste. This could be an inspiration in producing zero-waste as well. Numerous innovations are currently in use or being developed and hence will add to making the living buildings conceivable [66]. At a later stage, houses and workplaces which are intended to work like living creatures, particularly adjusted to their area and location, will be able to get their needs [67].

Today, maybe with a small impact of appeal of the environmentalist thought, a trend toward nature is tangible. Recent journals contain articles about activities demonstrating a natural character. This breeze of nature for the most part relies upon the technological improvement, contemporary construction methods, high-tech building components, materials, and efficient climate controlling systems. In addition, advanced computer, building information modeling (BIM), and IoT are considered the important inventions of the twenty-first century that provide accuracy to dwindle the gap. Technical information has extended and enhanced a great deal, where humans have more opportunity to make natural buildings and catch up with the efficiency of natural structures. Furthermore, nature is essential for our existence and for the development of innovative architecture and cities [68].

3.11 Conclusion

The chapter highlighted the purpose of why biomimicry has been presented and discussed. The approaches of biomimicry – design looking to biology and biology to design approach – were depicted. Furthermore, the levels of biomimicry, including organism level, behavior level, and ecosystem level, are presented. At the ecosystem level, it included mimicking the process strategies of the ecosystem and mimicking the function of ecosystem; moreover, levels of biomimetic information have been illustrated. Application types of biomimicry in architecture and and the sources of inspiration from nature – visual inspiration, conceptual inspiration, and computational inspiration – were highlighted. The future of biomimicry in general and in Egypt has been covered in this chapter.

Learning about biomimicry is profoundly important to find out how to use it and choose the appropriate level when designing a specific design of building. Also comprehending the approaches of biomimicry assists in taking the right decision in design phases to choose which approach will improve the quality of the design architecturally and environmentally. In addition, nature principles should be the architects' design specifications in order to have an environmentally friendly and sustainable architecture. Many iconic examples by legendary architects, such as Alvar Aalto, Sir Norman Foster, Zaha Hadid, Santiago Calatrava, and Stefano Boeri as well as Vincent Callebaut, have depicted biomimicry in architecture and created many innovative buildings to be followed by architects worldwide. Finally, a realization has been reached that there are no biomimetic architectural examples in Egypt till date; hence, this chapter aims to encourage considering biomimicry approaches in the design processes as a strategy to achieve sustainability, especially in the age of COVID-19.

This coverage will lead to the next chapter which will present the biomimicry's application in connection with sustainable elements and global architectural design examples. Moreover, these examples will be analyzed to find out which example is the best biomimetic example according to the criteria that were selected.

References

1. Acioly, C., Vigno, R., Jonsson, A., Amirtahmasebi, R., Vuova, Z., & Fox, E. (2020). *The new urban agenda* (pp. 67–144). United Nations Human Settlements Programme (UN-Habitat). Retrieved February 20, 2021, from https://unhabiat.org/sites/default/files/2020/12/nua_handbook_14dec2020_2.pdf
2. The Food and Agriculture Organization of the United Nations and Earthscan. (2011). *The state of the world's land and water resources for food and agriculture managing systems at risk.* Retrieved December 28, 2020, from http://www.fao.org/3/i1688e/i1688e.pdf
3. Global Status Report for Building and Construction. (2019). *World Green Building Council.* Retrieved February 22, 2021, from https://www.worldgbc.org/news-media/2019-global-status-report-buildings-and-construction

4. Pérez-Lombard, L., Ortiz, J., & Pout, C. (2008). A review on buildings energy consumption information. *Energy and Buildings, 2008*(40), 394–398. Retrieved December 28, 2020, from https://www.sciencedirect.com/science/article/abs/pii/S0378778807001016
5. Yang, L., Yan, H., & Lam, J. C. (2014). Thermal comfort and building energy consumption implications – A review. *Applied Energy, 115*, 164–173. Retrieved December 28, 2020, from https://www.sciencedirect.com/science/article/abs/pii/S0306261913008921
6. Hill, R. W., Wyse, G. A., & Anderson, M. (2012). *Animal physiology* (3rd ed.). Sinauer Associates, Inc.. Retrieved February 22, 2021, from https://www.dphu.org/uploads/attachements/books/books_1984_0.pdf
7. Satterthwaite, D. (2008). Cities' contribution to global warming: Notes on the allocation of greenhouse gas emissions. *Environment and Urbanization, 20*, 539–550. Retrieved 22 February 2021, from https://www.researchgate.net/publication/235732092_Cities%27_contribution_to_global_warming_Notes_on_the_allocation_of_greenhouse_gas_emissions
8. Ritchie, H., & Roser, M. (2018). *Urbanization – Our World in Data*. Retrieved February 22, 2021, from https://ourworldindata.org/urbanization
9. Ürge-Vorsatz, D., et al. (2014). Energy end-use: Buildings, chapter 10. In *Global energy assessment – Towards sustainable future* (pp. 649–760). Cambridge University Press and the International Institute for Applied Systems Analysis. Retrieved February 22, 2021, from at: http://3csep.ceu.hu/sites/default/files/publications/geachapter10buildingshires.pdf
10. Mallawarachchi, H., Rameezdeen, R., & De Silva, L. (2013). *Green buildings, resilience ability and the challenge of disaster risk*. International Conference on Building Resilience 2013. Retrieved February 15, 2021, from https://www.researchgate.net/publication/277892802_Green_buildings_resilience_ability_and_the_challenge_of_disaster_risk
11. Hill, R. W., Wyse, G. A., & Anderson, M. (2012). *Animal physiology* (3rd ed.). Sinauer Associates, Inc.
12. Del Grosso, A., & Basso, P. (2010). Adaptive building skin structures. *Smart Materials and Structures, 19*. Retrieved February 21, 2021, from https://iopscience.iop.org/article/10.1088/0964-1726/19/12/124011
13. John, G., Clements-Croome, D., & Jeronimidis, G. (2005). Sustainable building solutions: A review of lessons from the natural world. *Building and Environment, 40*, 319–328. https://www.sciencedirect.com/science/article/abs/pii/S0360132304001829
14. Bengani, V. (2015). Secondary case study on the Gherkin building, London. Retrieved February 23, 2021, from https://www.slideshare.net/VikramBengani/the-gherkin-case-study
15. Benyus, J. M. (2002). *Biomimicry. Innovation inspired by nature*. Harper Perennial. Retrieved January 2, 2021, from at: https://www.academia.edu/38300413/Janine_M_Benyus_Biomimicry_Innovation_Inspired_by_Nature_2002_Harper_Perennial_1_
16. UNEP. (2015). *Sustainable consumption and production global edition*. Unep.org. Retrieved February 18, 2021, from https://sustainabledevelopment.un.org/content/documents/1951Sustainable%20Consumption.pdf
17. MacDonald, A. (2007). How cradle to cradle design principles can be implemented into architectural construction methods and designing for the lifecycle of a building? (pp 10–19). Retrieved February 18, 2021, from https://intrepidarchitecture.files.wordpress.com/2012/12/cradle_to_cradle_architecture.pdf
18. Blok, V., & Gremmen, B. (2016). Ecological innovation: Biomimicry as a new way of thinking and acting ecologically. *Journal of Agricultural and Environmental Ethics*. https://www.researchgate.net/publication/289585359_Ecological_Innovation_Biomimicry_as_a_New_Way_of_Thinking_and_Acting_Ecologically
19. Rankouhi, A. R. (2012). Naturally inspired design investigation into the application of biomimicry. In *Architectural design*. Retrieved 2 January 2021, from https://issuu.com/azarch/docs/thesis
20. Adebisi, G. (2015). Assessment of biomimicry design concept adoption in architecture: Towards a sustainable built environment in Nigeria. *IOSR Journal of Environmental Science, Toxicology and Food Technology (IOSR-JESTFT)*, 41–46. https://www.academia.

edu/13030653/Assessment_of_Biomimcry_Design_Concept_Adoption_in_Architecture_Towards_a_Sustainable_Built_Environment_in_Nigeria
21. EL-Zeiny, R. M. (2012). *Biomimicry as a problem solving methodology in interior architecture*. ASEAN Conference on Environment-Behaviour Studies. Elsevier Ltd. Selection and peer-review under responsibility of the Centre for Environment. Retrieved January 2, 2021, from https://www.sciencedirect.com/science/article/pii/S1877042812031928
22. Pedersen- Zari, M. & Storey, J. (2007). An ecosystem based biomimitic theory for regenerative built environment. *Lisbon Sustainable Building Conference 07*. Retrieved January 2, 2021, from https://www.researchgate.net/publication/261477552_An_ecosystem_based_biomimetic_theory_for_a_regenerative_built_environment
23. EL-Zeiny, R. M. (2012). *Biomimicry as a problem solving methodology in interior architecture*. ASEAN Conference on Environment-Behavior Studies. Elsevier Ltd. Selection and peer-review under responsibility of the Centre for Environment. Retrieved January 2, 2021, from https://core.ac.uk/download/pdf/82730369.pdf
24. Mansour, H. (2010). *Biomimicry a 21st century design strategy integrating with nature as a sustainable way*. BUE. Retrieved January 2, 2021, from https://www.academia.edu/432839/BIOMIMICRY_A_21ST_CENTURY_DESIGN_STRATEGY_INTEGRATING_WITH_NATURE_IN_A_SUSTAINABLE_WAY
25. Sherif, M. S. (2016). Biomimicry as an approach for bio-inspired structure with the aid of computation. *Alexandria Engineering Journal, 55*, 707–714. https://www.researchgate.net/publication/284913393_Biomimicry_as_an_approach_for_bio-inspired_structure_with_the_aid_of_computation
26. Aman, A., Komari, C., Nimesh, M., Bhgat, S., & Bengani, V. (2015). *Secondary case study on the gherkin building*, London. Presented as part of the Structures Seminar for B.Arch. Programme at School of Planning & Architecture, New Delhi. Presentation. Retrieved January 2, 2021, from https://www.slideshare.net/VikramBengani/the-gherkin-case-study
27. Dash, S. (2018). Application of biomimicry in building design. *International Journal of Civil Engineering and Technology (IJCIET), 9*(2), 644–660. http://www.iaeme.com/MasterAdmin/UploadFolder/IJCIET_09_02_062-2-3/IJCIET_09_02_062-2-3.pdf
28. Ahmar, S. A. (2011). *Master thesis – Biomimicry as a tool for sustainable architectural design*. Faculty of Engineering, Alexandria University. Retrieved January 2, 2021, from https://www.academia.edu/36344867/BIOMIMICRY_AS_A_TOOL_FOR_SUSTAINABLE_ARCHITECTURAL_DESIGN_TOWARDS_MORPHOGENETIC_ARCHITECTURE
29. De Pauw, I. (2015). *Nature-inspired design strategies for sustainable product development (Ph.D)*. Delft University of Technology. Retrieved February 18, 2021, from https://www.ideal-co.nl/img/NatureInspired%20Design_Thesis_Pauw.pdf
30. Pedersen Zari, M., & Hecht, K. (2020). *Biomimicry for regenerative built environments: Mapping design strategies for producing ecosystem services*. Biomimetics. Retrieved February 18, 2021, from https://europepmc.org/article/med/32408595
31. Pedersen Zari, M. (2015). Mimicking ecosystems for bio-inspired intelligent urban built environments. *Intelligent Buildings International*. Retrieved February 18, 2021, from https://www.tandfonline.com/doi/abs/10.1080/17508975.2015.1007910?journalCode=tibi20
32. Pedersen-Zari, M. (2012). *Ecosystem services analysis for the design of regenerative urban built environments (Ph.D)*. Victoria University of Wellington. Retrieved February 21, 2021, from https://core.ac.uk/download/pdf/41337786.pdf
33. Portland Development Commission: Governance. (2021). *Structure and process*. Retrieved January 2, 2021, from https://core.ac.uk/download/pdf/37768765.pdf
34. Holden, E., Linnerud, K., & Banister, D. (2016). *The imperatives of sustainable development*. Wiley Online Library. Retrieved February 21, 2021, from https://www.researchgate.net/publication/308791774_The_Imperatives_of_Sustainable_Development
35. Pedersen-Zari, M. (2010). Biomimetic design for climate change adaptation and mitigation. *Architectural Science Review, 53*(2), 172–183. https://www.researchgate.net/publication/262857243_Biomimetic_design_for_climate_change_adaptation_and_mitigation

36. Arash, V, (2009). *Nature as a source of inspiration of architectural conceptual design*. Master thesis, Eastern Mediterranean University, Architecture – Gazima. Retrieved January 24, 2018, from http://i-rep.emu.edu.tr:8080/xmlui/handle/11129/185
37. City Hall. Foster + Partners. (2002). Foster+parnters.com. Retrieved February 21, 2021, from https://www.fosterandpartners.com/projects/city-hall/
38. London Assembly (2002). London.gov. Retrieved February 12, 2021, from: https://www.london.gov.uk/about-us/our-building-and-squares/about-our-building
39. Denver Art Museum. (2021). Denverartmuseum.org. Retrieved February 12, 2021, from https://www.denverartmuseum.org/en/edu/object/concept-model-denver-international-airport-passenger-terminal-denver-colorado-1991-95
40. Clark, E., & Gilpin, D. (2005). Selfridges, birmingham. *The Arup Journal*. Retrieved February 12, 2021, from http://www.formpig.com/pdf/formpig_ARUP%20selfridges_future%20systems.pdf
41. Nikolousi, M. (2021). *The selfridges building, Birmingham*. John Desmond Limited Architectural & Metal Finishes Specialist. Retrieved February 12, 2021, from https://www.johndesmond.com/blog/design/the-selfridges-building-birmingham/
42. Kaur, C. (2021). *20 incredible building facade designs that will make you rethink building envelope*. Rethinking The Future. Retrieved February 12, 2021, from https://www.johndesmond.com/blog/design/the-selfridges-building-birmingham/
43. Soumaya, M. (2021). *Modern in Denver, Colorado's design magazine*. Modern in Denver. Retrieved February 12, 2021, from https://www.modernindenver.com/2017/06/museo-soumaya-fr-ee-mexico-city-1994/
44. Cruz, D. (2021). *Soumaya Museum/FR-EE Fernando Romero Enterprise*. Archdaily. Retrieved February 21, 2021, from https://www.archdaily.com/452226/museo-soumaya-fr-ee-fernando-romero-enterprise
45. Escobar, M. (2016). City-performance: A countervisual interpretation of the tower from performative ethnography. URBS. *Revista De Estudios Urbanos Y Ciencias Sociales, 6*, 21–34. https://www.researchgate.net/publication/304405278_City-performance_a_countervisual_interpretation_of_the_tower_from_performative_ethnography
46. Glories, T. (2021). *Structurae, International Database and Gallery of Structure*. Retrieved February 12, 2021, from https://structurae.net/en/structures/torre-glories
47. Encircle photos explore the world daily. (2021). Encirclephotos.com. Retrieved February 12, 2021, from https://www.encirclephotos.com/image/torre-glories-in-sant-marti-district-in-barcelona-spain
48. Escalando la Torre Agbar. (2021) *Bcnhoy*. Retrieved February 12, 2021, from https://www.bcnhoy.com/escalando-la-torre-agbar.html
49. Olcayto, R. (2014). London Aquatic Centre by Zaha Hadid architects. *Architects Journal*. Retrieved February 12, 2021, from https://www.architectsjournal.co.uk.archive/london-aquatic-centre-by-zaha-hadid-architects
50. Moore, R. (2011). Olympics Aquatic Centre – Review: Zaha Hadid's London 2012 Aquatic Centre hasn't come cheap at £269m, but it is the Olympics' most majestic space. *The Guardian – The Observer*. Retrieved February 12, 2021, from https://www.theguardian.com/commentisfree/2014/mar/03/in-praise-london-aquatics-centre
51. Zaha Hadid Architects' London Aquatics Centre breaks record with 2.5M visitors since opening in 2014. (2017). Worldarchitecture.org. Retrieved February 12, 2021, from https://worldarchitecture.org/architecture-news/cvhzf/zaha-hadid-architects-london-aquatics-centre-breaks-record-with-2-5m-visitors-since-opening-in-2014.html/
52. Blankenbehler Be. (2015). Gare do Oriente train station, Lisbon, Portugal. Architecture revived. Retrieved February 12, 2021, from https://www.architecturerevived.com/gare-do-oriente-train-station-lisbon-portugal
53. Inexhibit. (2014). Inexhibit.com. Retrieved February 12, 2021, from https://inexhibit.com/case-studies/eoul-dongdeamun-design-plaza-by-zaha-hadid-architects/

54. Krichels, J. (2012). South Korea's Expo 2012 Pavilion: Active Façade Design, *The Architect's Newspaper*. Archpaper. Retrieved February 12, 2021, from https://archpaper.com/2012/06/south-koreas-expo-2012-pavilion-active-facade/ .
55. Lintula, K., Sirola, N. &Summanen, M. (2012). Kamppi chapel of silence. *Archilolvers*. Retrieved February 12, 2021, from https://www.archilovers.com/projects/51632/Kamppi-chapel-of-slience.html
56. Urban Cactus, Sky garden in the city. (2021). AEMSEN. Retrieved February 12, 2021, from https://www.aemsen.com/urban-cactus
57. Cub–shaped houses – The official website of the Netherlands Board of Tourism & Convention. Web.archive.org. (2011). Retrieved November 20, 2021, from https://web.archive.org/web/20110301155645/; http://www.holland.com/uk/cities/rotterdam/architecture/kubuswoningen.jsp
58. Morrison, J. (2018). *Building in 2018 is on the up*. THE TIMES: Arts, p.8, 29 January 2018. Retrieved February 12, 2021, from https://www.stefanoboeriarchitetti/en/news/the-times-building-in-2018-is-on-the-up/
59. Asia's first vertical forest may be the solution to all pollution problems. (2017) Economic times. Retrieved February 12, 2021, from https://m.economictimes.com/nation-world/asias-first-vertical-forest-may-be-the-solution-to-all-pollution-problems/go-green/slideshow/57056103.cms
60. Callebaut, D. (2021). NAUTILUS ECO-RESORT – Vincent Callebaut Architectures. VINCENT CALLEBAUT ARCHITECTURES. Retrieved December 23, 2021, from https://vincent.callebaut.org/object/170831_nautilusecoresort/nautilusecoresort/projects
61. Pavilion-expo. (2015). Nemesistudio. Retrieved April 10, 2019, from http://www.nemesistudio.it/en/projects/type/culture/item/714-italy-pavilion-expo-2015,-milan.html/
62. Life's Principles – Biomimicry Design Lens. (2016). Retrieved February 22, 2021, from https://biosis.net/wpcontent/uploads/2016/04/Biomimicry38_DesignLens_Lifes_Principles_WEB.pdf
63. Hooker, G. (2020). Collections Biomimicry 101. Asknature. Retrieved February 21, 2021, from https://asknature.org/collections/biomimicry-101/
64. Shahda, M. M., Elmokadem, A. A. F., & Abd Elhafeez, M. M. (2014). Biomimicry levels as an approach to the architectural sustainability. *Port Said Engineering Research Journal*. https://pserj.journals.ekb.eg/article_45298_b66878851668638814ecca068311d80.pdf
65. Nour ElDin, N., Abdou, A., & Abd ElGawad, I. (2016). Biomimetic potentials for building envelope adaptation in Egypt. *Procedia Environmental Sciences, 34*(375–386), Elsevier. https://www.sciencedirect.com/science/article/pii/S187802961630055X
66. Biomimicry Institute. (2021). Biomimicry.com. Retrieved January 24, 2018, from https://biomimicry.org/stories-field/stories-field-egy-osmo/
67. Gruber, P. (2011). Biomimitics in architecture. In P. Gruber, D. Bruckner, C. Hellmich, H. B. Schmied, & H. Stachelberger (Eds.), *Biological and medical physics, biomedical engineering* (pp. 128–148). Springer. Retrieved January 24, 2018, from https://www.researchgate.net/publication/225970908_Biomimetics_in_ArchitectureArchitekturbionik
68. Kennedy, E. B. (2016). *Biomimetic buildings: The emerging future of architecture*. GBER. Retrieved March 15, 2018, from http://globalbuiltenvironmentreview.co.uk/wp-content/uploads/2018/05/10.1-Kennedy_Revised-Biomimicry-MS_Clean-Copy_080916.pdf

Chapter 4
Sustainability and Global Applications of Biomimicry Approaches and Levels in Architecture

4.1 Introduction

Biomimetics is defined as "the interdisciplinary cooperation of biology and technology or other fields of innovation to solve practical problems through the functional analysis of biological systems, their abstraction into models, and the transfer of these models along with their applications into solutions" [1]. Lately, the applications of the biomimicry approach have been extended progressively to different fields. Such applications include urban and architectural design and contribute to the design and functionality of buildings through a long evolutionary process. Biomimicry approach contributes to a sustainable society at the conceptual level; in this chapter, trends in the application of biomimetics to architecture are investigated in order to identify potential issues and successes resulting from implementation [2]. In the application of biomimicry in architectural design (Figs. 4.1 and 4.2), parts of "organisms" such as form or surface structure are frequently mimicked, whereas in urban design or larger scale, the application of biomimicry is applied to mimic whole ecosystems [2]. Therefore, there is a need for a mindset shift as stated by Michael Pawlyn (Box 4.1).

Box 4.1: Need for Mindset Shift

What is needed is shift in mindset, perhaps even the way that we pay for things. (Michael Pawlyn)

M. Aboulnaga, S. E. Helmy, *Biomimetic Architecture and Its Role in Developing Sustainable, Regenerative, and Livable Cities*,
https://doi.org/10.1007/978-3-031-08292-4_4

Fig. 4.1 Biomimicry depicted in the exterior of the gardens by Marina Bay Sands in Singapore. (Image credit and source: https://upload.wikimedia.org/wikipedia/commons/0/0d/Marina_Bay_Sands_from_Gardens_By_The_Bay.jpg)

4.2 Biomimicry Application in Line with Sustainability

Biomimicry is becoming more important than ever due to the rapid increase of the world's population and the corresponding need to meet the basic services such as housing and buildings, especially with more than 55% of the world's population living in urban areas and expected to increase to 68% by 2050 as depicted in Box 4.2 [3]. More than 55% of the world's population are living in urban areas with a projected increase to reach 68% by 2050 as depicted in Box 4.2 [3]; therefore, biomimicry is becoming more important now than ever due to this rapid increase of the world's population along with the corresponding need to meet the basic services such as housing and buildings. The application of sustainability and biomimicry architecture contributes to solving such challenges. Recent success stories exist in terms of how biomimicry can be applied to buildings' design, while buildings serve to protect us from nature's extreme conditions; this doesn't mean that they do not have anything to learn from the biological world. In fact, nature regularly builds structures with functionality that human-built structures could usefully emulate, for example, the palm tree stem elastically moves left and right (swings) in order to withstand the severe wind force and stabilizes after the wind is calm again (Fig. 4.3). Biomimetic research, science, and applications continue to grow and are already influencing the next generation of building products and systems and building designs [4]. The core of the biomimetic approach is to engage the design with the

Fig. 4.2 The mushroom columns inspired from nature – the gardens by Marina Bay Sands in Singapore. (Image credit and source https://upload.wikimedia.org/wikipedia/commons/0/0d/Marina_Bay_Sands_from_Gardens_By_The_Bay.jpg)

Fig. 4.3 Biomimicry depicting palm trees swinging to withstand wind forces. (Image credit and source: Loz Blain, https://upload.wikimedia.org/wikipedia/commons/9/9e/Singapore%2C_Gardens_by_the_bay_-_panoramio_%286%29.jpg)

function, transferred by particular natural adaptation, which is the secret key to fruitful implementation of the biomimicry principles.

Box 4.2: The Increase in Global Population Highlights the Need for Biomimicry

For the first time in history, more than 55% of World's population lives in an urban area; by 2050 it is expected to increase to 68%. (Michael Pawlyn)

Architecture applications exist in biomimicry; they are reflected on many levels that need to be achieved in the design process, to mitigate climate change and the built environment adaptation [5]. Biomimicry plays a major role in attaining sustainability and the sustainable development goals (SDGs), mainly goal 11 – Sustainable Cities and Communities. The engagement of a biologist in the early design phase of projects' development is important to developing biomimicry architecture (Box 4.3) in line of sustainability. Figure 4.4 presents the six sustainability elements to achieve the biomimetic concept; these elements will be discussed in the next section.

Fig. 4.4 Sustainability elements. (Image source: Developed by authors)

Box 4.3: The Role of Biologists in Developing Biomimicry in Architecture

Having a biologist at the design table right from the early stage of a project could not only change this, but also throw up some major breakthroughs. (Michael Pawlyn)

4.2.1 Structural Efficiency

Structural efficiency examples of biomimicry architecture are clearly manifested in two iconic buildings: (a) Burj Khalifa, Dubai, UAE, and (b) ICD/ITKE Research Pavilion, Stuttgart, Germany

4.2.1.1 Burj Khalifa, Dubai, UAE

A vivid example of biomimicry architecture is depicted in Burj Khalifa's structural metaphor in Dubai, UAE, where the idea was inspired from the *Hymenocallis* or spider lily – a flower that is commonly cultivated in the desert of Dubai and the Gulf region (Fig. 4.5). Such flower has inspired the renowned American design office – Skidmore, Owing, and Miller (SOM) to create the structural design of the tallest building in the world by emulating the leaves of the lily flower as shown in Fig. 4.5a, which makes it an excellent example of biomimicry structural efficiency.

Fig. 4.5 Burj Khalifa, a vivid example of biomimicry in architecture depicting a desert flower in plan and structure. (**a**) Lily flower (**b**) Burj Khalifa with nature (**c**) The tallest building's structure inspired by *Hymenocallis* (lily flower) leaves. (Image credit and source: (**a**) https://commons.wikimedia.org/wiki/File:Atamasco_Lily.jpg (**b**) and (**c**) Mohsen Aboulnaga)

4.2.1.2 Institute of Computational Design: ICD/ITKE Research Pavilion 2016/2017

In nature "materials are expensive and shape is inexpensive." The ideal example of biomimicry is the adaptive structures, in light of the fact that it enables structures to adjust simply like living organisms do, by altering their behavior responding to the surrounding environmental conditions [5]. Nervi's capability to compete through lightness, domes and shells accomplishes factor 10 increment in efficiency, along with thin pressurized layers which take this significantly further to factor 100 increments in resource efficiency. This has been inspired by spider behavior [5]. Improving the scientific knowledge of the designers led to the creation of beautiful and efficient structures that are inspired from and existing in nature [5], as shown in the veins leaves and spider silk web of the Palazzetto dello Sport in Rome, Italy (Fig. 4.6). Such structures' forms are classified as a behavior level of the

Fig. 4.6 Palazzetto dello Sport stadium in Rome, Italy. (a) Veins' structure of a leaf (b) Reticulate veins in *Ficus carica* (c) Spider's silk of the web (d) Inner view of the Palazzetto dello Sport (e) Stadium's roof tensile structure. (Images' credit and source: (a) Vaibhav Madhav Naikare https://commons.m.wikimedia.org/wiki/File:Bottle_Green_Veins.jpg (b) Jon Richfield https://commons.m.wikimedia.org/wiki/File:Reticulate_veins_in_leaf_of_Ficus_carcia_IMG_9202s.jpg (c) https://commons.wikimedia.org/wiki/File:Spider_web_(Newark,_Ohio,_USA)_3_(31169062330).jpg (d) Flickr, https://www.flickr.com/photos/lifelog-it/38717931965/in/photolist-21Znwg8-ipbo8E-dz94hr-dzezhU-ipbitD-dzexmw-dzexUG-dt64v2-o9gNpB-dsWPWg-dtqD5Y-dtqbMU-dt9sf8-dt5X9E-2f1XVzw-dtqFFQ-dsWRBk-dtdNpT-28mWJNt-dsWQzx-28mWK1H-dsWNZH-dt9dut-dt9wXj-dtdCCM-dsWN4F-dtdRuS-dsWWuJ-dt6myK-dt5CLg-dt9sfS-dsWLQa-dtdJ5x-dt9v19-dt61tx-dte2vJ-dt6k1w-dt5FTn-dt6czt-dtdVVj-dt5EyX-dt6qJd-dt9KEC-dt66tH-dtdUhj-dtdLTx-dt6t4u-dt6pzD-dt5Rca-dtqtDS (e) Absurdicus, https://m.wikimedia.org/wiki/File:Palazzetto_dello_Sport.JPG)

biomimicry levels as presented in the ICD/ITKE Research Pavilion 2016–2017 in Stuttgart, Germany, which is portrayed in Fig. 4.7. The progress of installation mimicking spider behavior [5], an example where the spider constructs a flat sheet web, under which the air bubble is set, is consecutively fortified by various leveled plans of fibers from inside, as shown in Fig. 4.8.

Parametric design is a process based on algorithmic thinking that enables the expression of parameters and rules that collectively define, encode, and clarify the relationship between design intent and design response. Moreover, parametric design is an excellent way to engage in architecture to translate and apply the biomimicry thinking and bring it into reality [6]. In this context, fiber-reinforced structure is an inspiring structure from nature, and the geometrically formed structure mimics waves (Fig. 4.9). Examples that mimic nature are shown in Figs. 4.10 and 4.11. The Stuttgart Airport illustrates another inspiration from nature, where

(a) (b)

Fig. 4.7 The ICD/ITKE Research Pavilion 2016–2017 in Stuttgart, Germany. (**a**) Skelton of the pavilion (**b**) Close-up view of the fiber-reinforced structure. (Images' credit and source: (**a**) MSeses, https://commons.m.org/wiki/File:ICD_Research_Pavilion_2016_17_Stuttgart_01.jpg (**b**) MSeses, https://commons.m.org/wiki/File:ICD_Research_Pavilion_2016_17_Stuttgart_02.jpg)

(a) (b)

Fig. 4.8 Spider behavior. (**a**) Spider behavior in wetland forming a flat sheet web (**b**) A spider is forming the web. (Images' credit and source: (**a**) Tan Pek Nan, https://comoms.m.wikimedia.org/wiki/File:Araneidae_(Orb-weaver_Spider).jpg (**b**) Stephencdickson, https://comoms.m.wikimedia.org/wiki/File:A_classic_circular_form_spider%27s_web.jpg#)

Terminal 1 mimics tree branches in the form of columns as portrayed in Figs. 4.12, 4.13, and 4.14.

At the time when difficulties arise in finding solutions for cities' challenges and problems, specifically in the architecture and construction sectors, it is suggested to simplify complex issues – meaning to focus on basics and nature, yet learn from animals' behavior and how their bodies act and re-act to their surroundings. Thus, biomimicry is the way for inspiration to ensure sustainable architecture [6].

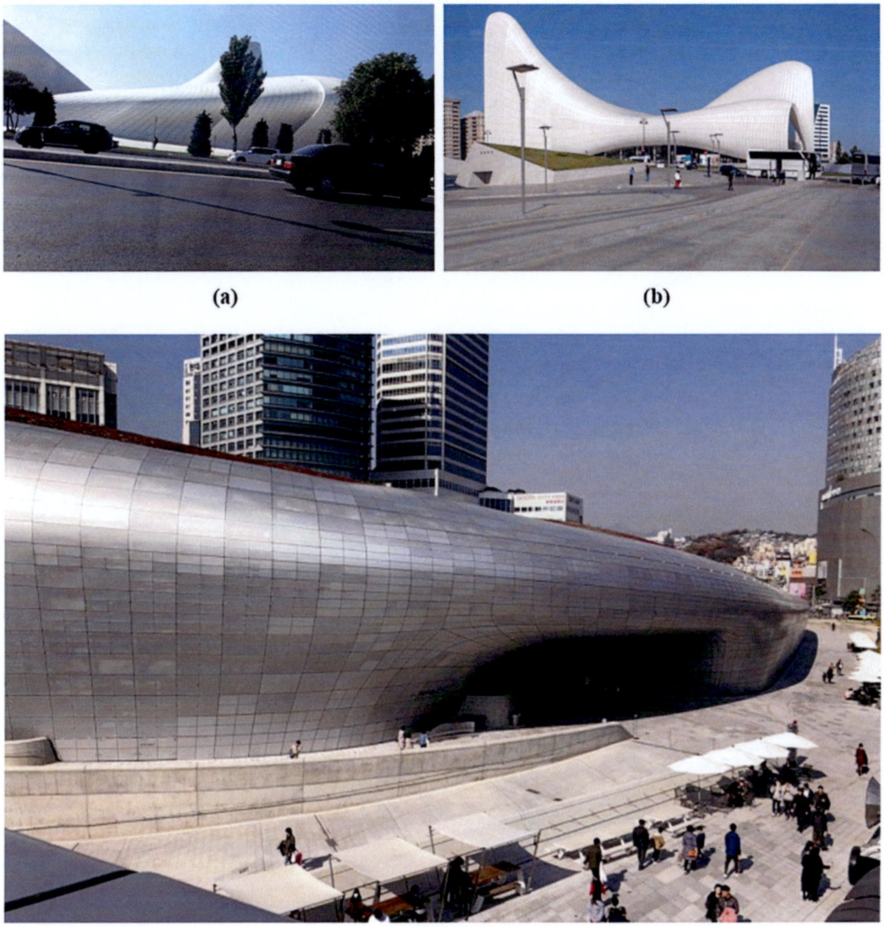

(a) (b)

(c)

Fig. 4.9 Fiber-reinforced structure examples in Baku, Azerbaijan and Seoul, South Korea. (**a**) View of Heydar Aliyev Center in Baku (**b**) Fiber-reinforced structure, Heydar Aliyev Center (**c**) Design center and exhibition using fiber-reinforced structure (FRS) in Seoul. (Images' credit and source: (**a**) Mohsen Aboulnaga (**b**) Matti Blume, https://commons.m.wikimedai/wiki/File:Heydar_Aliyev_Centre,_Baku_(P1090245-Pano).jpg (**c**) Mohsen Aboulnaga and Nada Nidal)

4.2.2 Material Manufacturing, ICD/ITKE Research Pavilion 2012

Emulating the efficiency of nature, in the man-made manufacturing processes, is a worthy goal in order to face the current and future challenges of the city including resource depletion, climate change, and above all the COVID-19 pandemic and crisis. All materials should undergo both biological and technological cycles, to solve the problem of waste production through natural system principles.

Fig. 4.10 Inspiring structure from the nature. (**a**) Termite mound structure (**b**) Inspiring structure from termite mounds (**c**) Birds' nest structure in nature (**d**) Beijing Olympic Stadium inspired from bird's nest structure. (Images' source: (**a**) https://upload.wikimedia.org/wikipedia/commons/a/a6/TermiteMound01LB.jpg (**b**) https://en.wikipedia.org/wiki/Organic_architecture#/media/File:R%C3%A5dhuset_metro_station_June_2015.jpg (**c**) https://commons.wikimedia.org/wiki/File:Birds_Nest_Nepal.jpg (**d**) https://commons.wikimedia.org/wiki/File:Beijing_national_stadium.jpg)

According to the rapid manufacturing of technologies, there is a scope for developing more cycled biological materials with much lower energy consumption than the technical cycle [5]. The excessive use of materials and the pressure on natural resources such as energy will lead to driving such usage from the technological cycle toward models of cyclical management that mimic natural systems as presented in Fig. 4.15. Hence, it is easily imaginable that manufacturing of a construction element, which uses cellulose, could achieve savings in energy compared to traditional approaches [5].

One of the most important processes that assist in realizing truly sustainable architecture is the shift from a linear, wasteful, and pollutant way of using resources to a closed loop model, which is shown in Fig. 4.16. The concepts of closed loop management of resources and biomimetic manufacturing are inextricably linked, as Julian Vincent neatly summarized [5]. The incorporation of the form generation techniques, the computational simulations, and automated manufacturing permitted the improvement of structure [5]. Such structure requires just a shell thickness of 4 millimeters of composite cover while spanning 8 m. Therefore, it is classified as an organism level in biomimicry levels; the focus is on the material and arthropods'

(a)

(b)

Fig. 4.11 Formed structure inspired by trees. (**a**) Lisbon Oriente Railway Station, Portugal (**b**) Tree structure of Stuttgart Airport in Germany. (Images' credit and source: (**a**) Joao Pimentel Ferreira, https://commons.m.wikimedia.org/wiki/File:Plataforma_ferroviaria_da_Gare_ do_ Oriente.jpg (**b**) Lasse Fuss, https://commons.wikimedia.org/wiki/File:Stuttgart_Flughafen_ Terminal_1.jpg#)

exoskeletons as illustrated in Fig. 4.17. Professor Achim Menges (ICD); Professor Jan Knippers (ITKE); IIGS University of Stuttgart, Germany; and the Center of Improvement automated creation process (robotic fabrication) developed inside the setting of the building industry the ICD/ITKE Research Pavilion 2012 in Germany. The pavilion structure is based on fiber or wire rolling of carbon and glass strand techniques.

Fig. 4.12 Interior of Terminal 1 of Stuttgart Airport at night showing the nature-shaped columns. (Image credit and source: CatalpaSpirit, https://commons.wikimedia.org/wiki/File:Stuttgart_Airport_Terminal_1.jpg)

Fig. 4.13 Geometry inspiration of Stuttgart airport – structure mimics tree branches. (**a**) General view of the interior of the airport (**b**) The structure of the passengers' hall mimicking tree branches (**c**) Close-up view of the structure of columns showing the tree stems (**d**) Detail of the column holding the roof. (Images' credit and source: Giftzwerg 88 (**a**) https://commons.m.wikimedia.org/wiki/File:Flughafen_Stuttgart_Interior_38.jpg#, (**b**) https://commons.m.wikimedia.org/wiki/File:Flughafen_Stuttgart_Interior_09.jpg#, (**c**) https://commons.m.wikimedia.org/wiki/File:EDDS-Terminal1-Bäume.jpg#, (**d**) https://commons.m.wikimedia.org/wiki/File:Flughafen_Stuttgart_Interior_03.jpg#)

Fig. 4.14 Interior of Stuttgart Airport in Germany – a structure inspired from tree branches. (Image credit and source: Giftzwerg 88 https://commons.m.wikimedia.org/wiki/File:Flughafen_Stuttgart_Interior_20.jpg#)

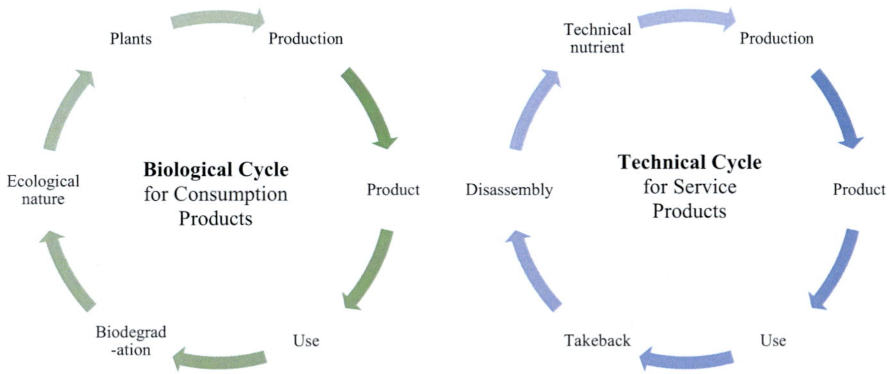

Fig. 4.15 Cradle to cradle waste recycling approach. (Image source: Developed by authors after https://epea.com/en/about-us/cradle-to-cradle)

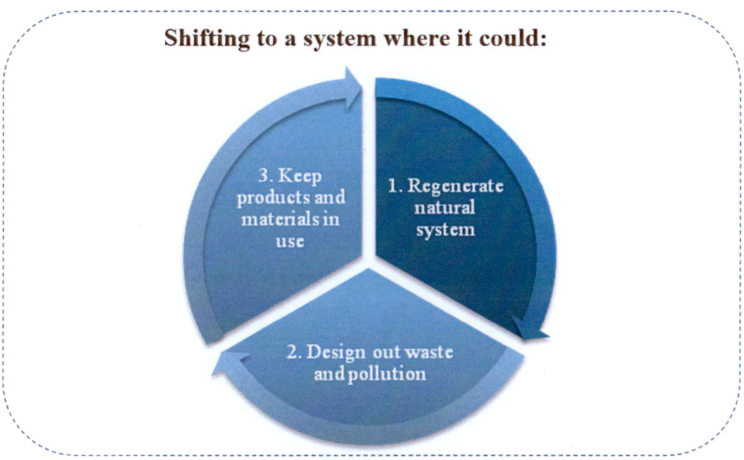

Fig. 4.16 The three targets in an efficient resource model. (Image source: Developed by authors)

4.2.3 Waste Management

Waste offers enormous potential due to the probability of generating more value from the same resources, which is achieved by moving toward zero-waste operations where a closed loop model in waste management is used instead of a linear loop model, as presented in Fig. 4.18.

A number of organizations already started to create networks that mimic an ecosystem, so that there is an increment in the amount of useful output, although it is from the same input. Achieving resource efficiency can be realized when reinterpreting nature and the function of energy and nutrients. This is the way in which ecosystems evolve for more efficient systems with less energy consumption [5]. If

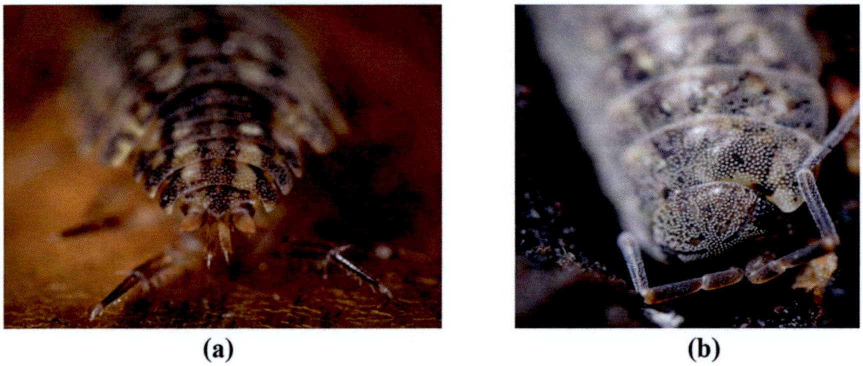

Fig. 4.17 Arthropods exoskeletons. (a) Lpidoniscus minutes – arthropods (b) Hard skeleton of arthropods. (Images' credit and source: Cassius Chaerea (a) https://commons.m.wikimedia.org/wiki/Category:Arthropoda#/media/File:3%ADSC1002-2.jpg (b) https://commons.m.wikimedia.org/wiki/File:Lepidoniscus_minutus.jpg)

Fig. 4.18 Closed loop cycle of waste materials. (Image source: Developed by authors after https://www.saveboard.nz/closed-loop.html/)

waste is seen as a nutrient resource, a new paradigm can be prominently created by consuming fewer resources. Significant examples of ecosystem thinking reflect values of being restorative to the environment and building local resilience. To prompt biomimicry, there is an urgent need for designers who work closely with industry and biology to generate a paradigm shift in the architecture field [5]. Apart from the hazard included in basing groups around monofunctional industries, models of

(a) (b) (c)

Fig. 4.19 Configuration of panels by raw materials. (**a**) Moss leaves on moist soil (**b**) Shape of the moss plant (**c**) Panels of bio-based raw materials (**d**) Pattern of Nylon. (Images' credit and source*:* (**a**) Pranesh Lokhande, https://commons.wikimedia.org/wiki/File:Leaves_of_moss.jpg (**b**) Isasza, https://commons.wikimedia.org/wiki/File:Sphaigne_de_roche.jpg# (**c**) Vigorini, https://commons.wikimedia.org/wiki/File:Particolare_di_calza_di_nylon.jpg# (**d**) Michael Ströck, https://commons.wikimedia.org/wiki/File:Nylon6_and_Nylon_66.png#)

ecosystem thinking include a variety of functions. At the behavioral level of biomimicry, the shape of the moss articulates matter as fertilizer and converts it into plant biomass, as illustrated in Fig. 4.19. In 1 cm, it has three mosses on a surface area of 0.17 m^2 [5]. The bio-based raw materials and building façade's panels are suitable without further coating for building's exteriors, thus reducing the supplementary additives, energy use, and CO_2 emissions. Moreover, the bio-based raw materials take different shapes according to the thermal condition in three dimensions [5].

4.2.4 Water Management

The main concepts in water management are: (a) water loss, (b) water storage, (c) water harvesting, (d) overabundance, and (e) wastewater treatment. In reducing water losses, organisms, especially those adapted to live in arid areas, have their methods to reduce water losses. This is often pursued by using nonliving matter to create shade, trap a layer of air next to its surface, reduce the evaporative process, and/or use a combination of both, for example, the cactus plant (Fig. 4.20). In wastewater treatment, it includes transformation of food production and water treatment systems following closed loop solutions rather than linear, wasteful, and polluting flows [5].

Another example is the Sahara Forest Project in Jordan and Qatar, where the characteristic of ecosystems, as a regenerative project, is an effective driver to create progressive solutions that went beyond sustainability for a restorative approach. Sahara Forest Project is a significant part of the inspiration for city challenge; it originated from studying the organisms that have officially adapted to the conditions in the desert [5]. It is considered a behavior level of biomimicry, since the seawater cooled the greenhouse in this project. Similarly, this

Fig. 4.20 Cactus adaptation in arid climate conditions. (**a**) A family of cactus (**b**) Old cactus plants adapting to arid climate (**c**) Cactus plant in Cairo. (Images' credit and source: (**a**) https://upload.wikimedia.org/wikipedia/commons/a/af/Mammillaria_compress_-_Compressed_Cactus.jpg (**b**) https://upload.wikimedia.org/wikipedia/commons/9/9b/Two_old_cactus_plant_on_cactus_garden.jpg (**c**) Mohsen Aboulnaga)

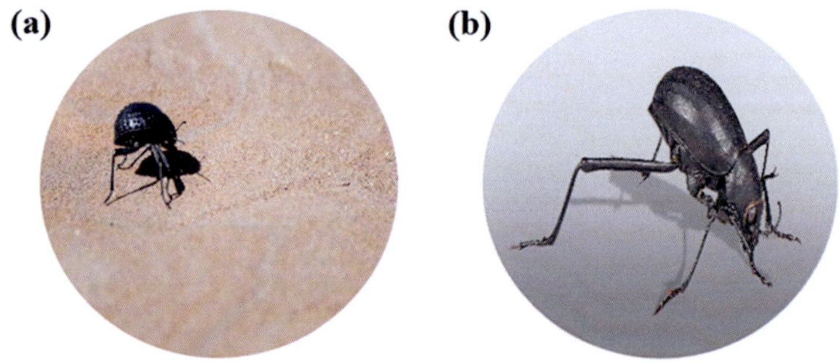

Fig. 4.21 Namibian fog-basking beetle as a model for water conservation. (**a**) The Namibian fog-basking beetle in desert areas, (**b**) Close-up view of the beetle's shell. (Images' source: (**a**) https://upload.wikimedia.org/wikipedia/commons/f/f0/Namib_desert_beetle.jpg (**b**) https://en.wikipedia.org/wiki/Onymacris_unguicularis#/media/File:Onymacris_unguicularis_MHNT_(cropped).jpg)

mechanism is mimicking the behavior of the Namibian fog-basking beetle harvesting water in the desert (Fig. 4.21).

4.2.5 Thermal Control and Self-Regulating Homeostatic Façade System

Homeostasis, the trend for an organism to sustain enduring conditions whether to keep warm or cool, is one of the applications that exhibit the interface between building and biology. In biology, the homeostasis in regard to thermal control is obvious in birds' feathers and bears' fur (Fig. 4.22). Similarly, the energy implications and thermoregulations in the construction are colossal. For organisms, keeping warm means the two fundamental sources of heat, based on solar energy, are direct solar gain and indirect solar gain. In contrast, keeping cool means the heat is moved in four ways (radiation, evaporation, convection, and conduction). Therefore, organisms that exist in hot regions go to incredible length avoiding grabbing heat. Thus,

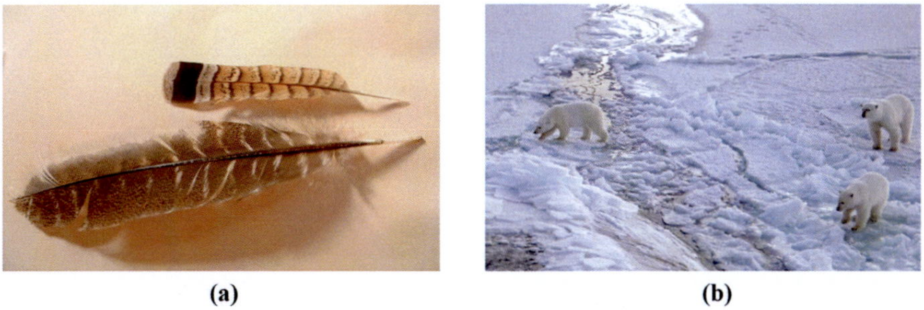

Fig. 4.22 Homeostatic adaptations in birds' feathers and polar bears' fur. (a) Birds' feathers for biomimicry (b) Bears' fur for biomimicry inspiration. (Images' credit and source: (a) Greudin, https://commons.m.wikimedia.org/wiki/File:2_feathers.jpg (b)Yeoman Alphonso, https://commons.m.wikimedia.org/wiki/File:Polar_bears_near_north_pole.jpg#)

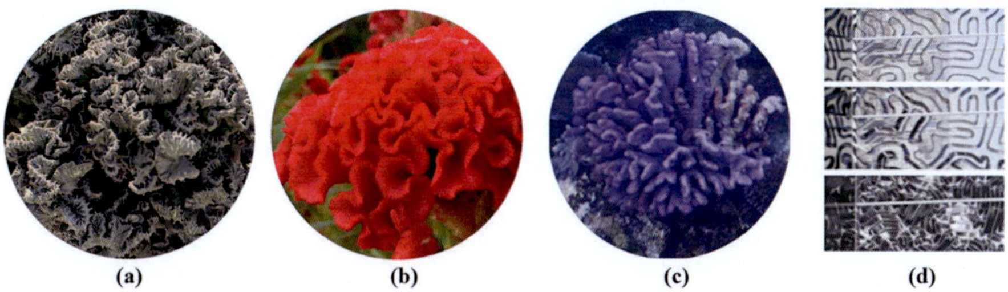

Fig. 4.23 Material patterns looking on the building's glass façade. (a) Cactus plant's pattern (b) Red cockscomb flowers' pattern (c) Coral reefs' pattern (d) Self-regulated homeostatic glass. (Images' source: (a) Mohsen Aboulnaga (b) Amore Mio, https://www.commons.m.wikimedia.org/wiki/File:Red_Cockscomb_ flowers.JPG# (c, d) https://issuu.com/ioanaleordean/docs/thesis_book_online-compressed)

avoid irradiative accumulation by remaining away from the sun or by minimizing the absorption of heat through conduction. Developing adaptive buildings that could adjust and acclimatize with changing contexts and conditions – such as in biology – is extremely needed to achieve sustainability and minimize the use of natural resources [5].

4.2.5.1 Self-Regulating Homeostatic Façade System

The self-regulating homeostatic façade is a glass system, which is exploited for huge buildings that open and close, that is, dynamic ones that respond to the indoor temperature in buildings. This inspiration was emulated from the nature and coral reefs as presented in Fig. 4.23. The self-regulating homeostatic façade system was designed in 2011 by architect Decker Yeadon, who got inspired from muscles and homeostasis in biological systems [6, 7]. Homeostasis is a state maintained by organisms that allow them to regulate a building's climate by automatically responding to the environmental conditions. Such a system enables for adjusting to surrounding conditions and exploiting viable and local materials. The system also opens and closes itself in reaction to the amount of light that enters the building, hence saving energy (Figs. 4.24 and 4.25). When the sunlight impinged on the silver-coated elastomers – a natural or synthetic polymer having elastic properties – an electric current is travelled across the surface, which brings about the material to deform or bend. As the amount of sunlight increased on the surface, such material inflates to provide shading (Fig. 4.26).

The Green Cast mixed-use building (shop, clinic, office, and private residence) in Odawara-shi, Kanagawa prefecture in Japan and designed by KENGO KUMA & Associates has its living façades with homeostatic system coupled with greenery, where the metal elements have different sizes due to the expansion and deformation from the heat as shown in Figs. 4.24a–h. The façade of the building – an extension

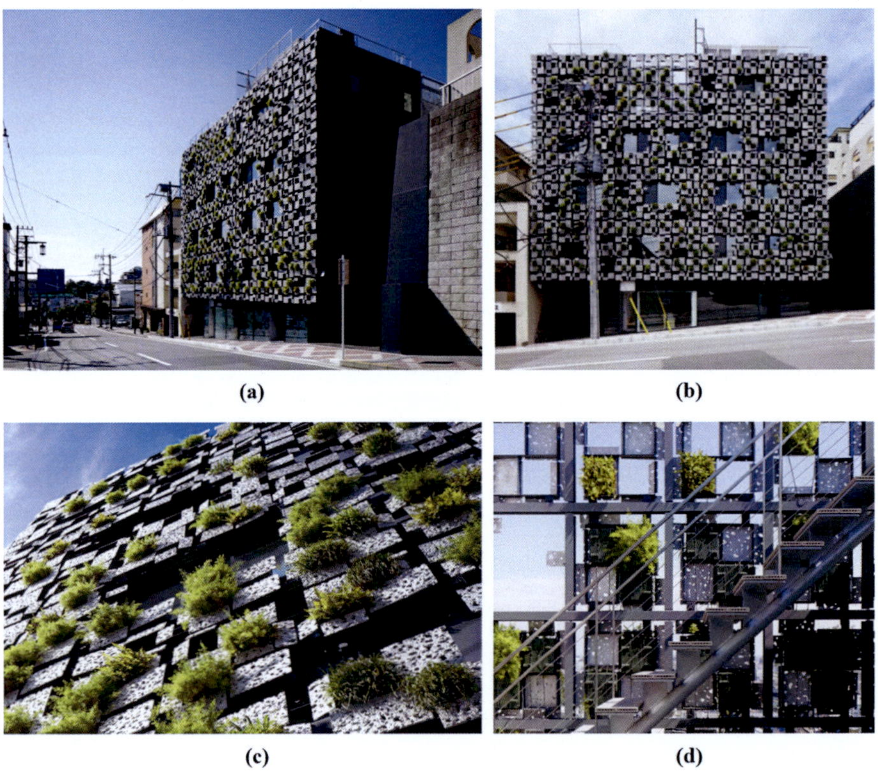

Fig. 4.24 Homeostatic living façades of a mixed-use building in Odawara-shi, Japan. (**a**) The Green Cast building's main façade (**b**) A front view of the living façade (**c**) Aluminum die-cast panels act as vertical planters (**d**) The monobloc casting slanted panels (**e**) Elevation of the die-cast iron panels acting as vertical planters (**f**) Plans, a section of the building, and detailed façade section depicting the casting slanted panels (**g**) View of the panels on the glazed façade (**h**) An interior showing the slanted panels. (Images' credit and source: Kengo Kuma Associates, https://kkaa.co.jp/works/ architecture/green-cast)

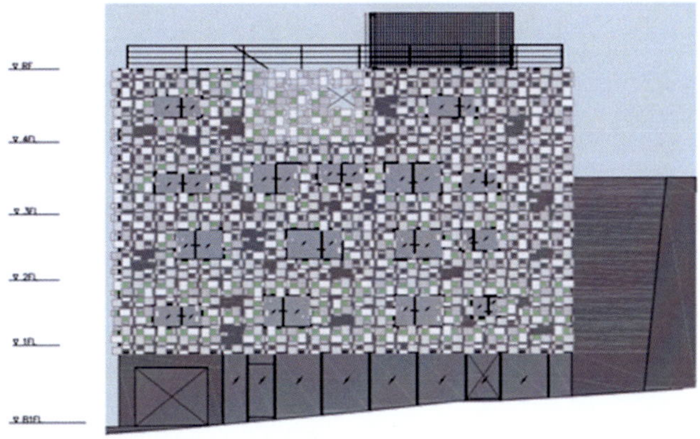

(e)

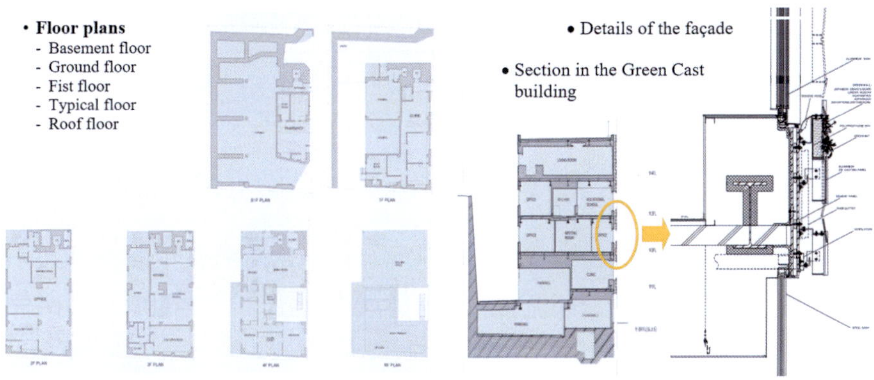

(f)

(g)

(h)

Fig. 4.24 (continued)

of an existing building, an underground, and four floors above it – is covered with planters made of aluminum die-cast panels, which provides space for facilities. The three (up to six) aluminum panels, which also form planters, are made in monobloc casting. Each panel is slanted, and its surface appears to be organic, of which cast comes from decayed styrene foam. Equipment such as watering hose, air reservoir for ventilation, and downpipes are installed behind the panels so that the façade can accommodate a comprehensive system for the building [8].

In a reverse situation, when the building's surface is not exposed to large amounts of natural light, the material shrinks or dwindles to furnish more light into the building's spaces. Thus, such façade system actually activates the building to be a living object that is expressed as environmentally responsiveness and sustainable [5, 6, 8].

The same technologies are used in the façade that depicts biomimicry architecture in the corporate office for Hispasat Technology Center in Madrid, Spain (Fig. 4.25). The details of the double-layered skin façade are presented in Fig. 4.25a, b, c. In addition, it is noticeable from Fig. 4.25 that the recycled plates are less dense on the northern façade to allow for better view toward the surrounding landscape (Fig. 4.25a, b), and they are more dense on the southern facades (Fig. 4.25c) to eliminate solar radiation, heat, and glare in summer.

Professor Juan Herreros – Estudio Herreros in Madrid, Spain – stated that this projects in itself a "*correction of an existing unsustainable circular building by splitting of the original massive and inoperative façade into a double layered light skin.*" He also added that "*the recycled aluminum plates are coated with different but similar colors to recreate the vibration of the landscape.*"

Also, the Center for Physical Sciences and Technology – Vilnius University in Vilnius, Lithuania – has living façades that incorporate dielectric elastomers and fiber-reinforced cladding as illustrated in Fig. 4.26.

In addition, the circular building homeostatic façade, with triangular lattice accentuates, is strongly in dialogue with nature to filter the light into indoor spaces and control solar radiation [9]. As far as the building's environmental performance is concerned, the lattice provides three various densities of openings, sweeping the solar spectrum in three sectors with different solar radiation.

The double-skin façade has the lattice away 1 m from the building's glazing; hence, it offers a shaded and ventilated chamber effect, which guarantees a decrease in heat gain as well as heat loss [9]. More examples of living façades using lattice and metal elements are depicted in the UK pavilion "Beehive" at EXPO 2015 Milan and the cylindrical building's façades in Berlin, Germany, as shown in Figs. 4.27 and 4.28, respectively.

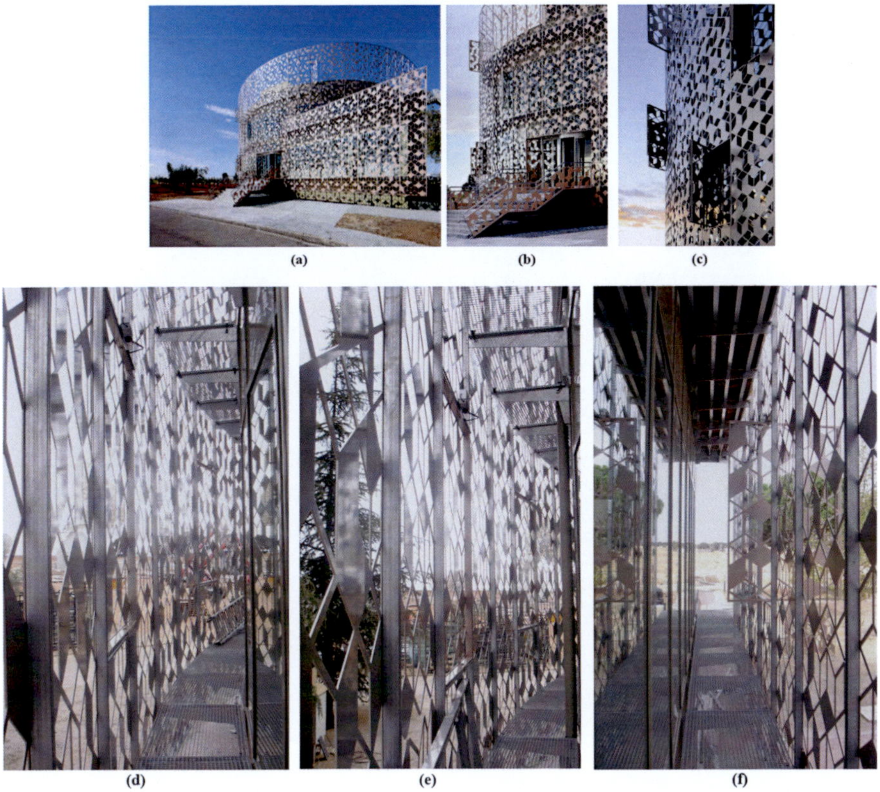

Fig. 4.25 The Corporate Office of Hispasat Technology Center with a double-layered light skin façade, Madrid, Spain. (**a**) The circular façade with triangular lattice accentuates in dialogue with surrounding nature (**b**) Double-skin façade to control light and heat (**c**) Recycled coated plates to control solar radiation (**d**) A double-layered light-skin that was created to split the original massive and inoperative façade (**e**) Recycled aluminum plates to allow for light and view through inner glazing to recreate vibration of the landscape (**f**) 3-D sectional elevation of the inner glazing of the double-skin façade behind the color plates to allow for ventilation and funnel air. (Images' credit: José Hevia and source: Estudio Herreros – Madrid, https://estudioherreros.com/en/project/satellite-control-center)

4.2.6 Renewable Energies: Strawscraper Building

Cities including buildings and transport as well as human needs for energy enforced them to try finding and creating more clean energy by several methods, rather than finding a way that could develop solutions as nature behaves, using less energy as a priority. Due to the growing manifestation of climate change impact and inappropriate strategic planning and actions to respond to this global threat, energy becomes one of the most priority challenges. Hence, it is important to reduce resource consumption, ensure sustainable development, and enhance the performance of buildings and cities. In this context, biomimicry approach includes principles for energy

Fig. 4.26 Center for Physical Sciences and Technology with living façades incorporating dielectric elastomers and fiber-reinforced cladding – Vilnius University in Vilnius, Lithuania. (Image's credit and source: Edgaras Kurauskas, https://commons.m.wikimedia.org/wiki/File:Nacionalinis_fizinių_ir_technologijios_mokslų_centras.jpg)

planning, which leads to a "solar economy." This term means that energy needs can be met depending solely on renewable energy (RE) sources such as solar photovoltaic and solar powers as shown in Fig. 4.29. Forms of RE include also wind, geothermal, biomass, and waves and tidal. Despite the contrast between man-made and nature's system, biomimicry proposes many interventions in the energy sector, which involves the following four principles as presented in Fig. 4.30 [5]. In conclusion, biomimetic solutions are resilient, nontoxic, restorative, unlimited energy source, and above all sustainable. According to David Mackay's detailed study, it is proved that the creation of solar economy could be achievable, but it could be a long journey. Solar power installations, on a concentrated and large scale, are a significant and primary pillar in achieving such an economy [5].

4.2.6.1 Strawscraper Building

The strawscraper (an urban power plant) was planned as a high-rise tower as an extension of Söder Torn tower (planned for 40 floors, but built at 26 only) on Södermaalm in Stockholm, Sweden. The building was originally designed by

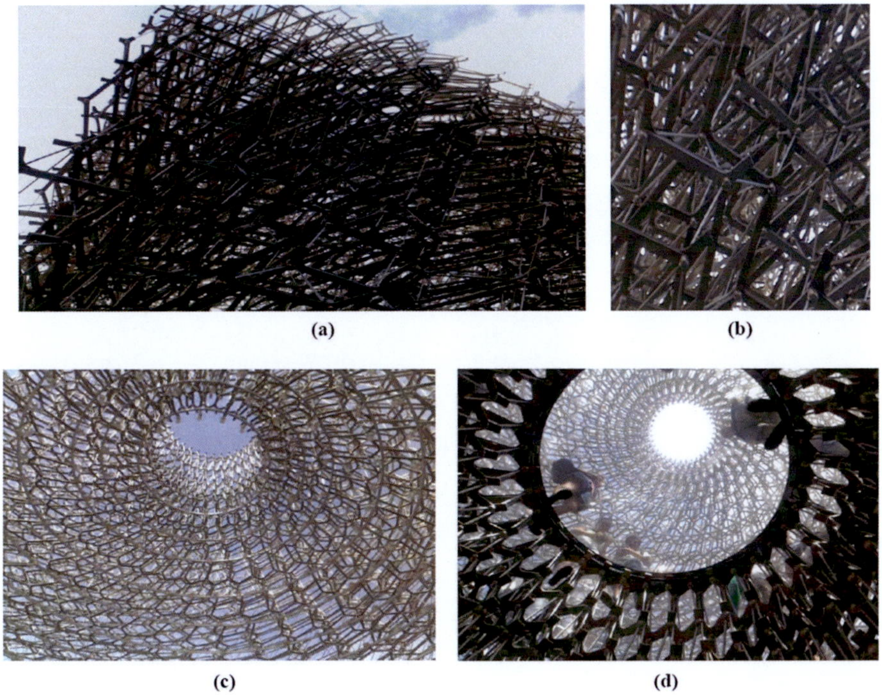

Fig. 4.27 The UK "Beehive" Pavilion in EXPO 2016 Milan. (**a**) The pavilion's façade with conical irregular lattice accentuates in dialogue with nature to filter the light into indoor spaces (**b**) Metallic screen façade to control solar radiation (**c**) A view from the pavilion's ground floor depicting the conical metallic lattice (**d**) Close-up view of the first floor where visitors move on the second floor of the living pavilion. (Image credit: Mohsen Aboulnaga)

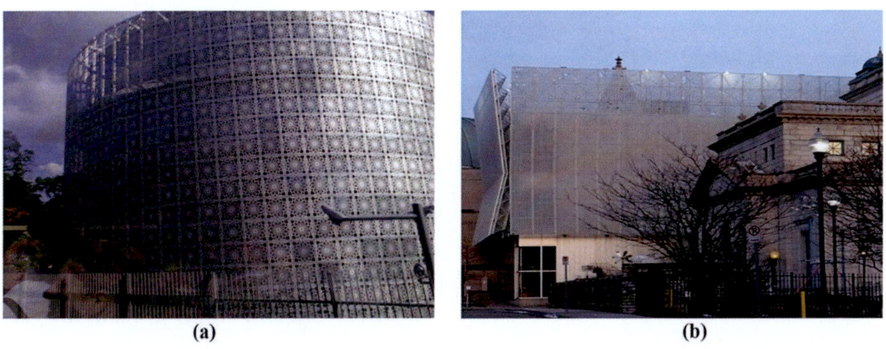

Fig. 4.28 A circular façade building with homeostatic metal elements in Berlin, Germany. (**a**) A cylindrical façade with homeostatic metal elements near Brien gate in Berlin, Germany (**b**) Homeostatic façade of Children's Museum in Pittsburgh – Pennsylvania, USA. (Image credit and source: (**a**) Mohsen Aboulnaga (**b**) Paul Sableman, https://commons.m.wikimedia.org/ wiki/ File:Childern%27s_Museum_-_Pittsburg_(25206956535)_(cropped).jpg#)

Fig. 4.29 Renewable energy resources. (**a**) Solar panels, (**b**) Wind turbines, (**c**) Water energy. (Images' source: (**a**) https://upload.wikimedia.org/wikipedia/ commons/3/37/Solar_Panels.jpg (**b**) https://commons.wikimedia.org/wiki/File:Paderborn _Germany_Turbines-of-Paderborn wind-farm-02.jpg (**c**) https://upload.wikimedia.org/wikipedia/commons/5/59/Water_fall_in_musanze.jpg)

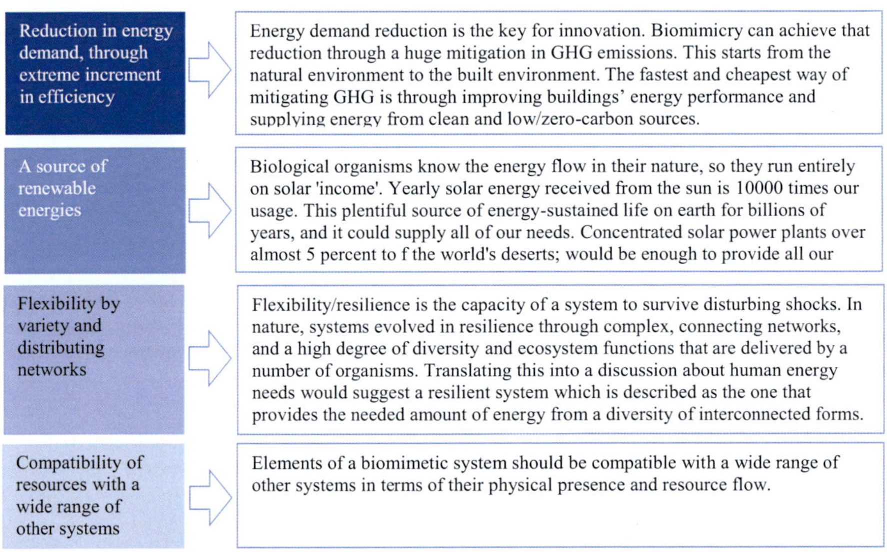

Fig. 4.30 Biomimicry suggested solutions for energy efficiency. (Source: Developed by authors)

Henning Larsen in 1997 [7], but Belatchew Arkitektur wished to develop the Söder Torn tower and used wind energy to generate electricity for the public use. The strawscraper is an extension of Söder Torn as shown in Fig. 4.31, a tower which produces energy using straws that can recover wind energy. Light breeze is enough for straws to function, because the tower can work in low wind velocity, and starts swaying generating energy [7]. Piezoelectric technology is a technique for creating a wise urban wind farm future, where a large number of thin straws with minimal wind movement can produce electricity. This new type of wind power plant opens up the possibilities for energy production by buildings generating structures [7]. Through this technology, the surfaces of old and new buildings can be transformed

Fig. 4.31 An example of utilizing renewable energy in Sweden. (**a**) Söder Torn building, Stockholm, Sweden (**b**) Strawscraper building – a Söder Torn extension. (Images' credit and source: (**a**) Holger Ellgaard, https://commons.m.wikimedia.org/wiki/File:Söder_torn_december_2014.jpg (**b**) Belatchew Arkitektur, https://www.archiscene.net/location/sweden/strawscraper-belatchew-arkitektur/)

into energy. The tower was not further developed due to the fact that Architect Henning Larsen has left the contract [7].

4.3 Architectural Design Examples: Biomimicry

Nature is an inspirational source of innovation due to the potential it offers to create a sustainable and regenerative built environment. Nature provides a large database for adaptive strategies that can be implemented in architectural design in general and the design of building's envelope in particular. Widespread and practical applications of biomimicry, as a design method, remain largely unrealized.

In the following section, thirty selected examples of biomimetic architecture, which depicted vivid biomimetic buildings inspired from nature, will be presented in order to clearly understand how biomimetic approaches have been applied in different countries through buildings' models worldwide. Furthermore, the levels of biomimicry applications in each example will be highlighted. Such examples are selected from Europe, Asia, Australia, Northern America, and Africa as presented in Fig. 4.32.

(a) **Examples in Europe.**

- Waterloo International Terminal in London, UK.
- The Eden Project in Cornwall, England, UK.
- Swiss Re – Gherkin Tower in London, England, UK.
- Selfridges Building in Birmingham, UK.
- The Munich Olympic Stadium, Munich, Germany.
- Tardigrada Botanica, Poland.
- Sagrada Familia Church in Barcelona, Spain.
- The Las Palmas' Water Theatre, Canary Islands, Spain.

Fig. 4.32 Selected global biomimetic examples depicting biomimicry from nature. (**a**) Examples of buildings mimicking nature in Europe – the United Kingdom, Germany, and Poland (**b**) Examples of European and Asian architecture following biomimicry concepts (**c**) Examples of Asian and Middle Eastern architecture following biomimicry concepts (**d**) Examples of Australian, American, and African architecture following biomimicry concepts (**e**) Examples of systems, hypothetical projects mimicking nature – the United States, China, Poland. (Source: Developed by authors)

(d)

(e)

Fig. 4.32 (continued)

- Tenerife Auditorium in Santa Cruz de Tenerife, Spain.
- The All Seasons Tent Tower in Yerevan, Armenia.

(b) **Examples in Asia.**

- Esplanade Theatre in Singapore, Singapore.
- Beijing Olympic Stadium and National Aquatic Center in Beijing, China.
- Omotesando Building in Tokyo, Japan.
- Residential Building in Songdo – Incheon, South Korea.
- The Lotus Temple in Delhi, India.
- Lavasa Township near Pune, Maharashtra, India.
- Aldar Headquarters in Abu Dhabi, UAE.
- The Sahara Forest project in Aqaba, Jordan.
- Doha Tower and Giant Cactus building in Doha, Qatar.

(c) **Examples in Australia.**

- Council House 2, Melbourne, Australia.

(d) **Examples in the Northern America.**

- Ultima Tower in San Francisco – CA, USA.
- Fennell Residence in Portland, Oregon, USA.
- Spiraling Chicago tower in Chicago, USA.

(e) **Examples in Africa.**

- Hydrological Center for the University of Namibia in Windhoek, Namibia.
- Eastgate Building in Harare, Zimbabwe.

(f) **General Examples: Systems.**

- Photovoltaic and Ferro systems.

(g) **Hypothetical Examples:** (i) Rain Collector in Nowak, Poland; (ii) Habitat 2020, China; (iii) Office Building of the Future project in the USA; and (iv) Lilypad Floating City, Belgium.

4.3.1 Waterloo International Terminal in London, UK

Organism level of biomimicry is manifested in London's Waterloo Terminal in London, United Kingdom. This example mimics the "pangolin" as shown in Fig. 4.33. The incoming and outgoing trains at the terminal generate changes in air pressure; therefore, the glazing system of the envelope is designed like "flexible scales" on the skin of the pangolin, allowing it to be adjusted based on the imposed forces of the air pressure as well as waistband shocks as illustrated in Fig. 4.33a, b, c [10]. The Waterloo Terminal is famous for its 400-meter-long curved glass roof, which is designed by architect Nicholas Grimshaw and Partners (Fig. 4.34). There is no doubt that advanced technology assists designers in implementing their idea and the applications from nature's systems. In this model, the organism's physical solutions have been mimicked at the level of form and function as shown in Figs. 4.35 and 4.36 [11]. The expectation of the increment percentage of the journeys in the terminal is expected to be 40% by 2043.

To ease congestion and create capacity for the future, Waterloo Station's upgrade involves modifying platforms one to four to suit longer trains (from eight cars to ten) and converting the platforms of the former Waterloo International Terminal (WIT) on the west side of the station for domestic services [11].

4.3.2 The Eden Project in Cornwall, England

The Eden Project in Cornwall, England, UK, is a building created according to biomimicry concept. It is considered one of the famous visitors' attractions in Cornwall, England. It was opened in March 2001 and was designed by Sir Nicholas

(a) (b) (c) (d)

Fig. 4.33 Waterloo International Terminal in London, UK. (**a**) The curved structure of Euro-train terminal (**b**) Side view of the roof mimicking a pangolin (**c**) Interior of the terminal showing the roof (**d**) The pangolin organism inspires the design. (Images' credit and source: (**a**) Maurits90, https://commons.m.wikimedia.org/wiki/File:London_Waterloo_former_Eurostar_station.JPG# (**b**) Timothy Baldwin, https://commons.m.wikimedia.org/wiki/File:Waterloo_International_Station_geograph.org.uk_-_111985.jpg.jpg# (**c**) Qwesy Qwesy, https://commons.m.wikimedia.org/wiki/File:Eurostar_-_panoramio.jpg# (**d**) David Brossard, https://commons.m.wikimedia.org/wiki/File:Manis_temminckii_(29645803646).jpg))

Grimshaw – Grimshaw Global. There are diverse types of plants that were collected from different climates and environments and placed inside the biomes. These biomes are a group of synthetic domes mimicking soap bubbles [12]. The Eden Project is gripped by neighboring domes, where each enclosure imitates a natural biome that accommodates thousands of types of plant species. The biome consists of hundreds of hexagonal and pentagonal puffed plastic cells supported by steel frames. Biome is a collection of plants and wildlife adapted to a certain climate, and it has characteristics' commonality for the environment they live in [13–15]. There are two biomes: the largest one mimics the environment of a rainforest, and the smaller one mimics a Mediterranean environment.

The outside botanical garden is considered one of the attractions in this project. Such botanical gardens house several plants, historical story of plants, and wildlife native to the United Kingdom in general (Fig. 4.37). The start was from the geodesic form, and it is formed by the American architect Buckminster Fuller. The geodesic form is selected for the lightweight and maximum surface area on the curve with strength [13]. The electricity green tariff is used in the buildings, where the energy comes from one of the wind turbines in Cornwall – among the first in Europe in

Fig. 4.34 Front view of the platform of Waterloo International Terminal illustrating the mimicked structure from the pangolin organism. (Image credit and source: Herbert Ortner, https://commons.m.wikimedia.org/wiki/File:Eurostar_3218_Waterloo.jpg#)

Fig. 4.35 Interior of the structure of Waterloo International Terminal in London. (Image credit and source: Smily Toerist, https://commons.m.wikimedia.org/wiki/File:Waterloo_International_1996.jpg#)

(a)

(b)

Fig. 4.36 Details of the mimicked roof of Waterloo International Terminal in London, UK. (**a**) The shell of the terminal mimicking a pangolin (Image credit and source: Cristian Bortes, https://commons.m.wikimedia.org/wiki/File:London_(8248551056).jpg#) (**b**) Close-up view of the covering steel truss covering the roof of the terminal. (Image credit and source: Piergiorgio Rossi, https://commons.m.wikimedia.org/wiki/File:Dettaglio_della_Waterlo_Station_Londra_2002).jpg#)

2010 permission received to build a geothermal electricity plant (GtEP). This GtEP generates 4 MW of electricity which is enough to supply about 5000 households with energy demand in Eden. The geometric panels on the domes are filled with ETFE (ethylene tetrafluoroethylene), and each pillow is connected to a web of interlocking steel tubes as shown in Figs. 4.37c, d and 4.38 [16].

4.3.3 Swiss Re: Gherkin Tower in London, England

The Gherkin Tower – designed by Sir Norman Foster – is forming a hexagonal building skin that acquired the inspiration from Venus flower under the water environment, which is known as a basket sponge (Fig. 4.39). This tower is a new translation for sustainability; it is integrated with nature, as nature is a source of enormous ideas and inspirations to designers and architects for creating sustainable architecture [17]. The building is inspired from the Venus flower basket sponge [16] as shown in Fig. 4.39b.

(a) (b) (c) (d)

Fig. 4.37 Eden Project inspired from nature in Cornwall – England, UK. (**a**) Eden biomes inspired from soap bubble (**b**) Inside the Eden Project (**c**) Eden Project structure elements (**d**) Roofs flailed with ETFE and the outer surfaces. (Images' source: (**a**) https://commons.wikimedia.org/wiki/File:Eden_Biomes_-_geograph.org.uk_-_217615.jpg (**b**) https://commons.wikimedia.org/wiki/File:Eden_Project_Panorama_2.jpg (**c**) https://commons.wikimedia.org/wiki/File:Eden_project2.jpg (**d**) https://commons.wikimedia.org/wiki/File:Biomes_of_Eden_Proyect.jpg)

Fig. 4.38 Ariel view of the biomes – Eco domes at Eden Project in Cornwall, England, UK. (Image credit and source: JuliesNotebook, https://commons.wikimedia.org/wiki/File:The_Eden_Project,_Cornwall.JPG)

The Gherkin Tower is fixed due to several approaches, including technical, architectural, social, and spatial approaches. The energy awareness is manifested by the form and structure of the building, with minimal space for maximizing light and views. It is designed with "light wells" that maximize daylight, which penetrates through the building and is controlled by sensors (light sensors and motion sensors) installed on the tower (Fig. 4.40). A pressure difference is generated by the aerodynamic shape of the building; such shape helps in providing natural ventilation when the wind comes through shafts of light, reduces the energy consumption of air conditioning, and, in turn, creates viability and a sustainable healthy indoor environment [16, 17]. Diagrid (diagonal + grid) is the structure of the building, where the cylindrical buildings' shape allows for the wind around the building to move, so that such shape deals with wind load as decreased buffeting, reduced vibrations, and diminished fluttering. In addition, the concept of the building was inspired by the shape of aircrafts and its streamlined bodies that interacts with the wind. Thinking about what makes this building unique? The answer is the diagrid. This iconic building is presenting the twenty-first century contemporary biomimicry architecture with its sustainability features.

The sustainability elements of the Gherkin are as follows:

(a) (b) (c)

Fig. 4.39 Gherkin Tower and Venus flower inspiration. (**a**) Façades of the Gherkin Tower, (**b**) Venus flower inspiring the design, (**c**) Underwater reef. (Images credit and source: (**a**) Karya Sendiri https://id.m.wikipedia.org/wiki/Berkas:London_UK_Gherkin-01.jpg (**b**) https://upload.wikimedia.org/wikipedia/commons/8/82/Venus_Flower_Basket_%28sponge-labelled%29.JPG (**c**) https://upload.wikimedia.org/wikipedia/commons/b/bd/Callyspongia_sp._%28Tube_sponge%29.jpg)

(a) (b) (c)

Fig. 4.40 Sustainability elements of the Gherkin Tower in London, UK. (**a**) Light travels down the six shafts and illuminates the six radial fingers of the Gherkin Tower. By creating a naturally lit building, the energy needed for lighting is drastically reduced. Thin films are also used to generate power for clean energy (**b**) The top roof with tinted glass panels to reduce both glare and solar heat gain yet provide a panoramic view (**c**) Tinted double-paned glass filled with argon reduces solar radiation that enters the building. These glass panels are either opened or fixed. The offices areas feature double glazed panels with low-E coating and an internal screen of laminated glass. (Images' credit and source: (**a**) Drroshdy, https://commons.m.wikimedia.org/wiki/File:The_Gherkin_London.jpg (**b**) Geekchic, https://commons.wikimedia.org/wiki/File:Top_floor_The_Gherkin.jpg (**c**) Samuel Zeller, https://commons.m.wikimedia.org/wiki/File:30_St_Mary_Axe_(The_Gherkin),_London, _United_Kingdom_(Unsplash).jpg)

- Wind load is reduced on the structure, by its aerodynamically designed shape, while the lower part attenuates and tapers which allows wind to wrap around the tower.
- Solar protection is 85%, which is achieved by advanced glazing technologies in the façade's design, ventilated cavities, and blinds screening.
- The main fuel used is natural gas (less carbon).
- The composition of the façade consists of two layers of glass panels; the outer skin is double glazed including a cavity for ventilation, with automated controlled curtains. An outer building weather sensors' systems are installed to monitor air and surface temperatures, wind speed, and sunlight level; hence closing blinds and opening window panels are performed only when needed.
- A group of gaps with six pipes made of natural ventilation system is installed on each floor, acting as a double glazing.
- The pipes are used for heating in winter and for cooling in summer, by taking warm air from the building. These pipes also ease the entry of light and consequently, and the cost of lighting is reduced.

- Energy savings are achieved by the systematic internal microclimate and solutions; this led to 50% reduction in energy use, and such solution is perfect for a building of this size.
- Lights' level and motion sensors are used for reducing unnecessary operation lighting.

Figure 4.40 illustrates the passivity of the light and the double-glazed façades of the building's skin. The materials, systems, and elements in this building include the following:

- Ten thousand tons of steel used to build Gherkin Tower.
- 24,000 square meters of Glass used on the exterior façade
- Recyclable materials aimed to be used whenever possible.
- The glazing system, specifically in the office area with a double glazing and single glazing for outer and inner screens, respectively.
- Solar control blinds intercepting heat gain before entering the office environment.

4.3.4 Selfridges Department Store Building in Birmingham, UK

Selfridges Department Store building in Birmingham, designed by architect Jan Kaplicky – founder of Future Systems – features a curvy space-age design that epitomizes what the aesthetic goal of bionic architecture is all about. His idea is centered on combining the organic and natural forms with high technology to achieve such optimization. Completed in 2003, the iconic building remains one of the leading forward-thinking biomimicry architecture [18]. The idea was inspired from nature – wasps' nests as shown in Fig. 4.41, where the organic natural form and structure are combined with high technology to accomplish optimization as illustrated in Fig. 4.41b, c [18]. The façades of Selfridges Building are covered by 1500 shiny aluminum discs, giving the building an almost splendid alien-like sheen (Fig. 4.41a) [19].

4.3.5 The Munich Olympic Stadium, Germany

The Olympic Stadium in Munich, Germany, is an example of using nature as a model for design. The stadium was designed by Architect Gunther Behnisch and Engineer Frei Otto as shown in Fig. 4.42 [10, 11]. The inspiration comes from dragonfly wings, and the wings' thickness of this fly is one three-thousandth of a millimeter. In spite of this very fine structure, however, these wings are extremely forceful (Fig. 4.43a, c). As the compartmental structure of the wings, which do not tear, these are able to withstand the pressure that forms during flight. The roof

(a) (b) (c)

Fig. 4.41 The Selfridges iconic building in Birmingham – England, UK. (**a**) Shiny façade, Selfridges Department Store (**b**) Wasp's Larva with eggs (**c**) Grub-like organisms. (Images' source: (**a**) https://commons.wikimedia.org/wiki/File:Selfridges_Store,_Bullring,_Birmingham_-_geograph.org.uk_-_983946.jpg (**b**) https://commons.wikimedia.org/wiki/File:Wasp_nest_3.jpg (**c**) https://commons.wikimedia.org/wiki/File:Wasp_Nest_03.jpg)

Fig. 4.42 The Munich Olympic Stadium in Germany. (Image credit and source: Amrei-Marie https://commons.wikimedia.org/wiki/File:Blick_vom_Olympiaberg_auf_das_Olympiastadion.jpg#)

design of the Munich Olympic Stadium was inspired and designed by a similar concept of compartmentalization as shown in Fig. 4.43d [10, 11].

Behnisch and Otto were responsible for designing and constructing the stadium, which hosted the Olympic Games in Munich in 1972. They uplifted a lightweight structure; the tensions are frustrated by a system of supports and cables (Fig. 4.43). They also developed the applications of mathematical calculations upon computational procedures to set the shape and behavior of the stadium's surface cover. Otto had to adapt the design to the preconceived form for the winner Günter Behnisch. They were keen to inspire the world through their design and emulate the motto of the games at that time; "The Happy Games" was their target and goal, as an

Fig. 4.43 Munich's Olympic Stadium and sources of inspiration. (**a**) The dragonfly inspiration (**b**) Structure imitation to dragonfly's wings (**c**) The Alps' mountains inspiring the design (**d**) Structure cover imitates the Alps. (Images' credit and source: (**a**) Ken Billington, https://commons.m.wikimedia.org/wiki/File:Emperor-Dragonfly-(8).JPG (**b**) Christian Benseler, https://commons.wikimedia.org/wiki/File:Muenchen_Olympiastadium_II_(10267906674).jpg# (**c**) Videcek, https://commons.wikimedia.org/wiki/File:Berghaus_Hintisberg.jpg# (**d**) Fred Romero, https://commons.m.wikimedia.org/wiki/File:München_-_Münchner_Olympiastadion.jpg#)

eccentric architecture mimicking the draping and the harmonious elevations of the Swiss Alps [10, 11], as illustrated in Fig. 4.43b. The whole Munich Park is presented in Fig. 4.44.

4.3.6 Tardigrada Botanica in Poland

Tardigrada Botanica is an example of biomimetic architecture in Poland (Fig. 4.45). It is an ecological architectural design that has the characteristics of a utopia, inspired by the form of the most resilient invertebrate Tardigrada (Box 4.4). The project is designed by architect Jaroslaw Kozakiewicz and built on the site of the former waste dump, where its biogases are used to heat the botanical garden inside

Box 4.4: Water Bear or Tardigrada Inspiring Architecture – Tardigrada Botanica

It is not the strongest of the species that survives, or the most intelligent, but the one most responsive to change. (Charles Darwin, 1809)

Fig. 4.44 Aerial view of Munich's Olympic Stadium and Park – a source of inspiration from nature. (Image credit and source: Tia Monto https://commons.wikimedia.org/wiki/File:Munich_-_Olympiapark_3.jpg#)

the structure [19]. The location of this building is situated in the village of Tumanek in Central Poland.

The structure of the building is inspired by an invertebrate called the water bear or Tardigrada. Such structure is shaped as a dome, which is composed of thousands of semitransparent material – ethylene tetrafluoroethylene (ETFE), a material that is air permeable and moist as shown in Fig. 4.45b. The plan of this building is mostly ellipse shape and 40-meter tall, and the widest diameter is 140 meters [20]. The garden inside this building contains tropical plants and many functions whether scientific or instructive as presented in Fig. 4.45c.

The microscopic invertebrate's existence is actually symbolic and of a critical meaning and magnitude for the whole design concept which relates to this microscopic creature in microscale [20]. It is clear from the microscopic photo, which shows that the surface of the bear has complicated structure, and Kozakiewicz copied it in macroscale, which is much bigger than humans.

According to the rules of biomimetics and similarly to the Tardigrada, the building is transparent. The creatures live in any condition, and the building shows the power of nature, which is dominant withstanding many ravages. The intriguing project is a tribute to the smallest elements of nature as well as nature in general. In that way, attention of Kozakiewicz focuses on the minute but at the same time reaches the mega dimension. In addition, biogas is emitted to warm the massive buildings; hence, even in the winter, there is an ability to grow tropical plants inside [20].

It exhibits a great example that summed experiments of sculptural and architectural artists. In this project, architecture is interpreted as the design of building that is mimicking forms of nature and/or caring for the use of renewable energy sources as well as promoting sustainable development. This diverse stylistically current is related to the degraded areas struggling for a new face and identity, such as the areas of postcommunist and postcolonial. The intention of the creator of a degraded area

(a) (b) (c)

Fig. 4.45 Tardigrada Botanica. (**a**) Invertebrate Tardigrada (**b**) The exterior of the Botanica (**c**) Interior of the garden with tropical plants. (Images' credit and source: (**a**) Schokraie, et al. 2012, https://commons.wikimedia.org/wiki/Category:Tardigrada#/media/File:SEM_image_of_Milnesium_tardigradum_in_active_state_-_journal.pone.0045682.g001-2.png (**b**) https://csw.torun.pl/sztuka/kolekcja-sztuki-csw/tardigrada-botanika-9937/attachment/kozakiewicz-jaroslaw-2/#!prettyPhoto/0/ (**c**) https://csw.torun.pl/sztuka/kolekcja-sztuki-csw/tardigrada-botanika-9937)

would be a place where the use of innovative solutions would generate positive changes both battered on earth and in the consciousness of the inhabitants [20].

4.3.7 Sagrada Familia Church in Barcelona, Spain

The Sagrada Familia, or Basílica i Temple Expiatori de la Sagrada Família, is a Roman Catholic Church located in Barcelona, Spain, as illustrated in Fig. 4.46. The iconic masterpiece is designed by renowned architect Antoni Gaudi, and the basilica remains under construction since its groundbreaking in 1882, with anticipated completion in 2026 [21]. However, it has been delayed due to the COVID-19 pandemic impact. The construction was interrupted during the Spanish Civil War of 1936, when many of the models Gaudi left after his death were destroyed. Gaudi's unique design combines traditional Gothic architecture with Art Nouveau forms to express Christian iconography; these composing 18 planned spires represent the 12 apostles, the four gospels, (the Virgin Mary and Jesus), the three grand façades for the Nativity, the Passion, and the Glory representing different aspects of Jesus's life and teachings [21, 22].

The inspiration for columns and towers was sourced from the tree branches in order to solve structural problems. The columns that replicated the branching canopies of trees solved the statics problems regarding the support of the vault shown in Figs. 4.47 and 4.48. To reduce the load of the roof structure and to let the light penetrate into the building, skylights have been designed in between the columns based on pieces of golden, green glass, and tiles. This is created to reflect daylight inside the church as shown in Figs. 4.49 and 4.50. All stained glass in the apse follows a plan of gradual tones [22, 23].

Nonetheless, the most notable example of biomimetic architecture in the Sagrada Familia is the branching columns of the nave which is the main, central area of a church. Evoking a sense of walking through a forest is the columns' branch, which provides greater support for the vault and the roof. The number of sides for each column depends on the load it needs to carry and increases as the column rises. Such design mimics the natural weight distribution of trees, allowing each column to bear a greater load than a traditional column as shown in Fig. 4.48 [24]. The branches transfer lateral loads efficiently to the nave columns, forgoing the need for the exterior flying buttresses of a typical Gothic cathedral. The interior columns support most of the weight and are capable of carrying a larger individual weight, compared to the outer buttresses which carry most of the load in traditional cathedrals. Such branches seek to attain equilibrium between the sections of the structure by minimizing the transfer of horizontal thrust (Figs. 4.49 and 4.52). The inclination of the branches is determined by the lateral thrusts and transfers those thrusts as vertically as possible. This equilibrium design minimizes the lateral loads by increasing the vertical load [23, 24].

The branches transfer lateral loads efficiently to the nave columns, forgoing the need for the exterior flying buttresses of a typical Gothic cathedral. The interior

Fig. 4.46 Main façade of the Sagrada Familia Church in Barcelona. (Image credit and source: Mohsen Aboulnaga)

Fig. 4.47 Interior of the Sagrada Familia. (Image source: https://commons.wikimedia.org/wiki/File:Sagrada_Fam%C3%ADlia,_Columns.jpg)

(a) (b)

Fig. 4.48 The columns of the church's nave are inspired from nature. (**a**) Tree's stem and branches (**b**) Branching columns in the interior of the church. (Images' credit and source: (**a**) Mohsen Aboulnaga (**b**) https://commons.wikimedia.org/wiki/File:Barcelona_panoramio_(263).jpg)

Fig. 4.49 Light coming through hyperboloid windows. (Image's source: https://commons.wikimedia.org/wiki/File:Window_Shadows_on_Ceiling_of_Sagrada_Fam%C3%ADlia_2010.JPG)

(a)

(b)

Fig. 4.50 Interior of the Sagrada Familia church in Barcelona. (**a**) Columns modeled branching canopies of a tree (**b**) The skylight behind the arcade. (Images' source: (**a**) https://upload.wikimedia.org/wikipedia/commons/0/0b/Sagrada_Familia_ceiling.jpg (**b**) https://commons.wikimedia.org/wiki/File:Sagrada_Familia_inside_Windows.jpg)

columns support most of the weight and are capable of carrying a larger individual weight, compared to the outer buttresses which carry most of the load in traditional cathedrals. The branches seek to attain equilibrium between the sections of the structure by minimizing the transfer of horizontal thrust (Figs. 4.47, 4.48, and 4.50). The inclination of the branches is determined by the lateral thrusts and transfers those thrusts as vertically as possible. This equilibrium design minimizes the lateral loads by increasing the vertical load [23, 24].

4.3.8 The Las Palmas Water Theatre in Canary Islands, Spain

Mimicry has a direct influence on inspiring architectural submissions. The Las Palmas Water Theatre "Teatro Del Aqua" project, designed by architect Nick Grimshaw, was part of a competition for the regeneration of the port of Las Palmas in the Canary Islands (Fig. 4.51). The design of the project was inspired by the properties of water collection from the fog collecting Namibian desert beetle (Fig. 4.51d). The Las Palmas Water Theatre gains from the sun and wind producing a large quantity of water for the town of Las Palmas [25].

The surroundings of Las Palmas provide the ideal conditions for such technology to sustain itself (Fig. 4.51a, b, c). Having a constant wind direction all year round and abundant sunshine assisted the project, in the sense that the basic solar thermal panels could be utilized to heat the seawater going into the evaporators, and this could ensure that moisture is collected by the wind going through [25].

Based on Sara Brown-Toganivalu, the destination is ideal since the Canary Islands are volcanic. The very steep sides below the seacut costs of constructing pipe systems in the deep-sea water. Among all the other ideal conditions, this proposal has the potential to create a very productive desalination (the removal of salt-water) plant, because at 1000 meters below the sea, the temperature is only 8 degrees centigrade, so the cold water could be used as a condenser.

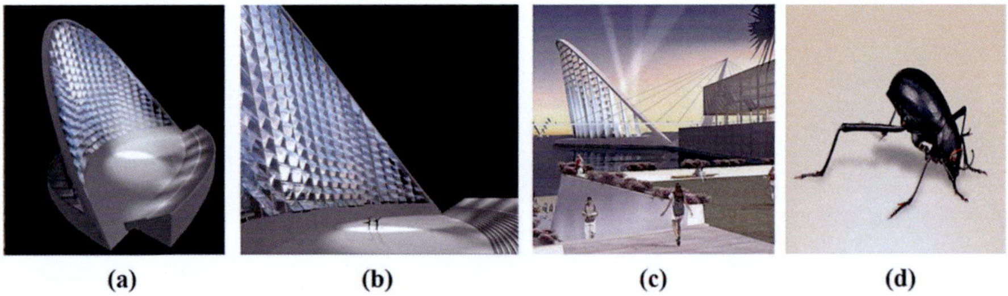

Fig. 4.51 Namibian fog-basking beetle and the theater with glass panels for collecting condensing seawater. (**a**) 3-D of the project (**b**) The main façade of Las Palmas Project (**c**) Exterior of Las Palmas Theater (**d**) Namibian beetle. (Images' source: (**a**) (**b**) (**c**) Grimshaw Global, http://uldevelop.egloos.com/ 1272040 (**d**) https://en.wikipedia.org/wiki/Onymacris_unguicularis#/media/File:Onymacris_unguicularis_MHNT_(cropped).jpg)

According to Grimshaw, *"The essence of the idea is to couple a series of evaporators and condensers such that airborne moisture from the evaporators is then collected from the condensers, which are cooled by deep seawater. This produces large quantities of distilled water"* [25]. The design includes a public venue for the performing arts and incorporates remarkable desalination methods [25].

Desalination of seawater and provision of a sustainable source of fresh water for arid urban areas could be pursued through a wide range of buildings. The design project has gained a lot of attention despite being a hypothetical building, because it is not only an eye-catching building but also a supplier of fresh water for a large part of the city of Las Palmas [25].

4.3.9 Tenerife Auditorium in Santa Cruz de Tenerife, Spain

The Tenerife Auditorium, situated between the Marine Park and the edge of the port, was designed by Grimshaw Global and opened in 2003 (Fig. 4.52). The distinctive concrete shell roof takes its inspiration directly from the landscape of Tenerife set against the Atlantic Ocean. The massive arc which is pouncing over the building like a wave depicting a symbolic power, serving not only the building that houses a 1558-seat auditorium for music and culture but also as a focal point for the islands [10].

4.3.10 The All Seasons Tent Tower in Yerevan, Armenia

The All Seasons Tent Tower in Yerevan, Armenia, is a public mixed building, which is designed by the OFIS Architecture. It is an international business center encompassing a museum and hotel with a total area of 82,130 m^2 [26]. The form of the tower symbolically represents the biblical mountain Ararat, and its inspired idea was created from the two terraced cylindrical towers connected at the ground floor (Fig. 4.53a). This tower provides several functions, such as a restaurant, pool, spa, fitness center, and bar in the higher tower, which will house a hotel and apartments along with a business center, shops, and a garage, and as for the lower, it is occupied with apartments. The connection between the two towers is a public area, including an exhibition, shopping areas, and restaurants connected to both the hotel lobby and the business entrance as well as a parking area which is located under the ground floor [26]. There is a part with fast parking and taxi or bus drop-off outside the plateau.

Figure 4.53 presents the two towers beside each other and the landscape. The terraces of the towers were not exposed, as the structural façade skin covers them, and this creates openings and full structural elements. The façade is a metal mesh, which illustrates the landscape element rising from the Plateau in Yerevan City, situated in a region that faces many earthquakes. For this reason, the priority when

(a) (b)

Fig. 4.52 Tenerife Auditorium, Santa Cruz de Tenerife, Spain. (**a**) The Auditorium inspired from a wave (**b**) A side view of the Auditorium. (Images' credit and source: (**a**) Diego Delso, https://commons.wikimedia.org/wiki/File:Auditorio_de_Tenerife,_ Santa_Cruz _ de_Tenerife,_España,_2012-12-15,_DD_02.jpg# (**b**) Wladyslaw, https://commons.wikimedia.org/wiki/File:Auditorio_de_Tenerife,_Sonnenuntergang.jpg#)

(a) (b)

Fig. 4.53 All Seasons Tent Towers in Yerevan, Armenia. (**a**) The metal mesh façade covering the two towers (**b**) A new Yerevan landscape. (Images' courtesy and source: (**a**) (**b**) OFIS – Slovenia/Paris, https://ofis.si/eng/projects/public/four_seasons_tent_tower.html)

choosing a structural system for a building is the safety and resilience during shocks of earthquakes, but this tall vertical structure resists the gravity load and forces resulting from earthquake action, and it has a reinforced concrete cores and composite columns [26].

The outer exteriors include a superior skin with adaptive outside shading devices to decrease sunlight gaining in summer. In addition, the cooling of inner spaces is accomplished primarily by use of typical slabs. In winter, the fresh air will be warmed inside the units and circulated into the rooms using the displacement ventilation [17]. In a nutshell, the two towers were inspired from nature and powered by solar energy and enveloped by a mesh fabric to filter sunlight and control temperature.

Fig. 4.54 Zvartnots Airport in Yerevan, Armenia. (Image credit and source: Armenia the World, https://commons.m.wikimedia.org/wiki/File:Zvartnots_airport.jpg)

Another example in Yerevan that partially mimics nature is Zvartnots Airport, where a curve structure is appearing in front of the building's main façade as shown in Fig. 4.54.

4.3.11 Esplanade Theaters in the City of Singapore, Singapore

The Esplanade Theater and Commercial district, located in the city-state of Singapore is one of the perfect biomimetic examples of art centers as a hub for performing arts (Fig. 4.55). It was completed in 2007 and designed by DP Architects and Michael Welford [27]. Its massive spine-covered peel or shell as illustrated in Figs. 4.55 and 4.56.

The main inspired behavior level from this plant is the "protection" as shown in Fig. 4.56a, and it uses its semirigid pressure skin to keep the seeds inside. Hence, this theater imitates the same behavior by its exterior as depicted in Fig. 4.56b. The dome of the Esplanade Theatre, details of the shell, and the Esplanade Theatres' park as well as the panoramic view of the complex at night are shown in Figs. 4.57, 4.58, 4.59, 4.60, and 4.61 respectively. Figure 4.62 also presents the interior of the theater, where the external envelope is vivid.

The criticism on this building was narrowed down to "the large area of glazing and openings." Yet another criticism stated that it is being insensitive and separated from the tropical climate of Singapore. Therefore, using a biomimetic approach was a decision taken after such criticisms; the building was designed in line with a

Fig. 4.55 Esplanade Theatres in the City of Singapore, Singapore. (Image credit and source: Bernard Spragg, https://commons.m.wikimedia.org/wiki/File:Esplanade_%E2%80%93_Theatres_on_the_Bay._Singapore_(12943310675).jpg)

biology approach. Thus, the target of the new design was the creation of a unique building that responds to its surrounding environments and culture [27–28].

The Esplanade Theatre contains an internal grid and bracing systems that combine the internal and external layers (Fig. 4.62). Along several areas in Esplanade Theatre, timber and stone were used as natural materials, and most of the floors are paved with stones as well (Fig. 4.62a). Sandstone cladding is used in the interior walls, and the triangular shades are made from insulating glass with aluminum fixtures cornering the intermediate points (Fig. 4.62b). This building is considered a landmark in Singapore, for dual reasons: (a) the dynamic sun shield and (**b**) the reflection of the Singaporean identity through its biomimetic inspiration.

The biomimetic approach solved the issues of concerns raised by the public. There are several outcomes from this project, including a convenient environment for the occupants, heat protection, natural light entering the building and interior protection from overheating, and lowering the HVAC usage, which mitigates GHG emissions [27–28].

(a) (b) (c) (d)

Fig. 4.56 Esplanade Theatres mimic nature – durian plant, Singapore. (**a**) Hanging durian plant on a tree (**b**) The shell of the durian fruit (**c**) Exterior of the Esplanade Theatres (**d**) Close view of the Esplanade Theatres. (Images' credit and source: (**a**) DXLINH, https://commons.m.wikimedia.org/wiki/File:Tender_fruit%27s_Durian_tree.JPG (**b**) Royleeyy, https://commons.m.wikimedia.org/wiki/File:D17_-_Durian.jpg# (**c**) Banku, https://commons.wikimedia.org/wiki/File:Esplanade_-_Theaters_on_the_Bay,_Singapore.jpg (**d**) Zairon, https://commons.wikimedia.org/wiki/File:Singapore_Esplanade_-_Theaters_by_the_Bay_4.jpg)

(a) (b)

Fig. 4.57 Dome of Esplanade Theatres, Singapore. (**a**) The dome mimicking durian shell (**b**) Triangular shades with insulating glass and aluminum fixture. (Image credit and source: (**a**) Basile Morin, https://commons.wikimedia.org/wiki/File:Dome_of_the_Esplanade_Theaters_on_the_Bay_Singapore.jpg (**b**) Martin Abegglen, https://commons.wikimedia.org/wiki/File:Theater_(1962336807).jpg#)

Fig. 4.58 Aerial view of Esplanade Theatres illustrating the Park inspired from nature. (Image credit and source: Zairon, https://commons.m.wikimedia.org/wiki/Category:Esplanade_Theaters_on_the_Bay#/media/File%3ASingapore_Esplanade_Theaters_by_the_Bay_viewed_from_UOB_Palaza_2.jpg)

Fig. 4.59 Front view of the Esplanade Theatres and park. (Image credit and source: Badener, https://commons.m.wikimedia.org/wiki/File:Singapore_Esplanade.jpg#)

Fig. 4.60 One of the two domes of Esplanade Theatres glittering and vividly appearing at night. (Image credit and source: Andrew Macmillan, https://commons.m.wikimedia.org/wiki/File:Dome_of_the_Esplanade_-_Theaters_on_the_Bay,_Singapore,_at_night.jpg)

Fig. 4.61 Glowing dome of the Esplanade Theatre at night. (Image credit and source: Basile Morin, https://commons.m.wikimedia.org/wiki/File:Esplanade_Theaters_on_the_Bay_Singapore_at_blue_hour.jpg)

(a) (b)

Fig. 4.62 Interior of Esplanade Theatre, Singapore. (**a**) The envelope from inside portrays the dynamic sun shield (**b**) Insulating glass from inside. (Images' credit and source: (**a**) Calvin Teo https://commons.m.wikimedia.org/wiki/File:Esplanade_concert_hall_reception.jpg (**b**) Calvin Teo https://commons.m.wikimedia.org/wiki/File:Esplanade_concert_hall_Circle_2_reception.jpg)

4.3.12 Beijing's Olympic Stadium and National Aquatic Center in China

Another biomimetic example is the iconic Beijing's National Stadium, which is known as "Bird's Nest" (Fig. 4.63). The sports complex, designed by Swiss architects Herzog and de Meuron is a clear and perfect example of applying biomimicry approach in modern architecture, based on its unique and innovative shape and creativity. Such biomimicry approach is not a new idea or new movement, but it is a revival of the early inspiration and relation to nature [11, 29].

The Beijing National Stadium was part of an international competition for the design of the main stadium for the 29th Olympic Games 2008, held by The Beijing Municipal Planning Commission [29]. The building has housed the opening and closing ceremonies of the Olympic Games. A set of criteria was included in the

(a) (b) (c)

Fig. 4.63 Façades' innovative structure of Beijing's National Stadium "Bird's Nest" in China. (**a**) Beijing National Stadium (**b**) Front view of the stadium structure (**c**) The structure of the stadium at night. (Images' source: (**a**) https://commons.m.wikimedia.org/wiki/File:Beijing_China_Beijing-National-Stadium-02.jpg (**b**) https://commons.wikimedia.org/wiki/File:Beijing_National_Stadium_ 20171117_043233.jpg (**c**) https://commons.wikimedia.org/wiki/File:Beijing_national_stadium_4.jpg)

program (Beijing, 2002): a stadium gauge of nearly 100,000 spectators during the games (to be reduced to about 80,000 after finishing the Olympic Games); a retractable or movable roof; a multipurpose and multifunctional design, for future efficiency and other uses; and an emphasis on green building and advanced technology [29].

The stadium is a straightforward inspiration from nature's ecosystem, where the straws of the bird's nest consist of many wooden membranes, and this was mimicked in the exposed exterior skeleton of the stadium building's structural elements. The aesthetic dimension through the motif itself and the minimal wasted material disguises the structure as shown in Fig. 4.64a, b. The material used to fill the stadium facades is made of ETFE (ethylene tetrafluoroethylene), and this draws the same analogy of the bird's nest. The nest in nature is isolated by stuffing small pieces of material between the twigs that make up the structure [30]. This semitransparent material protects the spectators inside the stadium and works as an acoustic insulation medium; meanwhile, it allows sunlight to penetrate and filter through its natural grass inside. Moreover, the panels are lighter than any other glass type or any aluminum panels, and that leads to reduction in the dead load supported by the roof. These panels are durable and self-cleaning too, hence helping in reducing the maintenance costs [29]. In addition, natural ventilation exists in the stadium by the openings in the stadium's façade and roof as shown in Fig. 4.64c, d.

(a) **(b)** **(c)** **(d)**

Fig. 4.64 Inspiration of Beijing National Stadium structure from a bird's nest. (**a**) Birds' nest structure inspiration (**b**) The façade of Beijing's stadium (**c**) Structural elements from inside (**d**) Stadium from inside with a void creating ventilation. (Image's credit and source: (**a**) Laslovarga, https://commons.m.wikimedia.org/wiki/File:American_Robin_Nest_with_Eggs.jpg (**b**) https://commons.wikimedia.org/wiki/File:Beijing_National_Stadium_2008_Summer_Paralympics.jpg (**c**) https://commons.wikimedia.org/wiki/File:Beijing_national_stadium_3.jpg (**d**) https://commons.wikimedia.org/wiki/File:Beijing_National_Stadium_Interior.jpg)

As another biomimicry concept, there is the National Aquatic Center in Beijing, China. Its idea comes from the honeycomb shape, which is regarded as perfect in structural balance. The honeycomb structure appears to be a pool of water from an aerial view as illustrated in Fig. 4.65. Furthermore, these structures were inspired from nature. The water cube's design depends on water bubbles in foam, while it may seem random, this structure is derived from standards of geometry and crystalline frameworks as presented in Fig. 4.66 [30, 31].

There is great and profound mathematics behind having such structures, and it is known for forming a dyad of bubbles; every three bubbles attached together intersect at 120 degrees. The more the bubbles, the more they are attached on many sides by other bubbles, creating flat faces as shown in Fig. 4.67. Ending up with watching the bubbles form polyhedral, this steel structure form engages three levels of biomimicry (form, process, and ecosystem) [32].

Both "the bird's nest" and "the water cube" used physics and chemistry science to create the structure and study the geometry of the foam bubbles and water molecules when transforming its concept into real steel structure. These two buildings are not only imitating forms from surrounding nature but also achieving sustainable architecture. In the water cube building, it uses solar energy for pool's water heating, reuses the gray water, filters it, and then returns water to the pools [32].

Fig. 4.65 Panoramic view of Beijing National Stadium and Aquatic Center. (Image source: https://commons.m.wikimedia.org/wiki/Category:Beijing_National_Stadium#/media/File%3AOlympic_stadium_of_Beijing.JPG)

(a) (b)

Fig. 4.66 National Aquatic Center in China. (**a**) Aerial view of the building (**b**) Both Stadium and Aquatic Center. (Images' source: (**a**) Leiem, https://commons.m.wikimedia.org/wiki/File:Shuilifang,_Beijing_Olympic_Park_(2019)_-_Blue_light.jpg# (**b**) Prinz W, https://commons.m.wikimedia.org/wiki/File:BNS%26BNA.jpg#)

(a) (b)

Fig. 4.67 National Aquatic Center in China and construction of bubble soap concept. (**a**) The envelope of the water cube (**b**) Interior of National Aquatic Center. (Image credit and source: (**a**) Hwanghsuhui, https://commons.m.wikimedia.org/wiki/File:Water_Cube_The_National_Aquatic_Center_Chaoyang_Beijing.jpg# (**b**) Arne Müseler, https://commons.m.wikimedia.org/wiki/File:Water_Cube_Ice_Cube_Beijing_2.jpg)

4.3.13 Omotesando Building, Tokyo, Japan

Architects use a wide range of design approaches to follow nature and exploit biomimicry as a source of inspiration, but nevertheless, this can certainly be achieved with the aid of computer modeling. The methods and the use of algorithms of generative modeling using special programmers can be improved by the study of computational models following natural processes and using their application to architectural design [33]. With the benefits of developed interlinkages between architecture, nanotechnology, and computer modeling, the new organic forms inspired by nature could be derived by exploiting computational programming/programmers; some of these examples have already been generated.

(a) (b)

Fig. 4.68 TOD's Omotesando building in Tokyo, Japan. (**a**) The façade of TOD's building (**b**) The emulated structure from nature at night. (Image credit and source: (**a**) Dconvertini https://commons.m.wikimedia.org/wiki/File:%E3%83%92%E3%83%A5%E3%83%BC%E3%82%B4%E3%83%9C%E3%82%B9%E8%A1%A8%E5%8F%82%E9%81%93%E5%BA%97_May_30,_2019.jpg%23mw-jump-to-license (**b**) d'n'c', https://commons.wikimedia.org/wiki/File:TOD%27S.jpg)

In Tokyo, Japan, examples of treelike façades in architecture were depicted in the TOD's Omotesando Building as shown in Fig. 4.68. The famous biomimetic building, completed in 2004, was designed by the Japanese architect Toyo Ito [11]. The curving, organic-looking shapes of the glass and concrete façades of TOD's building are inspired by the Zelkova trees that line the Omotesando Boulevard [33] as illustrated in Fig. 4.69.

4.3.14 Residential Building in New Songdo City, Incheon, South Korea

The residential buildings in Songdo smart city, Incheon, South Korea, are structurally profound towers that are generated by applying the biomimetic function of nature plants' properties such as the cactus plant (Fig. 4.70). This concept notably reduces the number of structural members used in its construction, which results in lessening the costs. As seen from Fig. 4.71a, two or three curved branches are created from the main stem of the cactus plant. In the three residential towers (Fig. 4.71b), the blue-glazed tower (left) is branching from the white tower, but in the other two blue towers, it is revered. This is a clear biomimetic approach inspired from nature. Figure 4.72 illustrates this concept of the towers from the lake.

(a) (b) (c)

Fig. 4.69 The concept of biomimicry of the famous TOD's Omotesando building in Tokyo inspired from Zelkova trees. (**a**) Dazzling TOD's main façade, (**b**) Side view of the façade, (**c**) Zelkova trees. (Images' credit and source: (**a**) Angaurits, https://commons.m.wikimedia.org/wiki/Category:Tod%27s_Omotesando_Building#/media/File:%3ATod's_at_Omotesando.jpg (**b**) https://upload.wikimedia.org/wikipedia/commons/thumb/c/cf/Omotesando_buildings.jpg/2560px-Omotesando_buildings.jpg (**c**) Mohsen Aboulnaga, Famartin, https://commons.m.wikimedia.org/wiki/File:2014-11-02_14_31_33_Zelkova_foliage_during_autumn_along_Woosamonsa_Road_in_Hopewell_Township,_New_Jersey.jpg#; Kici, https://en.m.wikimedia.org/wiki/Zelkova_serrata/mediaFile:Zelkova_serrata_japan-2006.jpg)

Another iconic example of the biomimicry approach is Songdo IBD Tri-bowl, which is located in the international business district in Songdo, Incheon, South Korea (Fig. 4.73). The complex is used as a multifunction such as exhibition, performance (play), and social gathering and recreation. The Tri-bowl design creates a free-curved floor under a flat ceiling [34]. The design concept of the Tri-bowl mimics nature in the form of Narcissus Geranium plants and its yellow or pink spring flower of the amaryllis family, Amaryllidaceous, as shown in Fig. 4.74. The Tri-bowl metal (Fig. 4.73a) emulates the segments of the flower's corolla, which has six petals that look like a cup (Fig. 4.74b, c).

Moreover, there are other examples in Songdo, Incheon, where the honeycomb structure has been inspired from nature in the design of a series of towers [34]. It is found that the honeycomb structure of these towers is able to maintain its own strength after twisting is applied. This concept notably reduces the number of structural members used in its construction, which results in lowering the building costs [35]. In addition, the analysis of the honeycomb arrangement was conducted by the Arup Engineering team [35].

Fig. 4.70 The twin towers inspired from nature in Songdo – Incheon, South Korea. (Image credit and source: Mohsen Aboulnaga and Nada Nidal 2019)

4.3.15 The Lotus Temple in Delhi, India

Bahai Temple in New Delhi, India, is one of the superior structures of the twentieth century. It is commonly known as "the Lotus Temple." The iconic biomimicry architecture is inspired by the lotus flower, as shown in Figs. 4.75, 4.76, 4.77, and 4.78. The Lotus Temple, which was built in 1986 and cost US$ 10 m (£16 m today)

(a) (b)

Fig. 4.71 The twin towers inspired from nature in Songdo – Incheon, South Korea. (**a**) Branches of a cactus plant (**b**) The inspired towers seen from Songdo Central Park. (Images' credit and source: (**a**) Mohsen Aboulnaga 2021 (**b**) Fleetham, https://commons.m.wikimedia.org/wiki/File:20141231_143713_HDR.jpg#)

Fig. 4.72 The towers inspired from nature in Songdo – Incheon, South Korea. (Image credit and source: Piotrus, https://commons.m.wikimedia.org/wiki/Category:Songdo_International_Business_District#/media/File:Songdo_International_Business_District_08.JPG)

Fig. 4.73 Songdo IBD Tri-bowl, central park in Songdo – Incheon, South Korea. (Image credit and source: Ken Eckert, https://en.m.wikipedia.org/wiki/File:Songdo_IBD_Tri-Bowl.jpg)

Fig. 4.74 Songdo IBD Tri Bowl mimicking Narcissus Geranium flower, Incheon, South Korea. (**a**) One of the tri-bowl in the form of a cup (**b**) Narcissus Geranium plant (**c**) Petals look like a cup. (Image credit and source: (**a**) Ken Eckert, https://en.m.wikipedia.org/wiki/File:Songdo_IBD_Tri-Bowl.jpg (**b**) and (**c**) KENPEI, https://commons.m.wikimedia.org/wiki/File:Narcissus_tazetta_var_chinensis1.jpg#)

and accommodates near to 2500 people, is a place for meditation and prayer, yet it has become a source of tourist attraction in India [36]. The temple has nine doors for the central hall and is made from white marble, which added an aesthetic dimension to the temple. There is also a pool with an area of 40 square meters to make the temple look like it is floating like a lily flower [36]. The appreciation of this temple originated from the integration between India's history and new amazing architecture. It is considered one of the most visited destinations worldwide, with the

(a) (b) (c)

Fig. 4.75 Lotus flower inspiring the structure of lotus temple. (**a**) The Lotus flower opening (**b**) Lotus flower inspiring the design (**c**) Model of the Lotus temple's structure. (Image's source: (**a**) Bohemianvegan https://en.m.wikipedia.org/wiki/File:White_Lotus_Blossom.JPG (**b**) Evak, https://commons.m.wikimedia.org/wiki/File:Nymphaea_tetragona.jpg (**c**) Wiki-uk, https://commons.m.wikimedia.org/wiki/File:104-Lotus-Temple-Information-Centre.JPG#)

(a) (b)

Fig. 4.76 Lotus temple in Delhi, India, a vivid biomimicry example from nature. (**a**) The lotus flower inspires the design, (**b**) Front view of Lotus Temple with five entrances. (Image source: (**a**) Jasmine Calypso, https://commons.wikimedia.org/wiki/File:Lotus_Flower_taken_at_Kenilworth_Garden.jpg, (**b**) Vinayaraj, https://commons.m.wikimedia.org/wiki/File:Lotus_Temple_-_Delhi,_various_views_(15).JPG#)

number of visitors reaching 100 million in 2014 [36]; this number exceeds the number of visitors at the Eiffel Tower in Paris and the Taj Mahal in India.

A special feature of this temple is that it is the first temple in India fitted with solar panels, generating 20% of its electricity from clean energy [36]. The Lotus Temple is creating a deeper meaning due to the structure, symbolic, and the perfect flower, which in turn sets up a feeling of unnaturalness and unearthly, especially when you are inside the temple [37] (Figs. 4.75 and 4.76) . The Lotus construction took 10 years to be completed, and it integrated the value of both aesthetic features and technology and their impact on architectural art. The Lotus Temple consists of three sets of leaves or petals, each created from thin concrete shells and covered by marble. The outer leaf consists of nine petals and is called the entrance leaves and opens outward forming the nine entrances around the outer hall as shown in Figs. 4.75b, 4.76, and 4.77, whereas the next set, called the outer leaves, points introvert. The outer hall is covered by the entrance and the outer leaves. The third

Fig. 4.77 Panoramic view of the Lotus Temple mimicking a lotus flower. (Image credit and source: ARwkmedia, https://commons.wikimedia.org/wiki/File:Lotus_Temple.png)

set of nine petals is called the inner leaves. Only the tips open out, somewhat like a partly open bud. Such a set rises above all, creating the main structure and including the central hall. The radial beams give protection from the rain outside and allow natural light to penetrate into the auditorium when the lotus opens [37, 38]. A front view of the Lotus Temple at night is shown in Fig. 4.78.

- **Problem definition:** In this project, the design itself was a problem. The project is to design the Baha "I" temple, which should act as a temple for all religions.
- **Reframe the problem:** The challenge for the architect, Fariborz Sahba, was to have a concept that would be acceptable to people of all religions and fit into the Indian context [38].
- **Biological solution research:** Based on research, the architect came up with the concept of the lotus flower for the design due to its importance to the Hindu religion and also Buddhist, and the lotus flower was chosen as the national flower of India as well [38].
- **Define the biological solution:** Flowers are conceptually symbolizing purity and holiness. The temple appears as a lotus flower surrounded by leaves afloat on water. The building is also in the form of a lotus flower, which has three layers of petals with nine petals in each layer [37].
- **Practical extraction:** The flower structure has nine inner and nine outer petals with 27 petals in total, surrounded by gardens and nine water bodies making it looks like floating on water.

Fig. 4.78 Front view of Delhi Lotus Temple at night. (Image credit and source: Vandelizer, https://commons.m.wikimedia.org/wiki/File:LotusDelhi.jpg#)

- **Principle application:** Similar to the form, the structure of the temple was inspired by nature, whereas the structure of the building is motivated by the shells [38].

Moreover, the inner nine petals form the nine entrances, which allow to enter the worship hall. These petals are separated at the top, which slightly looks like a bud. The opening is covered with glass to allow for natural light entry. The shells of the inner petals are supported by nine radial beams, mainly resting on the arches [38] as presented in Fig. 4.79. The structure consists of 27 giant marble petals and is connected through nine bridges, which run over the nine ponds that lead to the nine entrances [36, 38] as illustrated in Fig. 4.80.

4.3.16 Lavasa Township, India

Lavasa is a new Indian hill city developed by Hindustan Construction Company (HCC), where people can live, work, learn, and play in harmony with nature. An inclusive city, based on the principles of new urbanism principles, is developed over 18,000 acres of land, and it is a three-hour drive from Mumbai (Fig. 4.81). Lavasa Township depends on the biomimicry approach as a guide and mimics the banyan fig leaf by looking into its drip-tip system that allows water to run off while simultaneously cleaning its surface [24]. Lavasa hosts a complement of global leaders in hospitality, tourism, and education as well as healthcare, business research, and

Fig. 4.79 Interior of the Lotus Temple showing the nine petals. (Image credit and source: Vinayaraj, https://commons.m.wikimedia.org/wiki/File:Lotus_Temple_-_Delhi,_various_views_(3).JPG#)

(a) **(b)**

Fig. 4.80 Lotus temple showing the nine entrances running over nine water ponds. (**a**) One of the Lotus Temple's petals runs over a water pond leading to the entrance, (**b**) Half view of the temple's nine entrances over one of the nine ponds. (Images' credit and source: (**a**) Yatri Dani, https://commons.m.wikimedia.org/wiki/File:House_of_worship.jpg# (**b**) Abonzertv, https://en.m.wikipedia.org/wiki/File:Lotus_Temple_-_Temple_-_Half_View.png)

Fig. 4.81 Lavasa township new community rooted to nature. (**a**) Aerial view of Dasve Township, Lavasa site overlooking the river, Pune in India (**b**) Aerial view of Lavasa hills (**c**) Mountains engulf Dasve Town, Lavasa (**d**) The first phase of Dasve Valley with its new buildings, (**e**) Lakeside of Dasve Valley with its new buildings – Phase two (**f**) Lavasa city inside view – Phase two (**g**) The Center of Dasve Town, Lavasa (**h**) General view of the Convention Center at Dasve, Lavasa, (**i**) Hotel courtyard with nature, (**j**) The Convention Center at Dasve, Lavasa at night. (Images' credit and source: (**a**) Atulswaroop, https://commons.m.wikimedia.org/wiki/File:Lavasa1.jpg# (**b**) Arjun Singh Kulkarni, https://commons.m.wikimedia.org/wiki/File:Lavasahills.jpg (**c**) Cryongen, https://commons.m.wikimedia.org/wiki/File:Dasve_Town.jpg# (**d**) Trusharm512, https://commons.m.wikimedia.org/wiki/File:Lavasa4.jpg# (**e**) Cryongen, https://commons.m.wikimedia.org/wiki/File:Lakeside_View_at_Dasve_-_Lavasa.JPG# (**f**) https://commons.m.wikimedia.org/wiki/File:Planned_City_Lavasa_Inside_View.jpg# (**g**) Mayur239, https://commons.m.wikimedia.org/wiki/File:Lavasa_City.JPG# (**h**) Cryongen, https://commons.m.wikimedia.org/wiki/File:Lavasa_Convention_Centre.jpg# (**i**) Ankur P, https://commons.m.wikimedia.org/wiki/Category:Lavasa#/File%3A:Lavasa_Fortune_Hotel_courtyard%2C_Lavasa%2C_Maharashtra.jpg (**j**) Yoursamrut, https://commons.m.wikimedia.org/wiki/File:Lavasa_Nightscape.jpg#)

(f) (g) (h) (i) (j)

Fig. 4.81 (continued)

industry [39]. The Dasve Township, Lavasa site overlooking the river, Pune in India, is shown in Fig. 4.81a, b, c. The hill city provides contemporary yet timeless architectural designs and multiple options for housing like rental housing, starter homes, apartments, and villas (Fig. 4.81d, e, f, g). With the state-of-the-art infrastructure and amenities, it enables people to live their lives fully [39]. Lavasa Township includes five urban villages with 30,000–50,000 inhabitants. Also, the newly developed urban areas increase the tourism sector according to the visit of millions each year.

Traditional patterns of India's town planning, as well as vernacular forms of buildings, are the source of inspiration for the planning of the city (Fig. 4.81d, e, f, g). The new township imitates principles with culture, which have proven sustainable many years ago. To drive future landscape performance and renew deforested areas, a landscape master plan has been developed. Taking the environment into consideration is already a main part of the plan. In addition, Lavasa Township includes three self-sustaining towns (Dasve, Gadle, and Dhamanohol). Phase one of Dasve Valley (1700 acres) is completed and expected that 70% of deforested land will be restored by proper landscaping, reforestation, and slope greening (Fig. 4.81b, c, d, e); besides the reduction in carbon emissions, water consumption and waste will be around 30%, 65%, and 95%, respectively [39]. The Dasve Valley, Lavasa City inside view (stage two), and the Convention Center and hotel of Dasve town are shown in Fig. 4.81h, i, while Fig. 4.81j presents the Dasve Center at night.

4.3.17 Aldar Headquarters, Abu Dhabi, UAE

An additional biomimetic example in the Gulf region is the iconic commercial building of the Aldar Headquarters in Abu Dhabi, UAE (Fig. 4.82). The 121-meter high building accommodates the headquarters of the property developer Aldar. The landmark building is designed by MZ and Partners in 2005 and in collaboration with Arup, who developed the cores and completed it in 1 year depicting the first building engineering concept. The iconic structure allows for natural daylighting and a wide view of the beach on one side and the city on the other side [40–42]. The

(a) (b) (c)

Fig. 4.82 Aldar Headquarters' shell building in Abu Dhabi, UAE. (**a**) Shell form inspiration (**b**) Façade of the circular tower (**c**) Side view of Aldar tower. (Images' source: (**a**) https://www.flickr.com/photos/volvob12b/8837357438/in/photolist-esVLtG-5nM54x-4GhFnY-byx7eP-kVTiPk-WQRhs2-eW1SFx-dRPskz-hL6PL2-LJgoAJ-CQNGh2-SL825T-34STho-8z2ho6-7uBu1t-7uBtqa-4KuD1E-5HMRtL-7uBuzx-2k66ct7-9BxZyx-4KuBhU-4KuCpu-dPTEzJ-rZUtt-4KuCwh-BX6t2i-CSboUW-EpWTpY-4KqkVz-2iAp9aV-4Kqmyk-RvBeGR-4KuCRL-4KuBEE-4KuAsU-4KuCbs-4KuPB5-4KuBNj-4KuCCC-4KuBoq-5Qiufc-9h7RXc-fQip5w-28tc4BY-ezJiV9-aZpvU2-7KKUJ4-zwycX-2jNvtkq (**b**) FritzDaCat, https://en.wikipedia.org/wiki/File:Aldar_Headquarters_Building.jpg (**c**) https://www.flickr.com/photos/paolo_rosa/5507209368/in/photolist-Ldo1XL-DikTj9-6VD2Ev-bNmM7D-85yqdj-24zFPLc-cYrkdm-Eee7YB-ei6tq6-6ehQQo-c1yuV3-D4ZRqC-9oDTtE-qKyY9b-pNDfmL) tower offers 62,000 square meters of office spaces over 23 floors and is occupied by more than 120,000 people [42]. The simple structure of the building is centered on a biological solution, and the design concept is inspired by the clam shell, which has a deep meaning of maritime heritage of Abu Dhabi (Fig. 4.82a). For principle extraction, the architect imagined two giant curved circular walls joined together, representing the shell [40–42] as presented in Fig. 4.82b. The shell-emulated design powerfully fits into the iconic skyscrapers of Abu Dhabi. The biomimetic building is visible from Dubai to Abu Dhabi highway and is located in a privileged location, which is surrounded by the city, the sea, and the canal as shown in Fig. 4.82c [37, 40–42]. The design is based on a solution-based approach, where architects created a design that relates to Abu Dhabi water nature, being an island, and historically famous for pearls. The project was developed following the LEED principles of the US-Green Building Council (US-GBC).

4.3.18 The Sahara Forest Project in Aqaba, Jordan, and Qatar

Another biomimicry project is the Sahara Forest Project in Aqaba – Jordan and Qatar – which is a greenhouse design inspired by the Namibian fog-basking beetle. This beetle has found its own way to evolve its particular fresh water in the desert [43], yet it regulates its body temperature by accumulating heat by day and collecting the water droplets formed on its wings by the fog as shown in Fig. 4.83. The greenhouse design emulates the beetle to combat climate change in an arid climate. The exploration carried out by an architectural design team strived to create an ecosystem that was not only sustainable but also regenerative. Three mechanisms drove the design of the greenhouse: the main one is the creation of a saltwater-cooled greenhouse (SWCGH), concentrated solar power (CSP), and technologies for desert vegetation [16, 22, 44, 45].

(a) (b) (c)

Fig. 4.83 Sahara project and source of inspiration. (**a**) The Sahara Forest Project (**b**) Biomimicry concept (**c**) Namibian fog-basking beetle. (Images' credit and source: (**a**) https://en.wikipedia.org/wiki/File:The_Sahara_Forest_Project_pilot_facility_in_Qatar.jpg (**b**) Moongateclimber, https://commons.m.wikimedia.org/wiki/File:Namib_desert_beetle.jpg (**c**) https://www.flickr.com/photos/harryandrowena/8066930540/in/photolist-dhR8pj-7Zt8X8-5nh777-dQDGT3-6ZhKC-oZjzW-48N5Px-rYRCoT-2kYiYNg-e8MVgF-24qXZEv-24qXY3T)

The saltwater-cooled greenhouses (SWCGH) present suitable conditions for year-round cultivation. To create an SWCGH, Charlie Paton created a system to provide evaporative cooling and humidification in the greenhouse structure using saltwater [24, 45, 46]. The evaporated air condenses to fresh water allowing the greenhouse to remain heated at night. This system produces more water than the interior plants need; hence, the excess is spewed out for the surrounding plants to grow. The salt which is extracted from the evaporation process is crystallized at different stages to extract various elements. By crystallizing out the evaporated saltwater, calcium carbonate followed by sodium chlorite is created; these elements can both be compressed into building blocks [24, 45, 46].

Many other elements that exist in seawater are reused in other ways to minimize waste [24, 45, 46]. In addition, concentrated solar power (CSP) involves concentrating the heat from the sun with solar tracking mirrors to create steam that drives conventional turbines; this produces zero-carbon electricity, which is twice as effective as photovoltaics.

The Sahara Forest Project offers a solution to restoring forests and creating a solution for the vegetation of deserts as a source of food, water, and energy in an otherwise resource-constrained area of the world [24, 45, 46].

> The Sahara Forest project is a model for how we can create zero-carbon food, abundant renewable energy in some of the most water stressed parts of the of the planet as well as reversing desertification in certain areas. ([44])

4.3.19 Doha Tower and Giant Cactus Building

The cactus building is another vivid example of biomimicry. As mentioned earlier, the organism level in biomimicry's levels is defined as:

> The mimicking of an organism like a plant or animal and could involve emulating part of or the whole organism's form, material, function and process. ([43])

The design of the new building of the Minister of Municipal Affairs and Agriculture (MoMAA) in Doha, Qatar, was inspired by the cactus plant [43–47]. The staggering and stunning office building and the adjoining vegetative dome are perfect architectural examples in the biomimicry field. The building was designed by a Bangkok-based firm (Aesthetics Architects GO Group) in Thailand. The firm exploits the natural mechanism in the cactus and emulates the desert plant to create a unique sustainable solution to a complex problem of a very hot region in summer. The climate, which is classified as a hot-arid, has approximately 3.2 inches of precipitation annually [44]. The building was inspired by the shape of the hardy cactus plant knowing its ability to grow in hard desert climates. The cactus form and behavior were investigated to inform the building solution [47]. This building named "The Cactus Project" has been designed with several sustainability elements that have been taken from the organism level and behavior level of the cactus plant, including energy efficiency and sun shades (opened or closed as well as swinging up and down automatically to reach the prevailing temperature), depending on the preferred interior temperature. For heat and sunlight regulation, it is a mimicking of the cactus' behavior that proceeds transpiration at night instead of during the day to retain water. Thus, mimicking such behavior led to innovative solutions [47].

In addition, the dome, located at the base of the tower, has a botanical garden – an extra green point, including an edible garden and uses plants to clean up waste water. Moreover, it provides a sustainable food source irrigated from gray and black water treatment [47]. Furthermore, the cactus spine was one of the elements that inspired the shading devices incorporated in the MoMAA building. The spines' function is to provide shading for the plants from the intense sunlight in summer. Having many spines on the exterior skin shades the plants sufficiently to keep the internal temperature low enough to avoid water evaporation from the plants. This is a way of successfully surviving in such a hot climate. It is imperative to say that such technologies influence the design of buildings.

Another example of biomimicry in architecture is the Doha Tower, which was designed by Jean Nouvel [48]. The tower shape is created as a cylinder, like the Arizona cactus plant (Fig. 4.84). The tower is 231.5 m tall and comprises 46 floors

Fig. 4.84 Doha Tower inspired from nature. (**a**) Multiple cactus plants (**b**) Cylindrical cactus plant (**c**) Doha's cylinder tower. (Images' credit and source: (**a**) Forest & Kim Starr, https://commons.m.wikimedia.org/wiki/File:Starr_070320-5799_Echinopsis_pachanoi.jpg (**b**) WClarke, https://commons.m.wikimedia.org/wiki/File:Carnegiea_gigantea_in_ Suguaro_Park _ near_Tuscan,_Arizona_during_November_(58).jpg# (**c**) Ed Parsons, https://commons.m. wikimedia.org/wiki/File:P1000015_-_panoramio.jpg#)

in the shape of a cylinder volume that measures 45 in diameter [48]. The facades utilize a double-skin system which is composed of four butterfly aluminum elements for sun protection as illustrated in Fig. 4.84c. The concept has been a replica of Barcelona's Torre Agbar tower (Figs. 3.38 and 3.39 in Chap. 3).

4.3.20 Council House 2 (HC2) in Melbourne, Australia

The Council House 2 (CH2) building, completed in 2005, is inspired by the fundamentals of nature. Architect Mick Pearce, who was educated at the Architectural Association (AA) in London, UK, created such an environmentally viable building [49]. The CH2 has an important and interactive role between the city and nature, yet it is considered a model that the designers should follow. This building confers factual and metaphorical expressions of environmental intentions (Fig. 4.85). The key secret in mimicking the behavior of termites from nature has been expressed and achieved in the façade of the building. It mimics the structure of the termites with its shafts and vents, so that the façades control the climate, ventilate fresh air, and penetrate natural light [49]. Thus, it is the first commercial office building that achieves such a rate, yet the first building in Melbourne, Australia, that uses the biomimicry approach. The city of Melbourne aims to achieve zero-carbon emissions for the municipality by 2020 [49, 50]. In this context, the reduction of energy consumption in the CH2 building by 50% is a major contribution to the strategy of Melbourne City. The design of the CH2 building is truly considered to be in line with the fundamental "laws of nature."

Deciding on the success or failure of such a building regarding its biomimicry approach, it is clear that it exhibits qualities as a visually resolved architectural form instead of a box full of eco-tech objects [49, 50]. As illustrated in Fig. 4.85a, b, c, the CH2 façades are composed of timber vertical slats covering a fully glazed wall. These slats pivot vertically, open and close in response to the time of day and the angle of the sun, hence providing shades.

For this building, the heating, ventilating, and cooling (HVAC) system is designed with strategies taken from a termite mound, where the cool wind travels down into the bottom of the mound by bouncing through the channels and then is cooled and stored using wet soil (Fig. 4.85d). When the air warms, it flows upward and out of the mound by vents [49, 50]. The wind turbines for bioclimatic design of the CH2 building are shown in Fig. 4.85d, e.

4.3.21 Ultima Tower, San Francesco, CA, USA

Like in Europe and Asia, biomimicry has also been adopted and implemented in many building examples in the United States (Fig. 4.86). A clear example is the Ultima Tower – a proposal of a super skyscraper, which was designed by Eugene

Fig. 4.85 The CH2 sustainable building in Melbourne, Australia. (**a**) Main wooden façade (**b**) Opened timber louvers (**c**) Eastern façade with green balconies (**d**) Green terraces and wind turbines (**e**) Roof terraces and wind turbines to generate clean energy. (Images' credit and source: (**a**) Roryrory, https://commons.m.wikimedia.org/wiki/File:CouncilHouse2.jpg# (**b**) Nick Carson, https://commons.m.wikimedia.org/wiki/File:Council_House_2_Melbourne_Shutter_detail.JPG# (**c**) Biatch, https://commons.m.wikimedia.org/wiki/File:CH2_from_Tivoli_Arcade_Melborune.jpg (**d**) Nick Carson, https://commons.m.wikimedia.org/wiki/Category:Council_House_2#/media/File:%3ACouncil_House_2_Melbourne_east_and_turbine_detail.JPG (**e**) Stephen Bian, https://commons.m.wikimedia.org/wiki/File:Council_House_2_roof_garden.jpg#)

Tsui, an American architect in 1991 [11]. The building includes 500 stories with a total height of 10,560 feet with a base diameter of 6000 feet as shown in Fig. 4.86b [11]. The problem in this project is that the architect wished to solve the hypothetical design of the skyscraper to address the global population rapid increase, and it is planned to accommodate over one million inhabitants in a 2-mile high environmentally friendly tower based on a 1-mile wide horizontal plateau. The cost of the project amounts to US$150 billion. Moreover, it consists of several attributes and features of an ecosystem, such as rivers, lakes, streams, hills, and valleys and

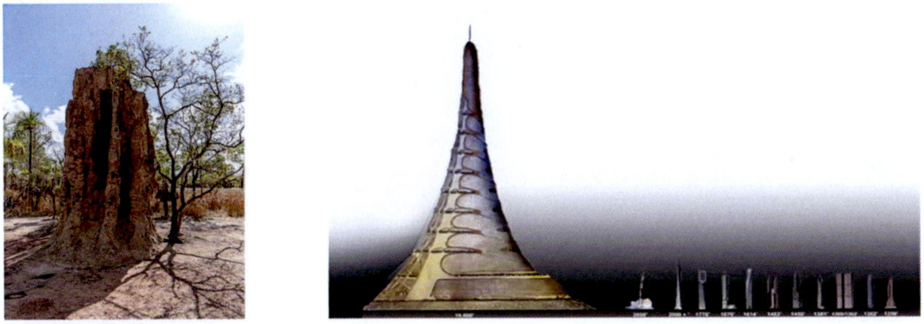

Fig. 4.86 Ultima Tower in USA and inspiration source of the form. (**a**) Form of termite inspiration (**b**) An impression of the Ultima Tower façade. (Image's source: (**a**) https://commons.wikimedia.org/wiki/File:Litchfield_National_Park_(AU),_Magnetic_Termite_Mounds_%2D%2D_2019_%2D%2D_3732.jpg (**b**) Eugene Tusi, https://www.tdrinc.com/ultima.html)

encompasses the soil and landscape on which buildings, for example, residential, retail, entertainment, commercials, and others, are designed [11].

In this project, the form of the termite mound was selected based on its ability to remove loads from the top to the bottom and for its effectiveness as an aerodynamic form that keeps stability in tall buildings as well (Fig. 4.86a). Hence, the aim of the building is to create an ecosystem for the people inside. Electricity generation will be through the difference in pressure between the bottom and the top of the tower. The building mimics the same visual mound of the termite mounds and that will be beneficial for the ventilation of the tower as illustrated in Fig. 4.86. A double-helix cable crossing network distributes tension across the outer covering, allowing the entire structure to absorb stresses rather than just one area. The architect has looked precisely into nature, especially the tree and how it behaves and how it moves water around their vertical lengths [11].

4.3.22 Fennell Residence, Portland, Oregon, USA

Another vivid example is the floating house, which is inspired by nature and developed by Robert Oshatz. The inspiration of the building is created from ocean waves, where the end result sits in harmony with its surroundings as presented in Fig. 4.87. Known for his curvaceous, Oshatz's swooping architecture is considered a unique approach for designing such a building.

This unique home is kept afloat by locally sourced 80-foot Douglas fir logs, and the exterior design takes its inspiration from ocean waves as shown in Fig. 4.87 [30]. The series of intersecting and overlapping organic forms creates a sense of constant movement that is clearly inspired by high waves (Fig. 4.87a) but likewise reflects the nature of the structure itself. More than just a clever aesthetic trick,

(a) (b)

Fig. 4.87 Fennell residence (floating house) and inspiration of sea wave. (**a**) Wave movements creating a water vortex (**b**) Mimicking wave vortex in the building design. (Images' credit and source: (**a**) NOAA, https://commons.m.wikimedia.org/wiki/File:Large_breaking_wave.jpg (**b**) Flickr, https://www.flickr.com/photos/45398879@N02/6279778617/in/photolist-ayVvp8-scV35b)

however, this repeated form serves the architectural functions from the interior volumes to allow natural light at both ends and along the roof line [51].

The house was designed as a series of curves that seem to flow over one another. A glue-laminated beam dives into the floor connecting the courtyard. The view draws the occupants through a door in the glass wall, where a patio on the water is located [51]. The glass fills the gaps in the structure, making the building feel light and transparent, with fine reflections that imply depth in addition to clerestory windows to allow for natural ventilation [52].

4.3.23 Spiraling Chicago Tower, Chicago, USA

The Chicago Spire has been called the world's most significant residential building, featuring 150 floors and 1194 residences. The slender structure rotates each floor at an average of 2.44 degrees to a total of 360 degrees from top to bottom, guaranteeing that no two views are the same [28]. It is the second tallest building in the world after Burj Khalifa [28]. The structure is replete with biomorphic elements where its idea was created from emulating Nautilus, and the tower helical form incorporates the golden mean and the Fibonacci sequence to closely mirror the spiraling growth of the nautilus shell as shown in Fig. 4.88. The sparkling tower super-structure incorporates a variety of sustainable elements, including a fluid façade shrouded in high-performance glass that was designed to protect migratory birds (Fig. 4.89). All landscaping is maintained via a rainwater recycling system, and a geothermal system draws river water to cool the building as well. The entire interior is outfitted with intelligent solutions for a smart building, including an energy management system and 15% higher energy efficiency standards [28].

Fig. 4.88 Nautilus shell inspiring the Spiraling Chicago tower. (Image credit and source: Lazo Wasem, https://commons.wikimedia.org/wiki/File:Nautilus_pompilius_(YPM_IZ_022690)_002.jpg)

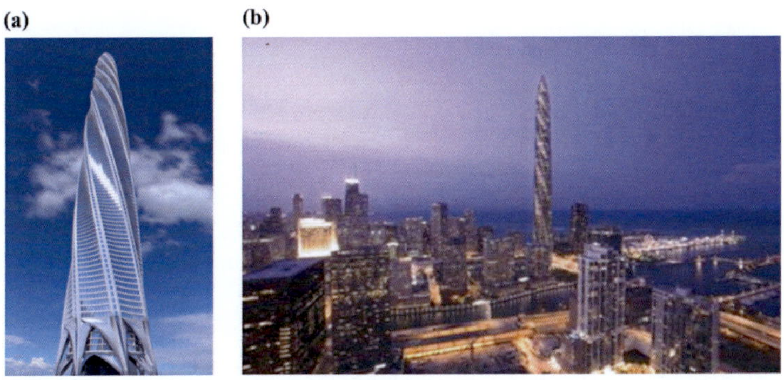

Fig. 4.89 Spiraling Chicago Spire design, Illinois, USA. (**a**) Front view of the Chicago Spire, (**b**) Aerial view of Chicago Spire. (Images' credit and source: (**a**) Milkomède, https://en.wikipedia.org/wiki/Chicago_Spire#/media/File:Chicago_Spire.png (**b**) Courtesy of Santiago Calatrava, https://en.wikipedia.org/wiki/Chicago_Spire.jpg)

4.3.24 Hydrological Center for the University of Namibia, Windhoek, Namibia

An example of biomimicry architecture developed in Africa is the Namibian Hydrological Centre. It is a hydrological center for the University of Namibia and is designed by KSS architect [11, 28, 53]. The inspired idea was based on the water collection techniques of an organism that has become more commonly known as the "Namibian Desert Beetle" and "The African Stenocara beetle." These are inspiring because the technique of this organism, referred to as "The Fog Catcher," explores sustainable development and water use in West Africa [11, 28, 53]. The architects applied biomimicry at the organism level, with a proposed fog catcher design and mimicry of a specific material as shown in Fig. 4.90.

Fig. 4.90 Mimicking the behavior of the beetle in the Hydrological Center Project – The University of Namibia. (**a**) Namibian desert beetle (**b**) View of the terminal. (Image's source: (**a**) https://commons.wikimedia.org/wiki/File: Stenocara_dentata.jpg (**b**) Image design and credit: Mathew Dean Parkes and image creation: Atlas Industries Ltd., source: https://www.thecadinsider.com/2010/11/)

This particular building was mimicked by the Namibia desert beetle, demonstrating biomimicry at an organism level. The Namibian Desert Beetle lives in an abandoned desert that scarcely observes any precipitation (Fig. 4.90a). Since the region is dry for most of the year, numerous creatures need to discover an alternative approach to survive and acquire a sufficient water source [11, 28, 53]. Despite there is less than one inch of precipitation every year, there is frequently fog in the morning; the beetle instinctually knows how to exploit such a situation. The unpredictable design of the beetle's shell provides the creepy-crawly with the needed nutrients and water to survive in such an environment. In the beetle's shell, there are pumps that work as hydrophilic (water-attracting) along with alternate parts which are hydrophobic that function as water-repelling. The water-attracting pump parts act like channels for water and moisture. The beetle is able to capture moisture from the swift-moving fog that moves over the desert by tilting its body into the wind [11, 28, 53].

4.3.25 Eastgate Building in Harare, Zimbabwe

The Eastgate Building in Harare, Zimbabwe, is another example of biomimicry architecture in Africa. Many organisms face environmental conditions in the same way humans do, and they need to solve problems they encounter. This organism

(a) (b) (c) (d)

Fig. 4.91 Different Australian and African termite mound. (**a**) Termite mound, Darwin (**b**) Termite mound, Ghana (**c**) Termite tower (**d**) Detail of a termite mound. (Images' credit and source: (**a**) Hansjoerg Morandell, https://commons.m.wikimedia.org/wiki/File:Litchfield_National_Park-03.jpg (**b**) Shawn Zamechek,https://commons.m.wikimedia.org/wiki/File:Termite_mound_-_Ghana,_West_Africa.jpg (**c**) Annette Teng, https://commons.m.wikimedia.org/wiki/File:Termite_Tower_-_panoramio.jpg#(**d**)Dinkum,https://commons.m.wikimedia.org/wiki/File:Termitière_gros_plan.jpg#)

works inside the environmental carrying capacity of a specific place with the impediment of energy and material availability. These points of confinement additionally create ecological niche adjustments in ecosystems, where there is not only continuous evolution by well-adapted organisms but also adjusted behaviors and relationships between organisms or species [12, 15, 16, 53].

In nature, organisms could do direct and indirect control of the resource flow to other species, which may cause changes in biotic (living) or abiotic (nonliving) materials or systems, and therefore, habitats are called ecosystem engineers [12, 53].

Knowing the optimization of an ecosystem, or thermal performance of spaces in nature, organisms could be done by having good internal air quality through a good ventilation system and gas exchange, and yet having reasonable comfort temperatures for the human occupation could be achieved and imitated by biomimicry [12, 16, 54].

Although many animal species can do this in an efficient way, the (termites) are classified as the masters of ventilation and gas exchange (Fig. 4.91). Comprehending the principles of how termites build their famous clay mounds, the macrotermitine termites (among many other termite species) construct mounds that maintain a constant internal temperature and achieve this only through their structure (Fig. 4.91) by using external energy sources [12, 15, 44, 54]. Each nest can house up to two million termites, all living, working, and breathing within the same closed volume of the nest. On top of this, the queen of the termites living in what is known as the "royal chamber" requires a constant temperature of 30 °C (suitable for the laying of her eggs), while outside temperatures can range from 2 °C to 40 °C [12, 44–54].

According to functionality, these outstanding mounds are not actually the residence for the colony, but rather serve as machines to capture the wind and power the active ventilation of the nest. Incorporation of these principles in buildings has been

(a) (b)

Fig. 4.92 Eastgate center in Harare, Zimbabwe. (**a**) Termite – an inspiration source (**b**) Exterior of the building. (Images' credit and source: (**a**) Dietmar Rabich, https://commons.wikimedia.org/wiki/File:Litchfield_National_Park_(AU),_Magnetic_Termite_Mounds_%2D%2D_2019_%2D%2D_3732.jpg (**b**) David Brazier, https://upload.wikimedia.org/wikipedia/commons/1/1e/Eastgate_Centre%2C_Harare%2C_Zimbabwe.jpg)

applied firstly by architect Mick Pearce, who designed Eastgate Center in Harare, Zimbabwe [15, 44–54] as shown in Fig. 4.92.

The shopping center and office building use approximately 35% of the energy required for temperature regulation in comparable traditional office buildings. This has inspired many other architects to follow its lead; the building and the inspiration source are shown in Fig. 4.92a. Such a complex design is ventilated and cooled naturally using biomimicry principles through examining the local mounds that are built by African termites [44, 54]. This food source needs a temperature of about 30 °C, although the exterior temperature fluctuates between about 2 °C and 40 °C between night and day, respectively, and the termites depend on their workers to oversee the temperature [35]. These workers are also replaced in the building by electrical energy, which is used to power the fans and then redirect the air. In this building, windows generally failed to behave as the openings that are in the termite mounds. As a result, humans rely on their inventions (electrical machines) to complete these tasks [35].

This building also comprises three sections: two outside structures and a glass center that connects them together. The glass space in the middle participates in normal convection and is typically open to the local breezes; amid the night cool air enters the building and goes through openings in the floor slabs. During the day, the heat rises; however, the thermal mass keeps it from rising extraordinarily. Toward the end of the day, warm air is vented via the central stacks both naturally and with the help of fans. This movement attracts the outside cool air, and this cycle continues [35]. The top of the open space is covered by a glass canopy, while the sides remain open for allowing natural airflow. Figure 4.92 presents the exterior of the Eastgate Building. This building looks like a termite mound due to its solid mass and appearance, intensive use of concrete, and similar openings in the façades (Fig. 4.92b). The biomimetic application addresses the ability of the building to get air inside and then warm or cool the air or the concrete [28, 29]. The first floor of each part has fans, which pull air from the open space and push it upward in vertical spines, so that it creates an air flow cycle; cool and fresh air constantly passes through the building downward, then upward, then exits from the chimney at the top, in addition to the circulation of air in each floor [15, 35].

4.3.26 General Examples

In addition to all the aforementioned biomimicry architecture examples, biomimetic applications have been adopted generally in examples of systems such as photovoltaic and Ferro systems. The following section highlights these vivid examples of biomimicry.

4.3.26.1 Photovoltaic System and Biomimicry

In the context of biomimicry, a question is why to use photovoltaic systems? Researchers created solar cells that more closely match nature. These cells are water gel-based ultimately synthetic leaves that couple plant chlorophyll with carbon materials, eventually resulting in a more flexible and cost-effective solar cell [55].

For harvesting solar energy, this could be considered mimicking the way that the leaf harvests energy as shown in Fig. 4.93. Such a solar panel product – called "Solar Ivy" and consists of synthetic leaves as photovoltaics – has been developed by SMIT (Sustainably Minded Interactive Technology) which is a company located in Brooklyn, New York, USA [55].

Solar Ivy can easily be mounted on a vertical wall due to its lightweight and flexibility and has doubled its functions, by creating aesthetic features and maximizing the area of power generation, as presented in Fig. 4.94. There are three types of

(a) (b) (c)

Fig. 4.93 Solar photovoltaic system. (**a**) *Viburnum lantana* leaf (**b**) Photovoltaic panels (**c**) Solar concentrated plants. (Images' source: (**a**) Krzysztof Golik https://commons.wikimedia.org/wiki/File:Leaf_of_Viburnum_lantana_01.jpg (**b**) Fernando Tomás, https://commons.wikimedia.org/wiki/File:Solar_Panels.jpg (**c**) NREL, https://commons.m.wikimedia/wiki/File:Solar_two.jpg)

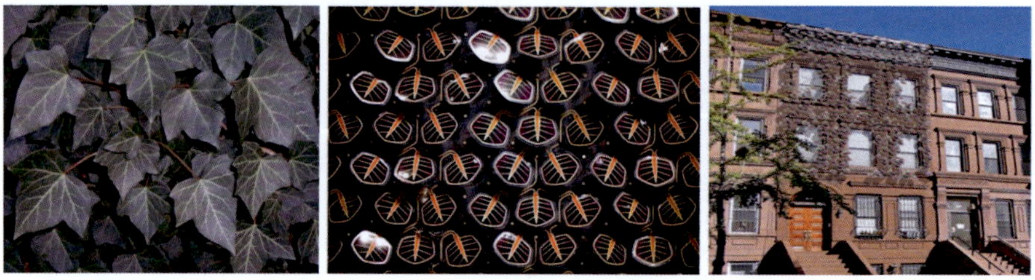

(a) (b) (c)

Fig. 4.94 Mimicking Ivy plant by "Solar Ivy" technology. (**a**) Ivy plant (**b**) Solar Ivy mimics the Ivy's growth (**c**) Ivy solar on façades. (Images' credit and source: (**a**) Uwe H. Friese, https://commons.m.wikimedia.org/wiki/File:Ivy_77uf_be.JPG (**b**) Kenny Louie, https://commons.wikimedia.org/wiki/File:GROW_(2507972766).jpg (**c**) Davey, http://www.syntheastwood.com/technology/solar-ivy)

these solar leaves – each leaf has a thin photovoltaic panel, the organic type contains nontoxic materials, and it costs approximately US$18 per leaf, to produce 0.5 watts. The amorphous silicon leaves cost US$23 per leaf each and produce 0.75 watts; however, the CIGS option is a more efficient type of thin film, which costs roughly US$21 per leaf and produces 4 watts of energy. Depending on the desired look, customers can select from a range of colors and shapes to suit their style. On average, 500 Solar Ivy leaves are recommended for a residential house, which would generate up to 250 watts of power [56].

This invention mimics the Ivy's natural growth on vertical surfaces in order to establish its artificial branching and tectonic qualities on such planes. Solar Ivy produces artificial photosynthesis without water. It also produces electrical energy instead of sugar [55]. Other types of solar panels don't work properly when installed on vertical walls, but solar Ivy performs and can do that, yet achievable by mimicking the real plant [56].

In addition, nature has designed the Ivy to populate vertical planes and still perform photosynthesis. Mimicking the natural mechanism of every leaf, in positioning itself in a certain way for capturing the rays of the sun, is one of the key elements that make SMIT's design successful [56]. Solar Ivy is not only capturing energy from sunlight but also creating an amusing visual aesthetic [56]. Moreover, solar Ivy could be placed on the facades of several traditional building types, such as residential or commercial where these places were one of the inspiration sources for this invention (Fig. 4.94).

It is clear from Fig. 4.94 that this solar Ivy utilizes and absorbs solar energy and works as a shade device as well. Furthermore, it moves toward the direction of the solar panel to maximize solar energy capture as illustrated in Fig. 4.95. The uniqueness of this idea has grown and turned into a fully functional business model, and it is a piece at the Museum of Modern Art's (MoMs) "Design and the Elastic Mind" exhibition in New York, USA [56].

Fig. 4.95 Leaves positioned to maximize solar energy capture. (Image credit and source: Kenny Louie, https://commons.wikimedia.org/wiki/File:GROW_(2507972766).jpg)

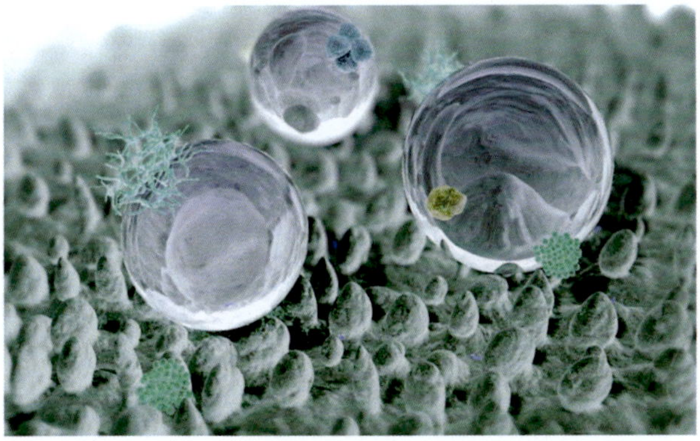

Fig. 4.96 The movement of water drop with lotus effect. (Image and credit source: William Thielicke, https://en.wikipedia.org/wiki/Lotus_effect#/media/File:Lotus3.jpg)

4.3.26.2 Ferro Systems and Biomimicry

Ferro system, a specialized glass manufacturing company, works especially in surface glass coating (Fig. 4.96). The Ferro glass company adopted a biomimetic approach, where it mimics the lotus plant that behaves in nature as a self-cleaning organism (Fig. 4.97). The researchers of the company found that ability of the lotus plant to remain clean, although it grows in a very muddy region (Fig. 4.97a, b) [35]. The secret is on the surface of each petal and the bumpy surface as presented in Fig. 4.97c.

Moreover, nanocoating of solar photovoltaics has increased the performance of the energy output. Improved performances after coating the panels **increase the** photovoltaic panels' energy intake by 10% as illustrated in Fig. 4.98.

(a) (b) (c)

Fig. 4.97 Self-cleaning behavior of the lotus flower. (**a**) A white lotus flower (**b**) Close-up view of the Lotus (**c**) Water drops on lotus surface – contact angles of about 147 degree. (Images' credit and source: (**a**) FroyAgta, https://commons.m.wikimedia.org/wiki/File:White_Lotus_Flower.jpg (**b**) FroyAgta, https://commons.m.wikimedia.org/wiki/File:White_Lotus_Flower_1.jpg (**c**) Na2jojon https://en.wikipedia.org/wiki/Lotus_effect#/media/File:DropConnectionAngel.jpg)

Fig. 4.98 Improved performances after coating the solar PV panels. (Image credit and source: Mohsen Aboulnaga)

4.3.27 Examples of Rainwater Collector and Low-Carbon Buildings

Moreover, hypothetical examples have portrayed models of biomimicry, which include six examples: the main collector in Nowak, Poland; The German Parliament, Berlin; Shanghai Skyscraper and Habitat 2020, China; the Office building of the Future in the United States; and Lilypad Floating City concept. The following section illustrates these examples.

4.3.27.1 Rain Collector in Nowak, Poland

This form of skyscraper is inspired by a leaf that emulates the mechanisms of plants' rainfall capturing and processing. The rooftop and building skin consist of systems of channels or gutters and targets to capture as much rainfall as possible to meet the daily needs of its inhabitants [57]. The revolutionary skyscraper design captures rainwater and reduces the building's water consumption by 40%. The lack of water problem is fundamental nowadays as the average water consumption is 150 liters/person, but 85 liters of it could be replaced by rain water [57]. Several factors have contributed to an increase in water consumption over the last 30 years, including, among others, an increase in the number of washing machines in homes, an increase in the number of garden irrigation water supplies, and the flushing of toilets. In nature, nevertheless, plants adapt themselves to water shortage or water surplus. Hence, architects, designers, and ecologists should correspondingly imitate these concepts and mechanisms for capturing rain water in future buildings' designs. The water tanks, as a large funnel, under the roof's top surface work as hydro-botanic water treatment units according to a chain of operations and treatments as well as transforming water into clean usable water, which is carried and transferred to apartments. A network of gutters on the skin of the building is designed for rainfall capture and flow down the building. As a result, the skyscraper's surface breathes and remains open to its surroundings, allowing for integration with the rest of the city [57].

The "Capture the Rain" skyscraper by polish architectural students, Ryszard Rychlicki and Agnieszka Nowak of H3AR, received a special mention for their proposal in the 2010 skyscraper competition [58]. The "Capture the Rain" skyscraper is a building and an external shell, which consists of gutters' systems that are aimed at capturing as much rainfall as possible to meet the daily needs of its inhabitants. Within the last 30 years, water consumption has significantly increased [58]. A third of the water being used in households in western countries is flushed in toilets. At present, an average American person uses five times more water than a citizen of developing countries due to the provision of a private swimming pool, lifestyle, and other improved living standards. On the other hand, a national hobby of the Danes is collecting rain water for washing and watering plants. In the last 10 years, the average use of pure water in Denmark has decreased by 40%, and residents of so-called eco-villages use one-third of the national average [58].

In view of such data, they decided to design a tower whose structure allows for the capturing and processing of as much rainfall as possible to provide water for its inhabitants. Millennia plants have been developing systems of capturing and processing rainfall [58]. Such systems help to deal with water deficits or surpluses. Initially, the design of the tower focused on shaping and modeling the surface of the roof to capture as much rainfall as possible. Under the roof's surface, there are water reservoirs in the form of a large funnel and reed fields, which serve as a hydro-botanic water treatment unit for processing water into usable water that is further transmitted to apartments. In addition, a network of gutters on the external surfaces of the building is designed to capture rainfall flowing down the building. Such flowing rainfall is transmitted to floors, and its surplus is stored in a reservoir under the building. The water captured and processed by the building may be used for flushing toilets, feeding washing machines, watering plants, cleaning floors, and other domestic applications. After analyzing rainfall in several large cities in developed countries, a formula was developed that shows what percentage of daily pure water consumption can be replaced by rainfall due to the technology used in these buildings [58].

4.3.27.2 Reichstag, New German Parliament in Berlin, Germany

The New Parliament building in Berlin features many sustainability elements (Figs. 4.99 and 4.100). It provides a model for sustainability through burning renewable biofuel-refined vegetable oil – in a co-generator to produce electricity, a clean energy source [59]. Such a system led to the reduction of CO_2 emission by 94% [59]. Soler collectors bring natural light in the center of the building, while automated solar shade louvers protect the interiors from undesired direct solar heat

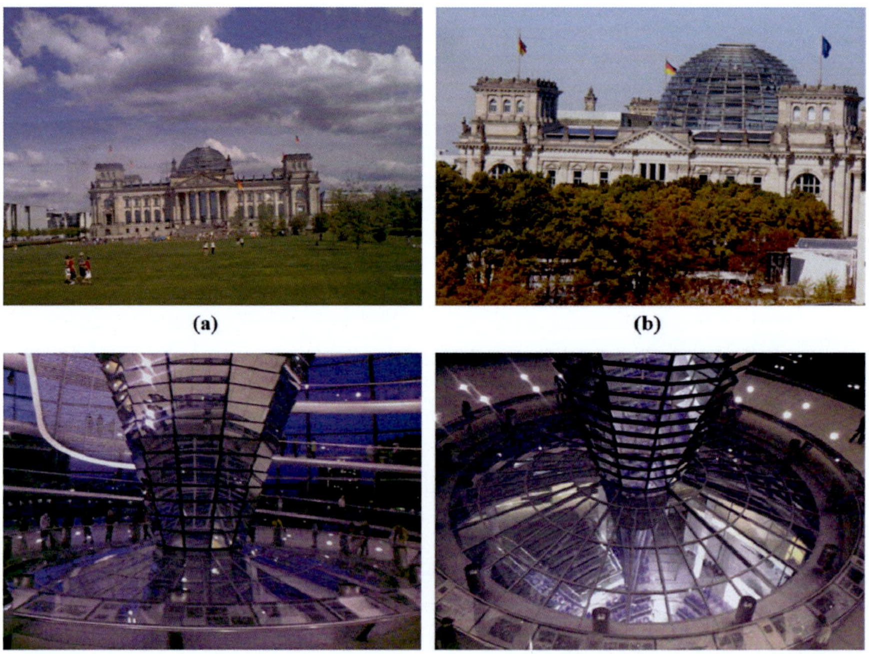

(a) (b) (c) (d)

Fig. 4.99 Parliament Building, Berlin, Germany. (**a**) General view of the New Parliament Building (**b**) Close-up view of the Reichstag glazed dome (**c**) The conical glass mirrors at night (**d**) Glazed atrium above the parliament's chamber. (Images' credit and source: (**a**) (**c**) (**d**) Mohsen Aboulnaga, (**b**) Colin Smith, https://commons.m.wikimedia.org/wiki/File:Berlin_-_Reichstag_-_geo.hlipp.de_-_29156.jpg#)

(a)

(b)

Fig. 4.100 Interior of the Parliament Building in Berlin, Germany. (**a**) The horizontal louvers inside the Reichstag building's dome to minimize sunlight and glare (**b**) Spiral ramps provide visitors with panoramic views of Berlin via the glazed dome (**c**) A view of the atrium where the cupola provides natural ventilation and the atrium increases daylight into the parliament's chamber (**d**) Interior showing the mirrors glass cone to enhance daylight effect and collect rainwater. (Images' credit and source: (**a**) Foto-AG Gymnasium Melle, https://commons.m.wikimedia.org/wiki/File:14-09-09-Bundestag-RalfR-104.jpg# (**b**) Rainer Strehl, https://commons.m.wikimedia.org/wiki/File:Mitte,_Berlin,_Germany_-_panoramio_(170).jpg# (**c**) Ole Neitzel, https://commons.m.wikimedia.org/wiki/File:190930_Blick_auf_die_Kuppel_aus_dem_Plenarsaal.jpg# (**d**) Foto-AG Gymnasium Melle, https://commons.m.wikimedia.org/wiki/File:14-09-09-Bundestag-RalfR-110.jpg#)

(Figs. 4.99b, 4.100a, b). The parliament's main chamber is naturally ventilated by virtue of the cupola, which is inspired from nature mimicking lily flowers as shown in Fig. 4.5a. Also, the building was created to optimize the use of passive systems while minimizing the use of active systems. The lighting and ventilation are monitored by central Building Management Systems (BMS), and a heat exchanger recovers waste energy (heat) from exhaust air [59].

(c)

(d)

Fig. 4.100 (continued)

According to Foster and partners, the performance of the building in use shows that the operational energy profile of the Reichstag Building indicates that it consumed 57% less primary energy than typical existing buildings and less 38% than the needed for new buildings. Such building utilizes locally produced biodiesel that aided in protecting and enhancing the local agriculture economy [59]. Figure 4.100c, d show that the mirror glass cone “the cupola” enhances daylight. The outer surfaces of the Reichstag Building glazed dome also collect rainwater for recycling and reusing it in landscape irrigation (Fig. 4.99b).

4.3.27.3 Shanghai Skyscraper, China

Another example that harvests rainwater is Shanghai skyscraper in China. The 632-meter high building, comprising 128 floors and an area of 420,000 m^2, is considered the tallest tower in Shanghai, China, and the second tallest building worldwide in 2020 according to Council of Tall Building and Urban Habitat (CTBUH) ranking [60–62]. The concept of this tall building was inspired from nature by mimicking twisting trees (Fig. 4.101a). Shanghai Tower also collects rainwater and reuses waste water and has combined cooling and heating power system and uses 40 other energy-saving measures cutting 34,000 metric tons from annual carbon footprint (Figs. 4.101 and 4.102). At the top of the tall building, there are 200 spinning wind turbines to generate clean energy, which is about 10% of the building’s electricity use (Fig. 4.103) [60]. The tower is also enfolded in two layers of glass to provide ventilation and natural cooling as shown in Fig. 4.101b). One-third of the site is made of green space encompassing 24 sky gardens sitting between the two skins and combined nature into architecture making it a green building [60].

4.3.27.4 Habitat 2020, China (Hypothetical)

For large or megacities, future sustainable housing is a top priority. The alliance for new ways of thinking to develop the future is vital. Nevertheless, the habitat is totally dependent on international energy and water grids. Conceiving strategies for a sustainable urban future could be achieved by looking at and understanding cities as dynamic and ever-evolving ecosystems [63]. Following nature through biomimicry is one of the most functional ways to reduce the ecological footprint of buildings. The Habitat 2020 building is envisioned for China’s futuristic example of biomimetic architecture that matches high-tech ideas with basic cellular functions in order to create a “living” building, especially the building’s skin that acts like a natural organism [17, 64] as shown in Fig. 4.104. The design concept fundamentally changes the current approach to buildings and habitats. This future habitat shifts from the current state, where the building surfaces are benign, inert “dumb” materials only used for construction and shielding purposes, to sensitive functional skins that are “alive” and act as membranes to harness energy as presented in Fig. 4.105. A membrane creates a strong link between the exterior and interior of the habitat,

(a) (b)

Fig. 4.101 Shanghai skyscraper in China inspired from nature. (**a**) Twisting trees (**b**) Shanghai tall building inspired from nature. (Images' credit and source: (**a**) Nerijus Navickas https://commons.m.wikimedia.org/wiki/File:Twisted_tree_-_panoramoi_(2).jpg# (**b**) Baycrest, https://commons.m.wikimedia.org/wiki/File:Shanghai_Tower_2015.jpg#)

used as a transporter to collect and channel air, water, and light from the outside to the inside space (Fig. 4.106), thus supplying the habitat with all necessary sources to enable living off the grid [63].

4.3.27.5 Office Building of the Future, USA (Hypothetical)

The Office Building of the Future is a proposed building whose inspiration is obtained from the tree as a living system. The system then applies it to a responsible way for a tower which has been designed by William McDonough. This project was developed by William McDonough + Partners in the United States for *Fortune* magazine to create such a design in 2006. William McDonough designed the first

Fig. 4.102 Façade of Shanghai Tower in China depicting the twisting glazed envelope. (Images' credit and source: Fredlyfish4, https://commons.m.wikimedia.org/wiki/File:Shanghai_Tower_September_2016_1.jpg#)

solar-powered house in Ireland, way back in 1977. In 1999, Time magazine recognized him as a "Hero for the Planet" [65]. In the Office Building of the Future, oxygen production is created through vertically distributed plant life, water distils, and energy production on a harmonious basis as well as adaptation to the varying

Fig. 4.103 Upward view of Shanghai Tower illustrating the double-skin glazed façade, rainwater gutters, and the 200 wind turbines on the top. (Images' credit and source: Ermell, https://commons.m.wikimedia.org/wiki/File:Shanghai_Shanghai_Tower_5166304.jpg#)

seasons for an overall efficient building, making this proposal similar to the tree as illustrated in Fig. 4.107 [65].

The design focuses on the possibilities of recent, for a future context, integrating green and advanced systems in a superefficient, forward-thinking architectural

Fig. 4.104 Habitat 2020, China. (**a**) Habitat building while raining (**b**) Tubes fixed on the building's skin. (Image's source: https://www.igreenspot.com/off-the-grid-sustainable-habitat-2020/)

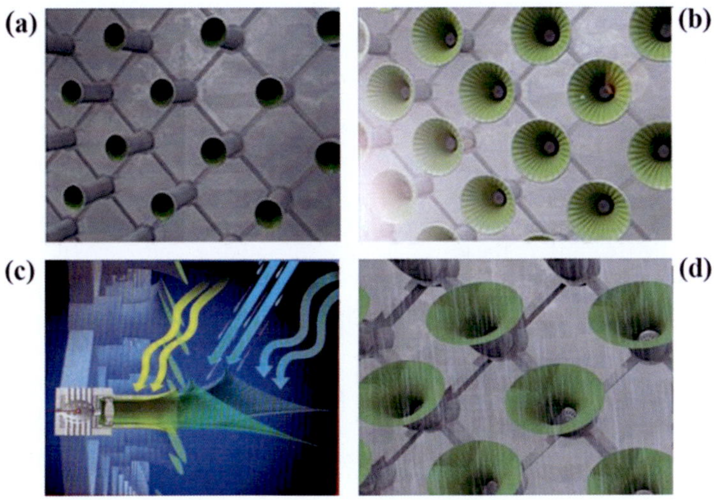

Fig. 4.105 Habitat 2020 living skin, China. (**a**) The "funnels" that create a membrane between interior and exterior when it is closed at night, (**b**) The "funnels" when it is tracking the brightness and angel of the sun, (**c**) Detail of the construction and functions of this membrane. The funnels change shape to make most efficient use of prevailing weather conditions, (**d**) Tubes harvesting rains. (Images' source: (**a, b, d**) https://www.igreenspot.com/off-the-grid-sustainable-habitat-2020/) (**c**) https://www.urbangardensweb.com/2009/09/06/off-the-grid-philips-sustainable-habitat-2020/)

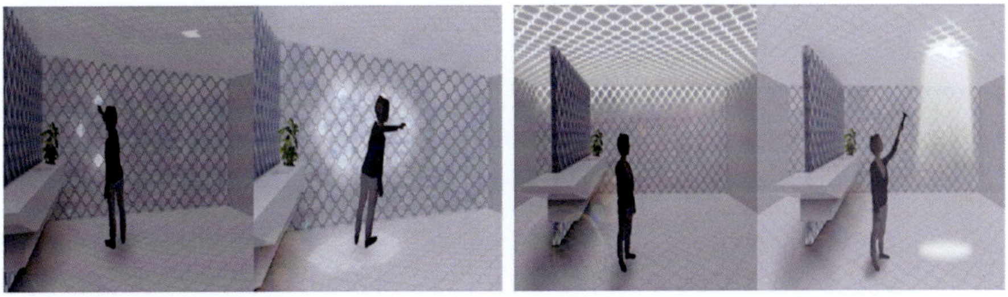

Fig. 4.106 Developing sustainable housing for a megalopolis in China, 2020. (Image's source https://www.igreenspot.com/off-the-grid-sustainable-habitat-2020/)

(a) (b)

Fig. 4.107 The Office Building of the Future Project, USA. (**a**) Exterior of the Office of the Future (**b**) Oxygen production via vertical plants. (Images' credit and source: William McDonough + Partners)

marvel. This tree building is a strong biomimicry application. Figure 4.108 presents the sustainability features of the tower, its exterior, and the landscape surrounding it.

According to the concept of Cradle to Cradle by William McDonough, who revealed the concept in 2006, he highlighted the sustainability elements of the tower as follows [65]:

(a) Form and function: It is well known that aerodynamic building's shape reduces the wind impact on it, and also a curved form reduces the amount of materials needed for construction, increases structural stability, and maximizes enclosed space. Plentiful flora, with a green rooftop, exists at the three-story atrium gardens.
(b) Tree tops: Asphalt and tar are the traditional rooftops, and these materials heat the surfaces that lead to the "urban heat island" effect of higher temperatures that can alter weather patterns and intensify smog. A layer of ground cover on this building's roof helps to regulate temperature, protects waterproof coatings, and absorbs and cleans stormwater.
(c) Soil and green: On the building's western facade, there are three-story atrium gardens. The greenery links outdoors and indoors together, providing a nature's breath.
(d) Water: It is recycled in this building more than one time. Greenhouses treat wastewater from bathtubs for reuse as irrigation for the gardens.
(e) Street smarts: Sun and shadows, the shape and orientation of the building are studied deeply and are fitted to the site. There is a park on the south facade that can capture maximum sun exposure, and its irregular form allows more daylight to reach the street. Gardens ring the base, contributing to the quality of life at street level.

Fig. 4.108 Sustainable elements of the Office Building of the Future Project. (Image credit and source: William McDonough + Partners)

(f) Solar power: a plant of 100,000 square feet of photovoltaic panels installed on the southern facade that convert solar energy into electricity collects sufficient energy to provide up to 40% of the building's needs. Costing at least 20% per kilowatt-hour several times as much as coal or natural gas, solar PV is expensive today. But the trends are good. Solar power is becoming more affordable, and the relative economics will improve as more states and countries regulate GHG emissions.

(g) Building skin: Transparency, weatherproofing, and insulation are the different materials' performance. These surfaces are becoming thinner, lighter, and smarter.

(h) Productive and smart workspace: Under-floor air appropriation enhances air quality.

Adaptable public spaces supplant settled individual stations. Seats and workstations are ergonomic. Smart monitors recognize the presence of individuals and change temperature, light, air, and sound as required. This enables people to control their conditions. Our adage: "We don't feel warmth or cool apparitions."

(i) Waste = food: In nature, supplements are cycled and reused perpetually. "Eco-viable design" tries to mimic the same cycles. All items, from building materials to furniture, are intended to return securely to the earth or to be reused like office seats that can be dismantled into parts and sent back to the manufacturer to wind up another item.

(j) Heating and cooling systems: These represent almost 30% of a building's energy use. A building can decrease energy use by exchanging heat between the building and the earth using a system that involves heat-absorbing fluid flowing through underground wells.

(k) A combined heat and power plant, energized by natural gas, works at up to 90% effectiveness and supplies the power that the sun-powered boards cannot.

4.3.27.6 Lilypad Floating City

The Lilypad Floating City is an amazing concept that has been created by Belgian architect Vincent Callebaut [17, 66]. The concept is a completely self-sufficient floating city that would accommodate up to 50,000 inhabitants. Callebaut apparently foresees his floating city as a place to live for people who have lost their homes due to sea level rising (SLR) resulting from climate change and could be located offshore as a major city as illustrated in Fig. 4.109 [67, 68]. The inspiration of the city is depicted in Box 4.5. At first glance, the city looks like something out of a starter's movie or like a couple of giant inflatable garden chairs that have washed out to sea, but according to the designer, it is the ultimate solution to rapidly SLR [69].

Major cities such as London, New York, and Tokyo are seen as being at huge risk from the oceans, which could rise by as much as three feet by the end of this century. Hence, the solution, by the award-winning Belgian architect Vincent Callebaut, is designed to be a new place to live for those whose homelands may be affected by climate change risks.

The "Floating Lilypad City" would float around the world as an independent and fully self-sustainable home. With a lake at its center to collect and purify rainwater, it would be accessed by three separate marinas and feature artificial mountains to offer the inhabitants a change of scenery from the seascape. Power for the central accommodation hub is provided through a series of renewable energy sources, including solar panels on the mountain sides, wind turbines, and a tidal power station that harnesses the energy of the waves [70–72].

The floating city will be accessed by three marinas and will feature three "mountains" to offer the inhabitants a change of scenery (Box 4.6). It will actually produce much more energy than it consumes and be an entirely "zero-emissions city" as all the carbon dioxide and the waste will be recycled [70]. Callebaut would like to see his Lilypad concept become a reality by 2100 [70, 72].

Box 4.5: Recommended Location of the Lilypad Floating City

The Lilypad floating city could be located close to land or set free to follow the ocean currents wherever they may lead. (Vincent Callebaut)

Box 4.6: Inspiration of the City

The design of the city is inspired by the shape of the great Amazonian Victoria Regia Lilypad

The goal is to create a harmonious coexistence of humans and nature. (Vincent Callebaut)

4.4 The Role of Robotic Strategies in Biomimetic Architecture

In the past few years, the use of robots in architecture has taken great steps, due to the advances achieved in the fields of digitalization, virtualization, and automation as well as artificial intelligence. Digital information is used not only for informing plans but also for creating materials that are needed to construct a building using architectural robots like (3-D printers and robotic fabrication) by mimicking nature. The 3-D printers assist architects to redesign their manufacturing pattern, while mimicking natural organisms like spiders – that creates the potential to print large structures, using advanced concrete or other advanced materials, to provide extra strength where needed and conserve material [10]. For example, the team of "Architectural Association grads" designed a concept called "Proto Home," which improves a new way of construction using the strengths of 3-D printing. The spindly structure is designed to imitate the way bones grow in humans [10]. Robotic production applications are designed to be applied from a micro scale to a mega scale. A new group of researchers, artists, designers, and fabricators have used robotic fabrication technology in architecture, for example, the ICD/ITKE Research Pavilions Project. It is designed to use glass and carbon fibers that are woven depending on the light steel structure to make each unique panel [10]. In such projects, biomimicry at this level is improved when influenced by biomimetic processes,

(a) (b)

Fig. 4.109 Lilypad Floating City inspired by Vincent Callebaut. (**a**) The floating city near the shore (**b**) Close-up view of the floating city. (Images' credit and source: (**a**) https://www.bing.com/images/search?view=detailV2&ccid=1M8H1tcV&id=46020312BC147829802EC8507CDA91FE090DA703&thid=OIP.1M8H1tcV3HL2BqkU94wfTAHaFa&mediaurl=https%3a%2f%2fth.bing.com%2fth%2fid%2fR.d4cf07d6d715dc72f606a914f78c1f4c%3frik%3dA6cNCf6R2nxQyA%26riu%3dhttp%253a%252f%252fwww.inhabitat.com%252fwp-content%252fuploads%252flilypad2.jpg%26ehk%3drAR8Uesk7Y6azY7TnA7pHjI3ha3yTssJzutYzOCCl6Y%253d%26risl%3d%26pid%3dImgRaw%26r%3d0&exph=393&expw=537&q=lilypad+floating+city&simid=608022933947816529&FORM=IRPRST&ck=3F1ABA4987798DBAF04CA6F5A4DBB60D&selectedIndex=9&qft=+filterui%3alicenseType-Any&ajaxhist=0&ajaxserp=0 (**b**) https://www.bing.com/images/search?view=detailV2&ccid=WwzSrLRO&id=20DAB41773A92960CAAB7E199000FA709FB36348&thid=OIP.WwzSrLROvRGLR9y3DunjggHaEx&mediaurl=https%3a%2f%2fth.bing.com%2fth%2fid%2fR.5b0cd2acb44ebd118b47dcb70ee9e382%3frik%3dSGOzn3D6AJAZfg%26riu%3dhttp%253a%252f%252fblog-imgs-37-origin.fc2.com%252fm%252fh%252ff%252f fmhfkillerloop%252fLILYPAD_01.jpg%26ehk%3dwd%252fPxVAYENPO624DYa78vk8VchtKHI9amvOLpILtaX4%253d%26risl%3d%26pid%3dImgRaw%26r%3d0&exph=512&expw=795&q=lilypad+floating+city&simid=608044348650121681&FORM=IRPRST&ck=DC516FB6D20E36E0F70990E36C4499A4&selectedIndex=3&qft=+filterui%3alicenseType-Any&ajaxhist=0&ajaxserp=0

via copying material organization strategies, which can be found in most natural constructions and play a vital role in their material efficiency [10].

4.5 Summary of Previous Global Biomimetic Examples

After highlighting and illustrating various examples of biomimetic architectural examples worldwide, it is clear that applying the biomimicry approach in architecture on many levels (organism level, behavior level, and ecosystem level) is difficult. All these examples lead to sustainable architecture and contribute to environmental and natural conservation. Nevertheless, there are many existing buildings that are already engaged in biomimicry concepts and other hypothetical projects that seek to achieve sustainability, energy conservation, and environmental protection in addition to resisting the impact of climate change. Many countries, including the United Kingdom, Germany, Poland, China, Japan, South Korea, India, Singapore, the United States, Australia, Namibia, and Zimbabwe, have used vivid examples of biomimicry in building design and architecture. This means these countries are aware of the importance of biomimicry concept in attaining sustainability and the SDGs, yet how biomimetic architecture entails a great impact on the environment. It is evident from the study that organism and behavior levels are the most applied and engaged levels in building design of the biomimicry approach in architecture. It is hoped that more nations could adopt this approach and engage biology in architecture since the early beginning of the design phase of any project. The following section provides a summary of all previous examples discussed earlier. These are listed in Table 4.1 in terms of six items: building's name, country, inspiration source, application in design, problem solved, and level of biomimicry (organism, behavior, and ecosystem).

4.6 Sustainability Comparative Analysis and Evaluation

According to the previous examples that depicted different levels and approaches in applying biomimicry thinking in architectural design, the analysis of the performance of these buildings and its ranking according to sustainable architecture evaluation elements are presented and discussed.

4.6.1 Comparative Evaluation of Biomimicry Architecture and Sustainability

Sustainability is an accepted term that has been in use since the early twentieth century in renewable resource management; in the building design professions, sustainable development (SD) has become a focal and main part of the agenda; such designs are inspired by nature as an inspiration source in terms of architecture, still devious.

Biomimicry is a potential approach that integrates ecological sustainability to architecture by understanding natural processes to understand its form and the environment within an ecosystem, which includes [72]: **sustainability,** achieves the needs of the present, but without exposing the ability of future generations to meet their needs too; **sustainability in architecture,** looks for reducing the negative environmental impact of buildings, through efficiency and modesty in using materials and energy; it uses conscious energy conservation and ecology in the design of the built environment; and **sustainable design,** reduces the impact on the environment and human health, consequently, improving the performance during a building's life cycle, and takes into consideration energy, water, building materials, and solid waste [73].

Table 4.1 Summary of all previous highlighted examples of biomimicry worldwide

	Building's name	Country	Inspiration source	Application in design	Problem solved	Level of biomimicry
1	Waterloo International Terminal	London, UK	Pangolin, conceptual inspiration	The glass panels in the facades mimics the flexible arrangement of the pangolin, so they are able to move in response to the imposed air pressure forces	Response to the imposed air pressure forces	Organism level
2	Eden Project	Cornwall, England, UK	Soap bubbles and pollen grains	Efficient structural design, by providing maximum strength with maximum volume and minimal surface	Design buildings that provide the environment to create different microclimates	Organism level
3	Swiss Re, Gherkin Tower	London, England, UK	Venus flower basket sponge	Double-glazing. The pipes are used for cooling in summer, removing hot air from the building, and for heating in winter. These also allow easier entry of light, with a consequent reduction of lighting costs	Energy efficiency	Organism level
4	Selfridges	Birmingham, UK	Nature wasps' nest	Organic natural form and structure are combined with high technology	Combine organic and natural forms with high technology	Organism level
5	The Munich Olympic Stadium	Munich, Germany	Dragonfly wings, mountains	The roof of the Munich Olympic Stadium was designed along the same principle of compartmentalization	Thin and very strong structure covering the material for the top of the stadium	Organism level
6	Tardigrada Botanica	Poland	Tardigrada	Biogas is emitted there to serve as heat for the colossal buildings, so that even in the winter you will be able to grow tropical plants inside	Biogas is emitted there to serve heat to the colossal buildings, so that even in the winter you will be able to grow tropical plants inside	Behavior level
7	Sagrada Familia Church	Barcelona, Spain	Tree branches and canopies	The unique set of the nine petals, known as the entrance leaves and the other layers of concrete shells resemble lotus flowers	Statics problems in supporting the vault	Organism level

	Building's name	Country	Inspiration source	Application in design	Problem solved	Level of biomimicry
8	The Las Palmas Water Theater	Spain	Namibian fog-basking beetle	Imitate the behavior of the Namibian beetle in having fresh water	Reduced structural members and construction costs	Organism level
9	Tenerife Auditorium	Spain	Atlantic Ocean	Concrete shell roof	Symbolic power, serving not only the building that houses a modern 1558-seat auditorium for music and culture but also as a focal point for the islands	Organism level
10	The All Seasons Tower	Yerevan, Armenia	Volcanos	The external facades will feature a high-performance skin with an adaptable external shading device to reduce solar gains in the summer	Solar energy is covered in a mesh skin that filters sunlight for temperature regulation and safety during earthquakes	Organism level
11	Esplanade Theatre	Singapore	Layered durian plant (a tropical fruit)	Protecting, it uses its semirigid pressured skin to protect and keep the seeds inside, so that this theater imitates the same behavior with its exterior	Sunlight penetration, and heat radiation	Behavior level
12	Beijing Olympic Stadium and National Aquatic Center	Beijing, China	Bird's nest and honeycomb and water bubbles	The form of the stadium and its structure are the same as a bird's nest The aquatic center form is inspired by the honeycomb shape structure Also inspiration for the skin of the building came from water bubbles, this structure is derived from principles of geometry and crystalline systems, and the bubble collects solar energy to heat the pool	Easily recyclable with little material wasted The ETFE panels protect the people inside and offer an acoustic insulation, and allow sunlight to be filtered through them Self-cleaning, durable, reduces maintenance cost	Behavior level

(continued)

Table 4.1 (continued)

	Building's name	Country	Inspiration source	Application in design	Problem solved	Level of biomimicry
13	Omotesando Building	Tokyo, Japan	Zelkova tree branches	Curving, organic-looking shapes of the glass and concrete facade	Computational models follow natural processes and use their application to architectural design	Organism level
14	Residential Building	Songdo, Incheon, South Korea	Human spine and honeycomb structure	The form and structure of the building are similar to a human spine and honeycomb	Reduction in structure elements and the building cost	Organism level
15	The Lotus Temple	Delhi, India	Lotus flower	The unique set of the nine petals, known as the entrance leaves and the other layers of concrete shells resemble lotus flowers	Structure system	Organism level
16	Lavasa Township	India	Fig leaf	Foundation stores water Drip tip system water to clean its surface	Responds to seasonal flooding Excess water removal	Ecosystem level
17	Aldar Headquarter	Abu Dhabi, UAE	Seashell	Two giant curved circular walls came together representing the shell	Clear goal to create an iconic structure. It mimics a shell with a pearl. The building is visibly seen from various *angles*. It fits into the skyscrapers skyline of Abu Dhabi	Organism level
18	The Sahara Forest Project	Jordan and Qatar	Namibian fog-basking beetle	The greenhouse design emulates this beetle to combat climate change in an arid climate zone	Creating a saltwater-cooled greenhouse, concentrated solar power (CSP) and technologies for desert vegetation	Behavior level
19	Qatar's Giant Cactus Building	Doha, Qatar	Cactus plant	Taking inspiration from the way these plants deal with the desert climate	Energy efficiency in the hot desert environment and the use sunshades	Behavior level

	Building's name	Country	Inspiration source	Application in design	Problem solved	Level of biomimicry
20	Ultima Tower	USA	Termite mounds	The design is extremely aerodynamically efficient, and resistant to earthquake shock waves. The structure is cooled in a similar way to a termite nest	Solve the global population crisis	Behavior level
21	Fennell Residence	Portland, Oregon	Ocean's wave	Series of intersecting and overlapping organic forms creates a sense of constant movement	Repeated form serves a series of architectural functions, from defining the interior to permitting exterior natural light at both ends and along the roof	Organism level
22	Spiraling Chicago Tower	Chicago, USA	Nautilus shell	Its helical form uses the golden mean and the Fibonacci sequence to closely mirror the spiraling growth of the nautilus shell	Energy efficiency	Organism level
23	Hydrological Center for the University of Namibia	Windhoek, Namibia	The Namibian desert beetle, known as the African *Stenocara* beetle	Collecting water, mimicking the techniques of the Namibian desert beetle	Turn water droplets collected from fog into usable water	Organism level
24	Eastgate Building	Harare, Zimbabwe	Termite mounds	Drawing air from the outside to inside, then it is warmed or cooled by the building's mass depending on which is hotter Fans and chimneys	Achieving good internal air quality via proper ventilation, maintaining satisfactory temperatures for human	Behavior level
25	Photovoltaic Systems "Ivy Technology"	USA	Photosynthesis process of plant's leaves	Mimicking plant's leaves and Ivy plant behavior in capturing sunlight to produce energy	Energy production	Behavior level

(continued)

Table 4.1 (continued)

	Building's name	Country	Inspiration source	Application in design	Problem solved	Level of biomimicry
26	Ferro Systems	Spain, Portugal, and Europe	Lotus plant self-cleaning process	Self-cleaning concrete that keeps buildings free from pollutants in the atmosphere	Self-surface cleaning	Behavior level
27	Rain Collector	Nowak, Poland	Leaf of the tree	Harvesting rainwater through the design of the elevation that imitate leaf behavior	Reduce building's water consumption	Behavior level
28	Habitat 2020	China		Breathing and acting façade	Looking to and understanding cities as dynamic and ever-evolving ecosystems	Behavior level and Ecosystem level
29	The Office Building of the Future	USA	The tree	The assumed design focuses on the possibilities of recent times, for a future context, integrating green and advanced systems in a super-efficient, forward thinking architectural marvel	Oxygen production through vertically distributed plant life, water distillation, energy production	Ecosystem level
30	Lilipad Floating City	Universal	Shape of the great Amazonian Victoria Regia Lilypad	Self-sufficient floating city, independent and fully self-sustainable home	Substituting homes due to sea level rising resulting from climate change	Ecosystem level

Biomimicry provides several ideas related to the functionality of materials at nanoscale; integrated intelligence as sensing and actuating at all scales; adaptive response as nano, micro, and macro scales; and architecture and additional functionality. Both fauna and flora respond to temperature, humidity, water scarcity, and surplus and other changes in their environment. By analyzing sustainability examples from nature and realizing the benefits and the values that have been discovered in different creatures, the appreciation toward nature increases. Knowing how organisms and nature, as a system, deal with sustainable buildings and how their adaptation can be mimicked in the built environment.

4.6.2 *Why Do We Need Sustainable Architecture?*

The concept of sustainability involves the dynamic and complex balance of environmental (man-made and natural), economic, and social dimensions from many earlier sources including the theory of the Third Ecology [69]. Sustainable development (SD) is vital for maintaining the environmental stability of our earth. Also, SD focuses on preserving and efficiently using natural resources. The definition of SD is based on two concepts (needs and limits). Due to huge material consumption and producing tons of waste from construction projects, architecture faces a unique challenge in the field of sustainability. Therefore, sustainable buildings aim at reducing the impact on the environment by saving energy and resource efficiency. This includes the following principles:

- Reducing nonrenewable resource consumption.
- Modifying the natural environment.
- Eliminating or minimizing the use of toxic materials.

In addition, sustainable buildings are the ones that have the least reverse impacts on the ecosystem, in terms of the building itself, the surroundings, and the broader regional and global setting as well as ensure suitable management of the buildings' stock in order to reduce energy use and improve environmental quality. Thus, sustainable building design should:

- Save water and energy needs on site.
- Evolve as surrounding conditions change and adapt to site climate.
- Operate with no pollution and do not generate waste.
- Elevate the health and well-being of all inhabitants, as a healthy ecosystem does.
- Encompass energy-efficient integrated systems, to maximize efficiency and comfort.
- Improve the health and diversity of the local ecosystem rather than degrade it.
- Inspire people to dream [70].

4.6.2.1 Environmental Benefits

Preserving the environment and limiting the expenditure of the earth's natural resources is the main purpose behind the sustainable building. Thus, it is necessary to keep and protect the ecosystem by the following:

- Enhance air and water quality.
- Reduce waste.
- Reduce emissions.
- Keep water.
- Control temperature.

4.6.2.2 Economic Benefits

There are many economic benefits that can be obtained by having the ability to use sustainable materials, reduce energy consumption, and improve water efficiency:

- Helps in the growth of the "Green Market".
- Engages the life cycle of the building.
- Improves occupants' attendance and productivity.
- Increases property value and reduces operating costs.

4.6.2.3 Social Benefits

Social aspects shouldn't be ignored as it is one of the main pillars of sustainability and sustainable buildings. So, the benefits that result from improving indoor environmental quality are:

- Comfort and health improvement for the occupants.
- Environment that has an aesthetics.
- Less tension on local infrastructure.
- Increment in occupants' morale and spirits.
- Improvement in the productivity of the worker.

The green movement is at an inflection point, as the benefits of sustainable building become more apparent over time [71].

4.6.3 The Analytical Study

Based on the above study, it is imperative to conduct an assessment of these examples. According to the aforementioned examples, these include an analytical study for 12 worldwide building examples that applied the biomimicry approaches in different levels and elements of the building through the building's skin, systems, and structure and form to achieve a reduction in energy consumption and focus on techniques and strategies applied which will be introduced below. They are the most buildings with environmental sustainability elements and biomimetic features at various levels of biomimicry, imitating not only the form but also the behavior and ecosystem of organisms. Tables 4.2a, 4.2b, 4.2c and 4.2d presents the analytical comparison for the selected 12 examples from several environmental aspects in order to find out the criteria to evaluate which biomimetic building is the most sustainable and saved energy. The evaluation is based on three levels (excellent, very good, and good). The "Excellent" evaluation rank is determined when it achieves all the sustainability components, the "Very Good" evaluation rank is reached when most of the sustainability components are attained, and the "Good" evaluation rank is reached when at least two elements of the sustainability comparative components are achieved. Table 4.3 lists the biomimetic architectural design examples that reach the three ranks.

According to the previous analysis, it was found that four buildings achieved many of the sustainability features and proved that biomimicry could contribute to having an environmentally sustainable architecture. The comparison between the four buildings is listed in Table 4.3 to decide on the optimum building that follows the biomimicry approach and achieves sustainability.

4.7 Conclusion

At the end of this chapter, biomimicry applications based on sustainability elements have been discussed. Examples exhibiting structural efficiency, material manufacturing, waste management, and water management as well as thermal control and renewable energy have been highlighted. In addition, several biomimetic architectural examples were assessed in terms of sustainability and sustainable design to illustrate their link to the approach of biomimicry in architecture. The study also revealed that the notion of why sustainable architecture in the built environment is highly and urgently needed. The comparative analysis between the 12 selected case studies of biomimicry approaches and their levels exhibited positive indicators to achieve optimum buildings that depict sustainability features related to the biomimicry approach and those linked to climate change (CC). Moreover, the study indicates how biomimicry could contribute to improve the built environment and mitigate climate change.

To have a deep understanding of biomimicry approach and comprehension, the next chapter will present the approaches of biomimicry that are linked to CC to find out if there is a concrete connectivity between biomimicry and mitigating greenhouse gas (GHG) emissions and climate change.

Table 4.2a Analytical comparison between 12 global biomimetic buildings

Building's name		Waterloo International Terminal in the United Kingdom	Eastgate Building in Zimbabwe	CH2 Council House 2 in Australia
Building's image				
Building type		Euro-star train terminal	Commercial office building	Commercial office building
Aims		Terminal that responds to pressure differences, in entering and departing the station Glass panels fixing make up the structure mimic Design an ecologically/environmentally friendly building	Energy efficiency Water recycle Waste management Designing for thermal control	Lighthouse environmental project As a result: it is 64% lower compared with even five green star building Electricity consumption reduced by 85% and GHGs by 87% and produced only 13% of emission
Energy efficiency	Energy saving	2 m £ saved in energy cost	Use 10% less of the energy used in similar sized conventional building	82%
	Natural light	√	√	Saved by 65%
	Natural ventilation	–	Use about 35% of the energy required for temperature regulation	Saved by 65%, provide nearby 100% fresh air
	Usage of PV panels	–	√	√
Water recycling		–	√	72% reduction in water use
Waste management		–	√	Recycling facilities for waste

Building's name	Waterloo International Terminal in the United Kingdom	Eastgate Building in Zimbabwe	CH2 Council House 2 in Australia
Material used	6–7 m £ has been saved by removing the need to build a new supporting structure, and 1480 tons of carbon have been saved by reusing existing piles	Glass and steel suspension bridges	All recycled timber Steel Concrete
Biomimetic level	Organism level	Behavior level	Organism and behavior level
Evaluation	●	**	**

● Very good example to present the impact of biomimicry in sustainable architecture
** Excellent example to present the impact of biomimicry in sustainable architecture

Table 4.2b Analytical comparison between 12 global biomimetic buildings

Building's name		Beijing Olympic Stadium	National Aquatic Center	Munich Olympic Stadium
Building's image				
Country		China	China	Germany
Building type		Olympic stadium	Olympic aquatic center	Olympic stadium
Aims		Structural biomimetic building is inspired by the bird's nest in nature, and has an iconic building in Beijing with sustainable environmental impact	Achieving several sustainability features: Water recycling Natural lighting	Build structure that emulates The game's motto "The happy game" – Designers created a structure, which constantly behaves along the site and mimics the harmonious elevation of the Swiss Alps
Energy efficiency	Energy saving	√	√ Reduction by 30% and capturing solar energy	–
	Natural lighting	√	√ Reduction of artificial lighting by 55%	√
	Natural ventilation	√	√	√
	Usage of photovoltaic panels	–	ETFE energy savings similar to savings when covering the roof with solar panels	–

Building's name	Beijing Olympic Stadium	National Aquatic Center	Munich Olympic Stadium
Water recycling	–	√	–
Waste management	–	–	–
Material used	Steel that is easily recyclable	Composed of energy-efficient ETFE	Tensile roofs of steel cable and acrylic panels
Biomimetic level	Organism level	Organism level	Organism level
Evaluation	●	**	**

● Very good example to present the impact of biomimicry in sustainable architecture
** Excellent example to present the impact of biomimicry in sustainable architecture

Table 4.2c Analytical comparison between 12 global biomimetic buildings

Building's name		Gherkin tower	Doha Tower	Esplanade theater
Building's image				
Country		United Kingdom	Qatar	Singapore
Building type		Commercial high rise building	Doha Tower	Arts Center
Aims		Wind resistant structures Materials anti split Make an architecture that achieve natural ventilation	Energy efficiency How it could survive in a hot dry environment by mimicking the cactus behavior Temperature regulation	Environmental goals such as, energy efficiency, insulated green house
Energy efficiency	Energy saving	√ The building freshen air as the same mechanism of glass sponge saving up to 50%	√	√ Energy reduction by 30%
	Natural lighting	√ Windows allow entering natural light and fresh air to the structure	√	√ 45% natural lighting
	Natural ventilation	–	–	–
	Usage of photovoltaic panels	–	–	–
Water recycling		–	√	–
Waste management		–	–	–

Building's name	Gherkin tower	Doha Tower	Esplanade theater
Material used	Steel and glass cladding Concrete piles for the foundation	Steel Glass Concrete	Aluminum, insulated glass, steel, and timber stone
Biomimetic level	Organism level	Behavior level	Organism level and behavior level
Evaluation	●	●	●

● Very good example to present the impact of biomimicry in sustainable architecture
** Excellent example to present the impact of biomimicry in sustainable architecture

Table 4.2d Analytical comparison between 12 global biomimetic buildings

Building's name		Office Building of the Future	Habitat 2020	The Eden project
Building's image			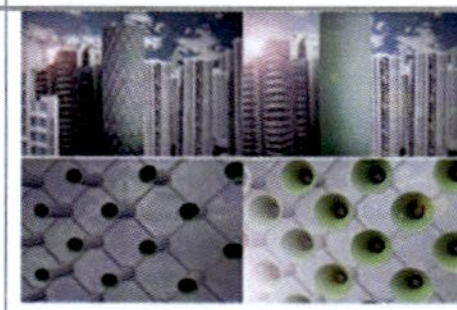	
Building		Proposal in the United States	Proposal in China	United Kingdom
Building type		High-rise office building	Residential or office building, not recognized	Multiple greenhouse complex
Aims		Create a building that imitates nature and the ecosystem; by making oxygen, distilling water and producing energy	Create "living" structures that operate like natural organisms	Sourced materials, energy efficient, and constructed with minimal waste
Energy efficiency	Energy saving	√	√	√
	Natural lighting	√	√	√
	Natural ventilation	√	√	–
	Usage of photovoltaic panels	√	√	–
Water recycling		√	√	√
Waste management		√	√	–

Building's name	Office Building of the Future	Habitat 2020	The Eden project
Material used	This hypothetical building, the materials that will be used is that perform different functions	A sensitive functional layer on the building's skin	Building's structure is created from double curved (glued laminated) timber beams environmental and social standards, all the way to Eden, recycled wood, plant-based flooring (clay sand (low carbon footprint)
Biomimetic level	Behavior level	Ecosystem level	Organism level
Evaluation	**	**	●

● Very good example to present the impact of biomimicry in sustainable architecture
** Excellent example to present the impact of biomimicry in sustainable architecture

Table 4.3 Comparison between the best four global biomimetic buildings cases

Building's name		Eastgate Building	CH2 Melbourne City Council House 2	Office Building of the Future	Habitat 2020
Building's image					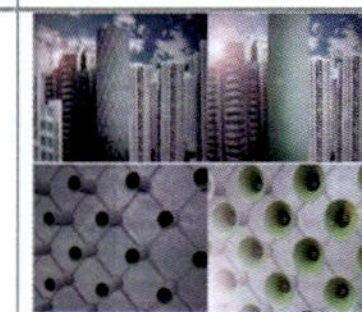
Architect		Mick Pearce	Advanced Environmental Concepts, Lincoln Scott, Design Inc.	William McDonough + Partners	Philips' Design
Country		Zimbabwe	Australia	USA	China
Building type		Commercial office building	Commercial office building	High-rise office building	Residential or office building, not realized
Existence		Exist	Exist	Proposal	Proposal
Energy efficiency	Energy saving	Use 10% less energy used in a similar sized conventional building	82%	√	√
	Natural lighting	√	Saved by 65%	√	√
	Natural ventilation	Use about 35% of energy required for temperature regulation	Saved by 65, and provide nearly 100% fresh air	√	√
	Usage of photovoltaic panels	√	√	√	√

Building's name	Eastgate Building	CH2 Melbourne City Council House 2	Office Building of the Future	Habitat 2020
Water recycling	√	72% reduction water consumption	√	√
Waste Management	√	Recycling facilities for office waste	√	√
Biomimetic level	Behavior level	Organism level, behavior level	Behavior level	Ecosystem level
Evaluation	**	**	**	**
Ranking	**2**	**1**	**3**	**4**

● Very good example to present the impact of biomimicry in sustainable architecture
** Excellent example to present the impact of biomimicry in sustainable architecture

References

1. ISO 18458. (2015). *Biomimetics—Terminology, concepts and methodology*. International Organization for Standardization. Retreived 23 February 2021, from https://www.iso.org/obp/ui/#iso:std:iso:18458:ed-1:v1:en
2. Uchiyama, Y., Blanco, E., & Kohsaka, R. (2020). Application of biomimetics to architectural and urban design: A review a cross scales. *Sustaniability, MDPI, 12*(23), 1–15. https://res.mdpi.com/d_attachment/sustainability/sustainability-12-09813/article_deploy/sustainability-12-09813-v2.pdf
3. *68% of the world population projected to live in urban areas by 2050, says UN | UN DESA | United Nations Department of Economic and Social Affairs*. (2021). Retrieved 14 May 2021, from https://www.un.org/development/desa/en/news/population/2018-revision-of-world-urbnaization-prospects.html
4. Manish, S., Achana, K., Anvitha, B., & Indira, D. (2017). Application of biomimicry towards sustainability of buildings. *International Journal of Science & Engineering Research, 8*. https://www.ijser.org/researchpaper/APPLICATION-OF-BIOMIMCRY-TOWARDS-SUSTAINABILITY-OF-BUILDINGS.pdf
5. Leordean, L. (2016), *Biomimicry in architecture – Mitigation and adaptation to climate change*. Politecnico di Milano. Retreived 20 Januay 2018, from https://issuu.com/ioanaleordean/docs/thesis_book_online-compressed
6. Begumsardan. (2019). *Learning from nature via Biomimicry*. IllustrArch. Retrieved 23 February 2021, from https://illustrarch.com/articles/827-learning-from-nature-via-biomimicry.html
7. *Strawscraper by Belatchew Arkitektur*. (2013). *Archiscene*. Retrived 25 February 2021, from https://www.archiscene.net/location/sweden/strawscraper-belatchew-arkitektur/
8. Kuma, K. (2022). *Green Cast*. Kengo Kuma Associates. Retrieved 7 Jaunary 2022 from https://kkaa.co.jp/works/architecture/green-cast
9. *Corporate Office Building for Hispasat Technology Center | Herreros Arquitectos*. (2013). Archdaily. Retrieved 10 May 2021, from https://www.archdaily.com/372080/corporate-office-building-for-hispasat-technology-center-herreros-arquitectos
10. Cuce, E., Nachan, Z., Cuce, P., Sher, F., & Neighbour, G. (2019). Strategies for ideal indoor environments toward low/zero carbon buildings through a biomimetic approach. *International Journal of Ambient Energy, 40*(1), 86–95. https://doi.org/10.1080/01430750.2017.1372807
11. Jalil, W., & Kahachi, H. (2019). The implementation of nano biomimicry for sustainability in architecture. *Journal of Enginnering and Sustainable Development, 23*(3), 25–41. https://doi.org/10.31272.jeasd.23.3.3
12. Adebisi, O., Onuwe, O., & Sani, M. (2015). Assessment of Biomimcry Design Concept Adoption in Architecture: Towards a Sustainable Built Environment in Nigeria. *IOSR Journal of Environmental Science, Toxicology & Food Technology (IOSR-JESTFT), 9*(4), 41–46. https://doi.org/10.9790/2402-09424146
13. Rust, C. (2011). *Blueprints from nature Investigating a biomimitic approach to architecture*. University of Cape Town. Retrieved 25 February 2021, from https://core.ac.uk/reader/185420200
14. Fisher, M. (2019). Terrestrial biomes. *Environmental Biology*. Retrieved 8 May 2021, from https://openoregon.pressbooks.pub/envirobiology/front-matter/introduction/
15. Pearce, M. (1996). *Passively cooled building inspired by termite mounds – Innovation – AskNature*. Asknature.org. Retrieved 12 May 2021, from https://asknature.org/idea/eastgate-centre/#.WkOIFlWWbIU
16. Nkandu, W., & Alibaba, H. (2018). Biomimicry as an alternative approach to sustainability. *Bulletin of the Polytechnic Institute of Jassy. CONSTRUCTIONs. ARCHITECTURE Section, 8*(1). Eastern Mediterranean University. https://doi.org/10.5923/j.arch.20180801.01/
17. Rao, R. (2014). Biomimicry in Architecture. *ISR Journals & Publications., 1*(3), 101–107. http://www.isrjournals.org/journals/civil_environmental_journals/biomimicryinarchitecture1401273613.pdf

18. Vahedi, A. (2009). *Nature as a source of inspiration of architectural conceptual design*. Master thesis, Eastern Mediterranean University. Architecture. Gazima. Retrieved 24 January 2021, from http://i-rep.emu.edu.tr:8080/xmlui/handle/11129/185
19. Kaur, C. (2021). 20 Incredible building facade designs that will make you rethink building envelope. *Rethinking the Future*. Retrieved 12 February 2021 from https://www.johndesmond.com/blog/design/the-selfridges-building-birmingham/
20. Worlowska, M., & Marko-Worlowska, M. (2011). Discover of the principle of nature-biomimitic architecture. In *Nature and culture symbiosis* (pp. 1285–1290). Retrieved 15 May 2021, from https://pdfcoffee.com/bio-9-pdf-free.html
21. Snaves. (2018). *Biomimetic architecture: Sagrada Familia. Steemit*. Retrieved 2 March 2021, from https://steemit.com/architecture/@snaves/biomimetic-architecture-sagrada-familia
22. *History of the Temple – Segrada familia – Segrada Familia*. Segradafamilia.org. (2021). Retrieved 15 May 2021, from https://sagradafamilia.org/en/history-of the-temple/
23. *Basilica De La Sagrada Familia – Segrada familia – Segrada Familia*. Segradafamilia.org. (2021). Retrieved 2 March 2021, from https://sagradafamilia.org/en/
24. Matai, V. (2018). *Biomimicry design inspiration from nature*. Retrieved 2 March 2021, from https://www.academia.edu/36288794/biomimicry_in_architecture_pdf
25. Florkowski, A., & Brown-Toganivalu, S. (n.d.). *Rethinking design. Bio-mimicary: Lighting design and Bio-mimicary: Water collection or treatment systems of water using systems research* [blog]. Retrieved 20 March 2021, from http://www.spatialdesign.info/blog/wp-content/uploads/2009/03/302_09_biomimetics_small3.pdf
26. Chin, A. (2010). *OFIS architects: All seasons tent tower*. Designboom. Retrieved 6 March 2021, from https://www.designboom.com/architecture/ofis-architects-all-seasons-tent-tower
27. Alwitra partner. (2016). *Case study esplanade-theatres on the bay*. Singapore. Retrieved 2 March 2021, from https://elmich.com/global/wp-content/uploads/2016/09/alwitra-ebroc-Esplanade1.pdf
28. Aithal, H. (2019). What is biomimetic architecture and why young Architects should know about it. *Rethinking the Future*. Retrieved 2 March 2021, from https://www.re-thinkingthefuture.com/fresh-perspectives/a539-what-is-biomimetic-architecture-and-why-young-architects-should-know-about-it/
29. Rogers, A., Yoon, B., & Malek, C. (2021). *Beijing Olympic Stadium 2008 as Biomimicry of a Bird's Nest*. Architectural Structure. ARCH 251. Retrieved 25 October 2020, from https://www.mcgill.ca/architecture/files/architecture/BiomimicrySSEFessay2007.pdf
30. Malani, S., Kshirsager, A., & Tiware, V. (2017). Biomimicry – Nature Inspired Building Structures. *International Journal of Engineering and Technology, 10*(1), 162–166. https://www.ripublication.com/irph/ijert_spl17/ijertv10n1spl_29.pdf
31. Herzog & De Meuron. (2010). *Beijing National Stadium (Bird's Nest): China's Behemoth Of Grandeur*. Archute. Retrieved 25 February 2021, from https://www.archute.com/beijing-national-stadium-birds-nest-chinas-behemoth-of-grandeur-by-herzog-de-meuron/
32. Rankouhi, R. (2012). *Naturally inspired design investigation into the application of biomimicry in architectural design*. Retrievd 25 Febraury 2021, from https://etda.libraries.psu.edu/files/final_submissions/534
33. *Tod's Omotesando, Where in Tokyo*. (2021). Wheretokyo.com. Retreived 6 March 2021, from https://whereintokyo.com/venues/25160.html
34. *Incheon Tri-bowl. IARC Architects*. ArchDaily. (2021). Retrieved 5 November 2021, from https://www.archdaily.com/190720/inchon-tri-bowl-iarc-architects
35. Spaho, K. (2011). *Biomimicry: Architecture that imitatees nature's functions, forms and parts*. Roger Williams University. Retrieved 2 March 2021, from https://docs.rwu.edu/archthese/72/
36. *Lotus Temple*. Institute of Civil Engineers (ICE). (2021). Retrieved 18 May 2021, from https://www.ice.org.uk/what-is-civil-engineering/what-do-civil-engineers-do/lotus-temple
37. Tirkey, A. (2018). *Manifestation of nature inspired architecture*, School of planning and architecture. New Delhi. Retrieved 2 March 2021, from https://issuu.com/archipreneuraloktirkey/docs/manifestation_of_nature_inspired_ar

38. All Answers Ltd. (2018). *Processes in achieving sustainable biomimetic design*. Retrieved from https://ukdiss.com/examples/sustainable-biomimetic-design.php?verf=1
39. Benyus, J. (2010). *Lavasa to become the world's first city to set standards using biomimicry*. Lavasa.com. Retrieved 6 June 2018, from http://lavasa.com/lavasaadmin/pdfs/Biomimicry%20 PC%20-%20press%20release%20-%20Final1.pdf
40. Aldar headquarters. (2017). *T & C Technology LLC*. Retrieved 2 March 2021, from http:// www.tandctech.com/en/projects/projects/aldar-headquarters
41. *Processes in Achieving Sustainable Biomimetic Design*. (2019). UKDiss. Retrieved 2 March 2021 from https://ukdiss.com/examples/sustainable-biomimetic-design.php#citethis
42. *Aldar Headquarters Building*. (2021). En.wikiarquitectura.com. Retrieved 4 March 2021, from https://en.wikiarquitectura.com/building/aldar-headquarters-building/https:// en.wikiarquitectura.com/building/aldar-headquarters-building/
43. *Levelsofbiomimicryanditsimportance*. (2019). Architecturever. Retrieved 25 February 2021, from https://architecturever.com/2019/09/07/levels-of-biomimecry-and-its-importantance-part3/
44. *Greening the sands with the sea | Atlas of the Future*. Atlas of the Future. (2021). Retrieved 14 May 2021, from https://atlasofthefuture.org/project/the-sahara-forest-project
45. Casalino, E. (2019). *The Sahara Forest Project: An Arid Ecosystem*. Biomimetic Science Institute. Retrived from https://biomimeticsciences.org/en/sahara-forest-project/
46. Caine, T. (2021). *The Sahara Forst Project |Smart Cities Dive*. Smartcitiesdive.com. Retrieved 14 May 2021, from https://smartcitiesdive.com/ex/sustainablecitiescollective/ sahara-forest-project/7635
47. Maglic, M. (2014). *Biomimicry: Using nature as a model for design*. University of Massachusetts Amherst. Retrieved from https://scholarworks.umass.edu/theses/871/
48. Nouvel, J. (2021). *Doha High Ride Office Tower – Ateliers Jean Nouvel*. Ateliers Jean Nouvel. Retireved 23 November 2021, from http://www.jeannouvel.com/en/projects/ doha-9-high-rise-office-tower/
49. Morris. R., (2007). *CH2*. Architectureau. Retrieved 6 March 2021, from https://architectureau. com/articles/ch2/
50. Hes, D., & Bayudi, R. (2005). Council House 2 (CH2), Melbourne CBD: A green building showcase in the making. In *Proccdings of sustainable building south east Asia*, Malysia. Retrieved 6 March 2021, from https://www.irbnet.de/daten/iconda/CIB_DC23463.pdf
51. *Donrob Staff, Organic 'Wood Wave' floating home*. (2021). Dornob.com. Retrieved 6 March 2021, from https://dornob.com/natural-curves-organic-wood-wave-floating-home-design/ https:/dornob.com/natural-curves-organic-wood-wave-floating-home-design/
52. Harvey, R. (2019). *Fennell Residence*. Archello. Retrieved 6 March 2021, from https://archello. com/project/fennell-residence
53. Kaya, L., Yucedag, C., & Asikkutlu, H. (2018). Reflection of biomimicry to spatial design. ECSAC. In *European conference on science, art and culture*. Antalya. Retrieved 17 May 2021, from https://www.researchgate.net/profile/Latif-Kaya/publication/328132089_REFLECTIONS_OF_BIOMIMICRY_TO_SPATIAL_ DESIGN/ links/5bba0231a6fdcc9552d56ce1/ REFLECTIONS-OF-BIOMIMICRY-TO-SPATIAL-DESIGN.pdf
54. Rohilla, P. (2017). *Biomimicry: Source of architecture innovation*. Slideshare. net. Retrieved 25 February 2021, from https://www.slideshare.net/axiompraveen/ biomimicry-source-of-architecture-innovation
55. Stecker, T. (2011). *Artificial leaf might provide easy, mobile energy*. Scientific American. Retrieved 15 November 2020, from https://www.scientificamerican.com/article/ artificial-leaf-might-provide-mobile-energy/
56. Borgobello, B. (2011). *Solar Ivy captures the sun's energy whilst creating a pleasing visual aesthetic*. New Atlas. Retrieved 15 November 2020, from https://newatlas.com/ solar-ivy-captures-suns-energy/19141/
57. Admin. (2010). *Rain collector skyscraper*. eVolo. Retrieved 8 March 2021, from https://www. evolo.us/rain-collector-skyscraper/

58. Naido, R. (2010). *Rain collector skyscraper*. Designboom. Retrieved 8 March 2021, from https://www.designboom.com/architecture/rain-collector-skyscraper/
59. www.fosterandpartners.com, F. (2021). *Reichstang, New German Parliament | Foster+Partners*. Fosterand partners.com. Retrieved 14 November 2021, from https://www.fosterandpartners.com/projects/reichstag-new-german-parliament
60. Roxburgh, H. (2016). Inside Shanghai Tower: China's tallest skyscraper claims to be the world's greenest. *The Guardian*. Retrieved 12 November, 2021, from https://www.theguardian.com/cities/2016/aug/23/Inside-shanghai-tower-china-tallest-building-green-skyscrapers
61. *Council of Tall Building and Urban Habitat*. Ctbuh.org. (2021). Retrieved 25 Nobemver 2021, from https://www.ctbuh.org
62. *100 Tallest Completed Buildings in the World – The Skyscrpers Center*. Skyscrapercenter.com (2021). Retrieved 25 Nobemver 2021, from https://www.skyscraperceter.com/buildings
63. Plaskoff, R. (2009). *Off the Grid: Philips sustainable Habitat 2020*. UrbanGardens. Retrieved 3 May 2021, from http://www.urbangardensweb.com/2009/09/06/off-the-grid-philips-sustainable-habitat-2020/
64. John, G., Clements-Croome, D., & Jeronimidis, G. (2005). Sustainable building solutions: A review of lessons from the natural world. *Building and Environment, 40*(3), 319–328. https://doi.org/10.1016/j.buildenv.2004.05.011
65. Foxlinadmin. (2010). *William Mcdonough's treescraper.* Reterived 10 May 2021, from https://foxlin.com/william-mcdonoughs-treescraper/
66. MGS Architecture. (2008). *Lilypad a floating City. NBM & CW Intraconstruction & Equipment Magazine.* Retrieved 15 March 2021, from https://www.nbmcw.com/articles-reports/building-construction-material/green-tall-construction/lilypad-a-floating-city.html
67. *Floating Lilypad City*. (2021). Aquatic Urbanism. Retrieved 15 March 2021, from https://aquaticurbanism.com/floating-lilypad-city/
68. DeAngelis, S. (2011). *Floating Cities*. Enterra solutions. Retrieved 15 March 2021, from https://enterrasolutions.com/blog/floating-cities/
69. Pinky, A. (2015). *Sustainable architecture*. Retrieved 10 March 2021, from https://www.slideshare.net/arundathipinky1/sustainable-architecture-55140947?next_slideshow=1
70. *Importance of sustainable architecture in 21st century.* (2017). Retrieved 10 December 2017, from http://www.architecture-student.com/sustainable-design/importance-of-sustainable-architecture-in-21st-century/
71. Design Build. (2011). *The benefits of sustainable building*. Thedesignbuildblog.wordpress. Retrieved 10 Decemeber 2017, from https://thedesignbuildblog.wordpress.com/2011/04/04/the-benefits-of-sustainable-building/.
72. Paul, R. (2008). *LILYPAD: Floating city for climate change refugees*. Inhabitat. Retrieved 15 March 2021, from https://inhabitat.com/lilypad-floating-cities-in-the-age-of-global-warming/
73. Gamage, A., & Hyde, R. (2012). A model based on Biomimicary to enhance ecologically sustainable design. *Architectural Science Review, 55*(3), 224–235. https://doi.org/10.1080/00038628.2012.709406

Chapter 5
Biomimicry in Architecture for Climate Change Mitigation and Adaptation: An Overview of Egypt, Italy, and Germany Actions towards Climate Change

> We need to turn the recovery into real opportunity to do things right for the future. (Antonio Guterres, United Nations Secretary General, April 22, 2021)

> With this restart, a window of hope and opportunity opens, an opportunity for nations to green their recovery packages and shape the 21st century economy in ways that are clean, green, health, safe and more resilient. (UN Climate Chief, April 22, 2021)

> We all agreed that climate change is one of the greatest and most daunting challenges of our age. We have a moral imperative to act and act now. (David Cameron, Former Prime Minister of the U.K.)

5.1 Introduction

It all starts with a question. Can the biomimicry approach to create and develop buildings be one of the solutions for climate change mitigation and adaptation? In fact, there could be a major role for biomimicry architecture in climate change mitigation and adaptation. Truly, climate change (CC) is manifested worldwide (Figs. 5.1 and 5.2) with growing cost of US$ 210 billion in damage in 2020, which is caused by disasters such as floods, heatwaves, and wildfire [1]. Based on an EU report 2005, the annual economic loss from weather- and climate-related events has increased significantly over the past 20 years from an average of less than US$ 11 billion per year [2]. According to Paul Watkiss et al. [3], the impact of global CC resulting from greenhouse gas emissions (GHGs) is varying, potentially dramatically increasing and probably representing the most serious long-term environmental issues the world is currently facing [3]. David King expressed the seriousness of global CC stating that it is the number one threat facing us and its severity overcomes the threat of terrorism. As the effects of CC increment, arrangements and activities must be applied without delaying to mitigate the emissions of GHGs [4, 5]. The urban built environment (UBE) is specifically in urgent need of such arrangement. The responsibility for almost the third of global GHG emissions relies upon the urban environment. Hence, the UBE should accommodate the process of adaptation to the CC impacts, as the main core of civilized life.

M. Aboulnaga, S. E. Helmy, *Biomimetic Architecture and Its Role in Developing Sustainable, Regenerative, and Livable Cities*,
https://doi.org/10.1007/978-3-031-08292-4_5

Fig. 5.1 Flooded areas due to highest tide in 50 years – climate change impacts in Venice, Italy. (Image credit and source: https://commons.wikimedia.org/wiki/Category:Floods_in_Venice#/media/File:Acqua_Alta_sur_la_place_Saint_Marc_%C3%A0_Venise_(8091188816).jpg)

With more than half of Earth's population living in urban areas spending most of their time indoors, building has a significant and leading role in mitigating GHG emissions with less expenses compared to other sectors, as stated by the Intergovernmental Panel on Climate Change (IPCC), but this goal requires drastic changes in the current practices [4, 5]. Thinking about living organisms and their way of adapting to similar problems such as CC and gas emissions will assist us to find a solution to overcome such problems. The adaptation and survival of some plants and animals in extreme weather or climate change severe events could provide a perspective on how buildings and constructions should or could work. Looking at the characteristics of biological systems that allow them to be flexible might likewise offer a potential model to consider [4, 5].

Box 5.1: The Role of Everyone in Improving the Quality of the Environment

The only way forward, if we are going to improve the quality of the environment, is to get everybody involved. (Richard Rogers – British Italian architect)

Fig. 5.2 Climate change impacts in cities – flooded areas of Lokoja in Kogi State in Nigeria. (Image source: Flickr, https://www.flickr.com/photos/condevcenter/25291918331/in/photolist-EwXGar-74eSzs-74aRG2-74eemN-9ptDm4-74eWMf-AEb2VP-zHhFLe-dEZ6vD-cbU9R3-dF6cz9-2hxHwE8-2akDcX3-dEZ8SH-dF8cs9-dEZc82-dF8x4b-dF5c59-dEYLAM-dF7KkQ-dF6u1f-74j238-2hSxUxK-2kCPFNH-dF8Kmo-e12rxc-dF2W6e-rgjz8T-dF8DGj-dF8zUq-dF7S3E-dF2XPc-dF3cwc-9ph6q2-dF2har-4xgrFs-kUas3V-9phpme-9phppg-dF81Zh-dF8GFj-dF7PHJ-dF35vz-dF8sfG-9KhXtc-ejjvhz-74dXBE-74ekj9-dF7jLW-dF84Tm/)

5.2 Climate Change in a Global Context

Climate is defined by the *Oxford Learner's Dictionary* as "the regular pattern of weather condition of a particular place." These weather conditions could be classified into mild, temperate, warm, and wet depending on season and/or location [5]. Conclusive evidence, however, shows a drift from the normal pattern of weather condition to a rather harmful and adverse trend as climate is now known to have a negative effect on the environment and invariably on the ecosystem, both animate and inanimate [5, 6]. Notable changes of the environment occurred because of worldwide climate change, for example, the reduction of the amount of glaciers, ice melting on rivers sooner than expected (Fig. 5.3), and the modifications in the times of growing some crops and plants. The expectation of scientists for the impacts and consequences of environmental changes is presently happening: ice cap melting, accelerated ocean level increase, and serious warm waves, among many others [6]. Almost all the scientists concerned with the studying of CC agreed on the fact that the primary cause of CC and global warming is due to human behaviors and activities. Global warming is happening due to the increase in the Earth's average surface temperature and its oceans due to GHGs, mainly carbon dioxide (CO_2) and methane resulting from burning fossil fuel – global trends 1900–2101 as shown in Fig. 5.4.

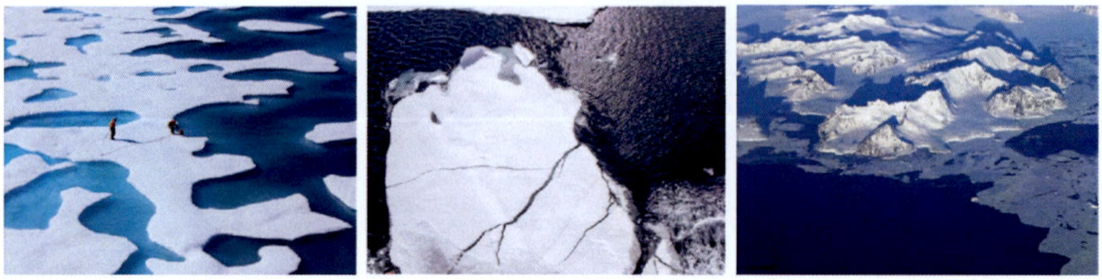

(a) (b) (c)

Fig. 5.3 Impact of climate change observed in the amount of glaciers and ice melting. (**a**) Melted ice (**b**) Ice cracking due to increase in temperature (**c**) Glacier ice blocks. (Image source: Flickr (**a**) https://www.flickr.com/photos/vintage_illustration/29025841378/in/photolist-LdV2p1-4eVwN3-gzktet-oZLiSC-2aRmVXQ-PC9d9m-69Lg8m-2aDALp6-9zeJJT-4UHncw-rfu7CA-28mMvqV-bYZTnE-6AP9XW-RwVZTQ-2dG5XDn-25AP6V3-73Whbq-2aRmVFh-2dBVpwK-2ix5Nxo-2ixETva-2iwPdXw-2ibcV2g-2iukV3i-PA1EB9-P8taKT-u9Umas-QzvkRt-6APcu3-2mebgwT-DPgTtU-2iYKsmD-2fe179m-oxjbvf-wxkuxP-eLuGZE-pP1PwM-2ixEsWu-2dxrC8C-bYZULL-RzvTnx-2dpiw3K-SWMe3s-2aRmYxu-2kcngJ-oiF6kH-2cvZSy9-2dxqRGy-ozWEVh, (**b**) and (**c**), https://www.flickr.com/photos/vintage_illustration/46351471211/in/photolist-2dBVpwK-2ix5Nxo-2ixETva-2iwPdXw-2ibcV2g-2iukV3i-PA1EB9-P8taKT-u9Umas-QzvkRt-6APcu3-2mebgwT-DPgTtU-2iYKsmD-2fe179m-oxjbvf-wxkuxP-eLuGZE-pP1PwM-2ixEsWu-2dxrC8C-bYZULL-RzvTnx-2dpiw3K-SWMe3s-2aRmYxu-2kcngJ-oiF6kH-2cvZSy9-2dxqRGy-ozWEVh-SWMdVJ-aguj6-2jDaTt-oBVrqF-31ndLV-QwFoC6-27hJu6s-23vsiiG-25tHyu3-NoNdrS-jVKMW-G5E7Vf-9uQ8tf-2jHDPX3-2jhC5cH-RbXLLG-2iBmiXe-25LJVEy-YXJ9mt)

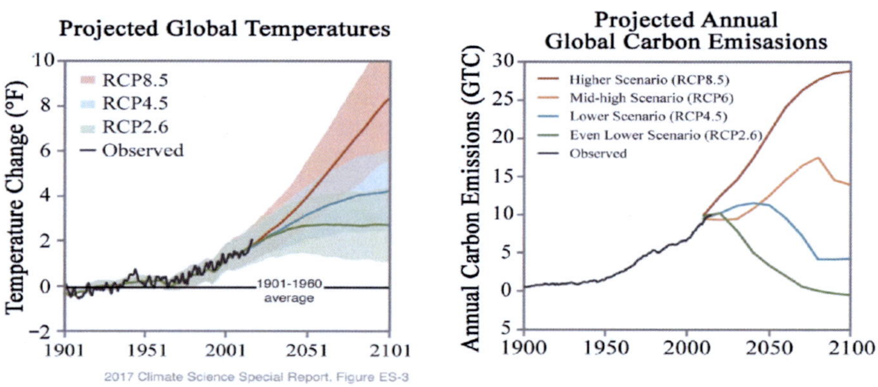

(a) **(b)**

Fig. 5.4 Global trends of temperature change and CO_2 from 1901–2101 to 1900–2100. (**a**) Global trends of temperature change (**b**) Annual global CO_2 emissions over time. (Image source: https://www.climate.gov/media/12886)

Such increase in the level of GHG emissions caused the average temperature of the world to increase [6]. Figure 5.5 presents the atmospheric CO_2 which has increased since the industrial revolution. Richard Rogers – a British Italian architect – emphasized on the role of everyone in enhancing the quality of the environment to mitigate CC (Box 5.1).

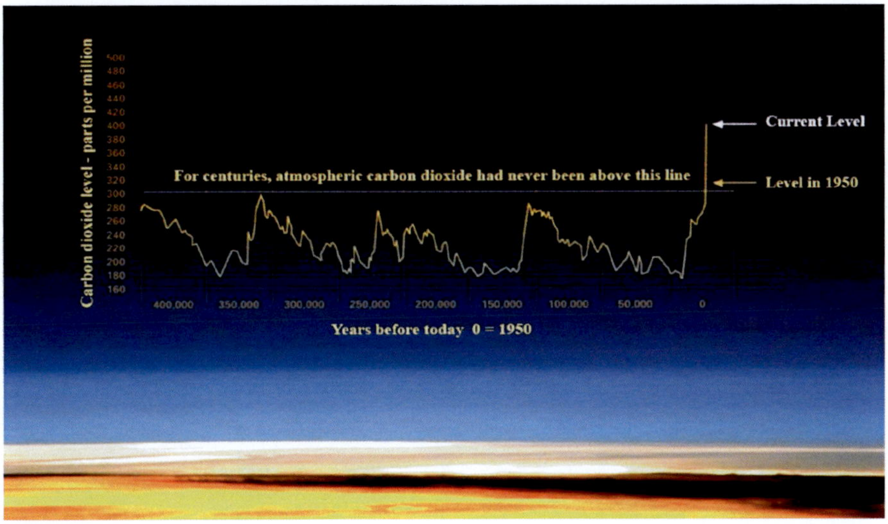

Fig. 5.5 Atmospheric CO_2 increase since the industrial revolution. (Image source: https://climate.nasa.gov/evidence/)

Information concerning temperature rise from four global science foundations shows global warming in previous decades. The most recent decade has been the hottest on record, notably the years 2018 and 2019 are the highest in record [7]. The below graphs present the correlation of air samples contained in ice cores and later direct estimations, and they also give proof that climatic CO_2 has expanded since the industrial revolution as shown in Fig. 5.5 [5].

Satellites orbiting the Earth and other innovative methods allow researchers to get a comprehensive view and collect extensive data about our planet and its atmosphere on a global scale. This compilation of information gathered over the years reveals signs of atmospheric evolution. Table 5.1, which demonstrates the proof for fast environmental change, is convincing. The current warming pattern is of specific noteworthiness on the grounds that the greater part of it is to be the consequence of human activities and behaviors since the mid-twentieth century and continuing at a rate that is remarkable over decades and centuries [5].

> Climate change is running faster today than the people trying to tackle it. We will be destroyed by Climate change. (Antonio Guterres, Secretary General of the United Nations – WEF 2020)

Table 5.1 Evidence for rapid climate change manifestation

Image	Evidence	Discussion
 Image source: https://climate.nasa.gov/evidence/	Temperature rising globally	A change in the planet's average temperature accrued, by increased carbon dioxide (CO_2) and other human-made emissions into the atmosphere. It has risen by 1.1 Celsius since the late nineteenth century
 Image source: https://climate.nasa.gov/evidence/	Warming oceans	The top 700 m of the oceans' surface have absorbed much of the heat
 Image source: https://climate.nasa.gov/evidence/	Shrinking ice sheets	Antarctic ice sheets reduced in mass. Loss of 150–250 Km^3 from Greenland
 Image source: https://climate.nasa.gov/evidence/	Glacial degradation	Glaciers are retracted almost everywhere around the world – Including the Alps, Himalayas, Andes, Rockies, Alaska, and Africa
 Image source: https://climate.nasa.gov/evidence/	Decreased snow cover	Satellites have detected that the amount of spring snow cover in the northern hemisphere has decreased over the past five decades and that the snow is melting earlier

(continued)

Table 5.1 (continued)

Image	Evidence	Discussion
 Image source: https://climate.nasa.gov/evidence/	Sea level rise (SLR)	The rate of the rising sea level in the last two decades is double of the last century since the rising number in inches for the past two decades
 Image source: https://climate.nasa.gov/evidence/	Decaying Arctic Sea ice	In the last decades, the thickness of Arctic Sea ice decreased rapidly
 Image source: https://climate.nasa.gov/evidence/	Extreme events	The high temperature record number in the USA has been increasing, and low temperature record decreased, since 1950. There is also an attended increment in the number of intense rainfall

Source: Developed by authors

5.3 Vital Signs

Scientists have projected that increasing global temperatures would cause a number of significant changes to the global climate system. Some of these changes, such as declining global snow and ice cover and sea level rising (SLR) are happening gradually as temperature rises. Other changes are a consequence of amplified climate instability, for example, the increasing frequency and intensity of extreme weather events such as heat waves, heavy precipitation, and storms. Below is a description of three of the expected changes of Earth systems caused by rising global temperatures: SLR, decreased snow and ice cover, and extreme weather events [8].

Global efforts have been manifested since the Paris Climate Agreement was signed in December 2015. The United Nations Climate Change Conference (COP21) culminated in a big success with the adoption of a new global climate change agreement, backed by 196 countries to mitigate climate change (CC) risks with the target yielded to keep the Earth's temperature at 1.5 °C [9]. In addition to the effort resulting during COP22 in Marrakech, Morocco, in 2016; COP 23 in Bonn, Germany, in 2017, as well as COP24 in Poland in 2018; and COP25 in

Madrid Spain in 2018, more is expected in COP26 in Glasgow in November 2021 [10–13]. Nonetheless, efforts toward climate change have been slowed down due to the impacts of the COVID-19 pandemic in 2020 and early 2021. Despite the uphill struggle with the COVID-19 crisis, COP26 will, however, be held in November 2021 in Glasgow, UK, to revive the effort toward mitigating climate change and discuss the global climate agenda [14]. Moreover, the USA rejoined the Paris Climate Agreement in February 2021 after the previous US administration pulled out in 2017. Despite the stumbling block, the USA invited 40 world leaders to the Climate Virtual Summit held on April 22, 2021, which highlights commitment toward CC mitigation as shown in Fig. 5.6 [15]. Such efforts were culminated and reaffirmed during COP26 (2021) in Glasgow, UK.

5.4 Causes and Effects of Climate Change

Climate change and global warming as terms are not only used and discussed extensively by world leaders, governments, and international organizations (the UN (United Nations), United Nations Framework Convention on Climate Change (UNFCCC), Intergovernmental Panel on Climate Change (IPCC), World Bank, Council of the European Union, Organization for Economic Cooperation and Development (OECD), C40, and International Council for Local Environmental Initiatives, ICLEI - Local Governments for Sustainability, etc.) but also dealt with interchangeably within scientific journals.

Global warming refers to surface temperature increase, while CC includes global warming and everything else that increases GHG amounts [16]. The IPCC defines CC as any change in climate over time, whether due to natural variability or as a result of human activity [16] as depicted in Box 5.2. Similarly, CC as a long-term alteration in global weather patterns, such as increase in temperature and storm

(a)

(b)

Fig. 5.6 The Climate Virtual Summit held on April 22, 2021. (**a**) UN secretary-general addressing world's leaders (**b**) US president addressing world's leader on CC. (Image source: (**a**) https://en.wikipedia.org/wiki/2021_Leaders_Summit_on_Climate#/media/File:President_Joe_Biden_at_the_Leaders_Summit_on_Climate_(02).jpg, (**b**) https://commons.wikimedia.org/wiki/Category:Joe_Biden_at_Leaders_Summit_on_Climate,_April_2021#/media/File:097-P20210422AS-1557_(51,145,710,189).jpg) activity, is often linked to the potential consequence of rising levels of heat-trapping gases known as GHG, which retain the radiant energy provided to the Earth by the sun in a process known as the greenhouse effect. Climate scientists believe that the Earth is currently facing a period of rapid warming brought on by rising levels of GHG in the atmosphere [16]. It is important to note that GHG occurs naturally and without them, the planet would be too cold to sustain life. However, since the beginning of the industrial revolution in the mid-1700s, human activities have added more and more of these gases into the atmosphere [16]. According to Izge and Maina, with more GHGs in the mix, the atmosphere acts like a thickening blanket and traps more heat [16]. In 1957, researchers at the Scripps Institution of Oceanography (SIO), based in San Diego, California, began monitoring carbon dioxide (CO_2) levels in the atmosphere from Hawaii's remote Mauna Loa Observatory located 3000 m above sea level. When the study commenced, CO_2 concentration in the Earth's atmosphere was 315 ppm, but the CO_2 concentration has increased to reach 323 ppm and 335 ppm by 1970 and 1980, respectively [17]. By 1988, atmospheric CO_2 had increased up to 350 ppm. Nonetheless, it reached 407 ppm in 2018; an increase of almost 29% in 61 years from the year 1957 level as presented in Fig. 5.7. In March 2020, it reached around 418 ppm [18] and it increased to 421 ppm in June 2022. Other researchers confirmed these findings; scientific interest in the accumulation of GHGs and their effect on the environment slowly began to grow. In 1988, the World Meteorological Organization (WMO) and the United Nations Environmental Program (UNEP) established the Intergovernmental Panel on Climate Change (IPCC) to be the panel dealing with CC [17].

Box 5.2: Climate Change

- Climate change is a complex multilayer evolution that can affect the global environmental climate in various ways.
- It is generally resulting from different types of emissions fostered by large corporations in their operation practices.
- Governments around the world have taken steps and measures to curb, monitor, and/or regulate large corporate practices to limit such emissions. (IPCC and United Nations)

The IPCC was the first international collaboration of scientists to access the scientific, technical and economic information related to the risk of human-induced

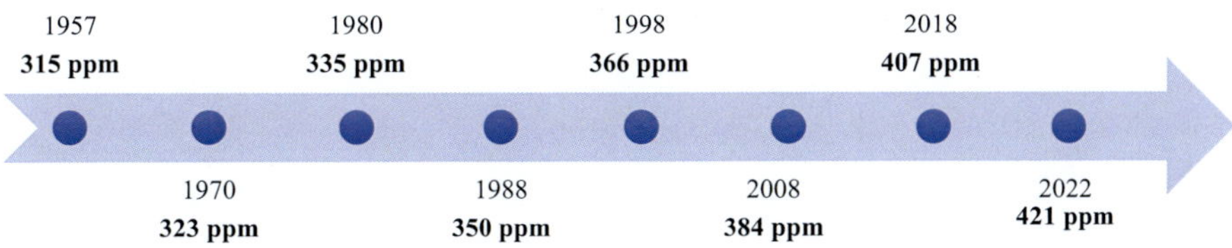

Fig. 5.7 Increase of CO_2 concentration in the Earth's atmosphere over the time between 1957 and 2018. (Image source: Developed by authors after Izge and Maina (2011) [16, 17])

CC. The IPCC created periodic assessment reports on advances in scientific understanding of the causes of CC and its potential impacts and strategies to control the GHG emissions. The IPCC also played a critical role in establishing the United Nations Framework Convention on Climate Change (UNFCCC), which provides an international policy framework for addressing CC issues. However, in this line, the United Nations General Assembly (UNGA) in Rio, Brazil, adopted this approach in 1992. Climate scientists around the world are currently monitoring atmospheric GHG concentrations and creating forecasts about their effects on global temperatures. These organizations gathered information about climate systems and used such information to create and test computer models that simulate how climate could change in response to changes on Earth and in the atmosphere. However, these models can only provide approximations, and some of the predictions based on them often spark controversy within the scientific community. Nevertheless, the basic concept of global warming is widely accepted by most climate scientists. The following session of the study aims at reviewing the causes and the impacts of climate change on agriculture in Africa and the adaptive measures that might be taken to ameliorate the negative impacts [17].

5.4.1 Causes of Climate Change

What are the circumstances and results of climate change? There are several factors contributing to CC. One of them is natural factors and other factors are caused by humans. The main cause of global warming stated by climate researchers is the increased levels of human expansion of "greenhouse effects," leading to trapping the heat in the atmosphere and preventing radiation from the Earth to the space [19]. Figure 5.8 shows a summary of the elements causing CC:

- CO_2, minor concentration. Carbon dioxide is emitted by human practices such as cutting forests and extensive use of fossil fuels as a source of energy in addition to natural causes such as the eruptions of volcanoes and breathing behavior. These causes and practices increased the concentrations of CO_2 by third, forcing the CC.

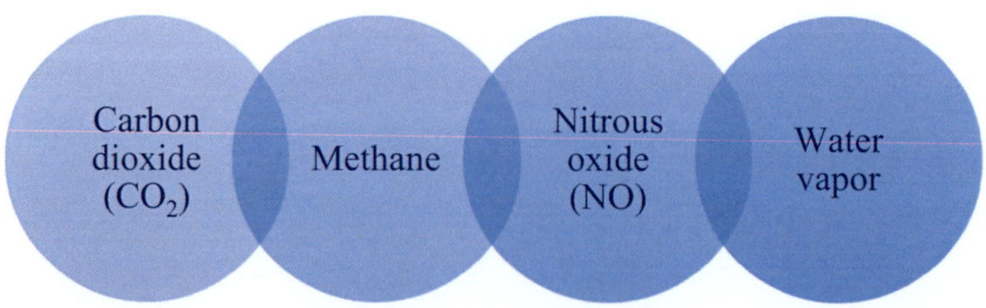

Fig. 5.8 Main elements causing climate change. (Source: Developed by authors after [19])

- Methane, a hydrocarbon gas generated both naturally and artificially. The causes include the waste decaying in landfills, farming especially rice farming, and management of cattle. On the molecular scale, it is more harmful.
- Nitrous oxide is an intense ozone-depleting substance created by soil farming activities, particularly the fertilization of soil, burning of fossil fuels, generation of nitric acid, and consumption of biomass.
- Water vapor is the most inexhaustible ozone-harming substance; however, critically, it is considered an indication of climate. The warmth of the Earth's atmosphere is increased by water vapor in addition to the clouds and precipitation, making them prime indicators [19].

Global warming is present due to the emissions of GHG, especially component of CO_2 as shown in Fig. 5.8. It is continuously generated by respiration, consumption of fossil fuels, and plant decomposition [19], so the breathing process resulting in CO_2 as the number one cause of CC makes the main contributor in emitting GHG emissions in human activities. On a daily basis, the heavy dependence on fossil fuels for generating energy is causing global warming. This nonrenewable type of energy generation is used not only for lighting homes, buildings, and streets, for cooling and heating of houses and other tertiary buildings, and for cooking using the oven but also for transport and transportation (vehicles, railways, or planes) as well as industries, trades, and agriculture. Figures 5.9 and 5.10 present the carbon cycle (storage and release) and the global GHG emissions (top emitting countries and sectors), respectively [19].

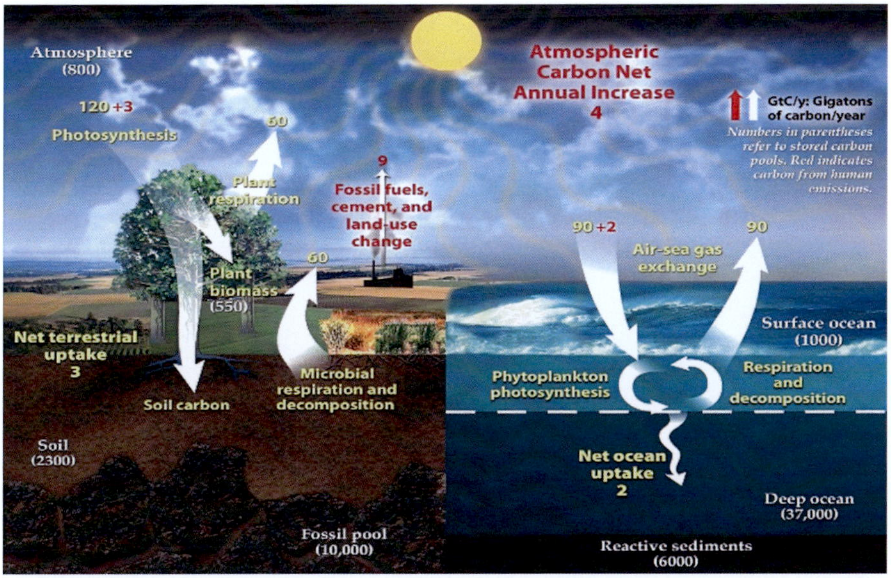

Fig. 5.9 Carbon cycle – storage and release. (**a**) Global highest GHG emissions from fossil fuels by country (**b**) Global GHG emissions by the economic sector. (Image source: https://en.wikipedia.org/wiki/Carbon_cycle#/media/File:Carbon_cycle.jpg)

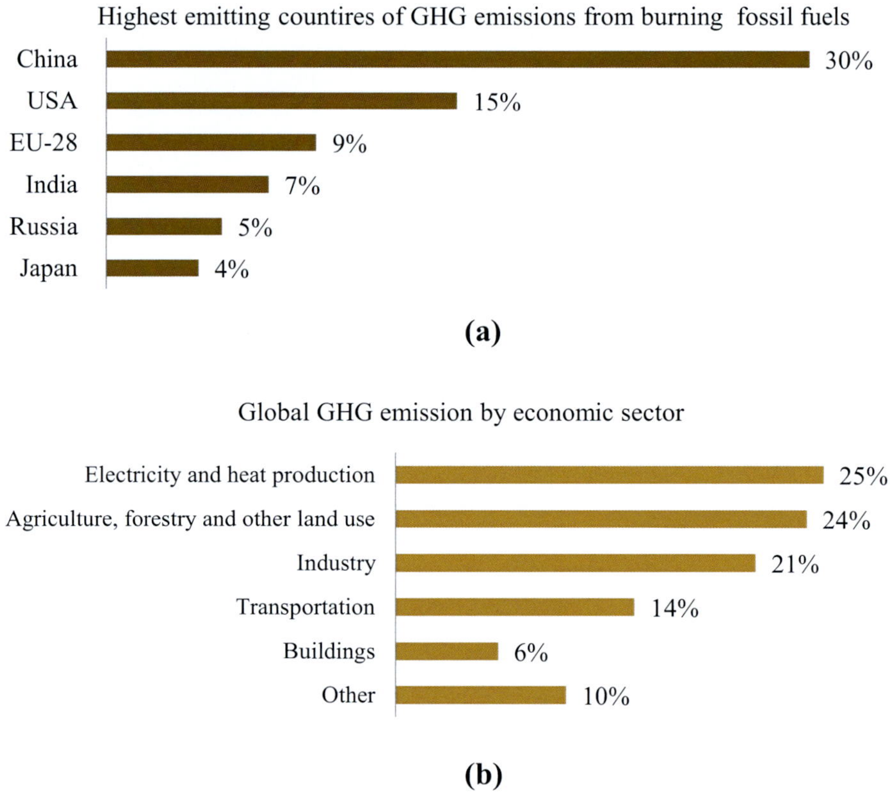

Fig. 5.10 Global GHG emissions 2017. (Image source: Developed by authors after EPA and Statista, https://www.statista.com/chart/9633/global-greenhouse-gas-emission-countries-and-industries/)

The percentage of CO_2 in the atmosphere increased in the last century due to consuming a high portion of fossil fuels. The process of burning fossil fuels generates CO_2 by combining carbon with oxygen present in air. Agriculture and farming, industrial activities, and others contributed to the increase of the GHG emission percentage globally [20].

Nature Contributions to Solar Irradiance

There is a reason behind the assumption that the changing energy output that comes from the sun could result in CC, due to the importance of solar energy affecting Earth's climate. Studies verified that solar energy changes affected the climate in previous times. For example, the little ice age occurred between the years 1650 and 1850. It was a result of the reduction in solar energy, when glaciers increased in the Alps and Greenland. Nevertheless, many evidences show that the current CC cannot be explained by the change in solar energy [20].

The CO_2 emissions from volcanoes are smaller than the emissions produced by human activities. NASA stated that the average emissions of volcanoes are 130 and 230 million tons of CO_2 per year. However, this amount is increased by 100 times upon using fossil fuels as a source of energy; about 26 billion tons of CO_2 are emitted to the atmosphere every year starting from 2005 [20].

Human Contributions

The IPCC, on its fifth evaluation report, stated that the contribution of human practices and behaviors in warming up the planet exceeded 95% [20]. It added that the human modern progress and development activities are 400 parts in million over the last 115 years. The report also indicated that a great deal of what is causing the Earth's temperature to increase is due to the human-generated GHG emissions [20]. The heavy dependence on fossil fuels in generating energy resources is among the reasons causing CC. Such dependence on fossil fuel energy to meet human needs globally is enormous and estimated at about 80%, which makes it very difficult to switch to other energy resources according to the Australian Greenhouse Office (AGO). This notion has changed as many countries enacted policies between 2020 and 2021 to ban vehicles that run by fossil fuel, such as the UK where only electrical cars will be allowed by 2030 [21].

Many other countries follow suit. Egypt's government moved in the same direction by exempting all imported electrical cars and busses from taxes from 2019 [22]. Table 5.2 shows the two types of causes of climate change [20].

5.4.2 Impacts of Climate Change

The futuristic impacts of CC worldwide involve more continuous rapidly spreading fires, longer times of dry seasons, heat waves in a few places and expanding, and length and power of typhoons and storms. Figures 5.11 and 5.12 illustrate a portion of the impacts of CC on the Earth. As previously illustrated, CC had notable impacts around the globe. The reduction of sea ice mass, the elevated sea levels, and the acute heat waves are among the impacts [20]. The latter has been manifested in Tokyo, Japan, in 2018, where 17 people were dead and more than 3000 were

Table 5.2 Causes of climate change

Anthropogenic	Natural
Chemical fertilizer	Sunspot and solar cycle
Deforestation	Ocean currents
Increased vehicles	Forest fire
Emissions of GHG	Volcanic eruption
Industries	Meteorites

Source: Developed by authors

Fig. 5.11 The effect of climate change on Earth. (Image credit and source: Christophe Licoppe, https://commons.m.wikimedia.org/wiki/File:visit_of_Ursula_von_der_Leyen,_President_of_the_European_Commission,_to_Rochefort_and_Pepinster_in_Belgium_12.jpg#mw-jump-to-licence)

admitted to hospitals due to heat-related problems as well as the severe heat wave that hit London, Paris, and the Netherlands on the same year [23]. In addition, the recent devastating floods due to CC impact in Belgium in July 2021 (Fig. 5.11 and Fig. 5.12). Therefore, the IPCC predicted an increased level of the average temperature of the Earth by 3 °C compared with that of 1990; it may be useful in some regions and not in others. The expenses will increase with the temperature warming up. The IPCC also stated that the expenses to repair the harms resulting from CC will be severe and will increase with time [20]. The harmful impacts of CC worldwide on the environmental and social life including the pollution of air and drinking water and the lack of food and housing are presented in Fig. 5.13. It is evaluated that the number of deaths will witness a great increase by 250,000 in the period of 2030–2050, caused by starvation, heat stress, and diseases as shown in Fig. 5.14. The expenses in health-related turfs are evaluated to be 2–4 billion US dollars per year by the year 2030. If we make better choices regarding our use of energy resources, it will result in improving health. Figure 5.15 shows the impact of CC on human health [6]. The human health is reversely affected by the warming up of the global climate due to the trapped GHGs. Therefore, many societies will suffer from this change in climate. Unfortunately, the impacts will affect people differently, depending on whether they have access to healthcare or not, with the most effect directed to those in sub-Saharan Africa. Unexpected changes will be harsh on youth or the elderly to adapt to it.

Fig. 5.12 Devastated impact of floods in Tilff, Liege region, Belgium, on July 16, 2021, caused by climate change. (Image credit and source: Régine Fabri, https://commons.m.wikimedia.org/wiki/File:floods_16_July_2021,_Belgium,_Tilff_1.jpg#mw-jump-to-licence)

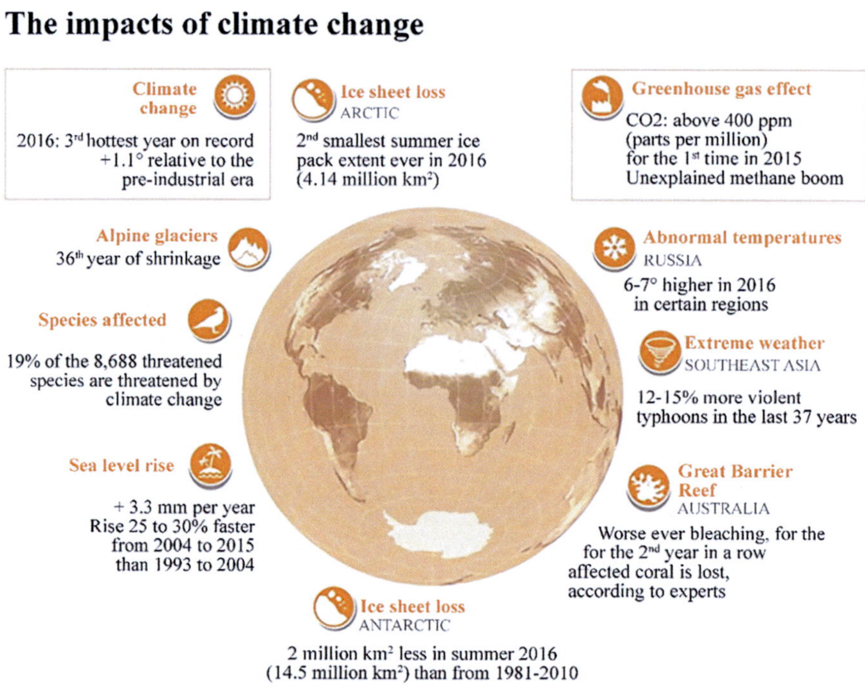

Fig. 5.13 The significant impacts of climate change worldwide. (Image credit and source: World Meteorological Organization (WMO), https://www.dawn.com/news/1336862)

Adverse nations will have less chance to adapt to the abrupt climate changes. The harsh conditions lead to great stress on the emergency systems and health [6], but nevertheless, this has been manifested not only by CC impacts but also by the coronavirus COVID-19 pandemic since March 2020 till the present – with the unprecedented case of India experiencing the amok of COVID-19 and the failure of the health system to cope with the spread of the pandemic, where the cases amounted to 400,000 cases and more than 3000 dead cases on April 30, 2021 [24].

Box 5.3: The Value of the Environment According to a Japanese Scientist, Dr. David Suzuki

The environment is so fundamental tour to continue existence that it must transcend politics and become a central value of all members of society. (David Suzuki)

Hence, it is imperative to reduce the production of GHG emissions and adapt to the new changes in order to mitigate the impacts on health and economies. A strain has been noticed on the economies affected by flooding, loss of yields of crops, and increased illnesses. According to the UN report, it is indicated that more flooding,

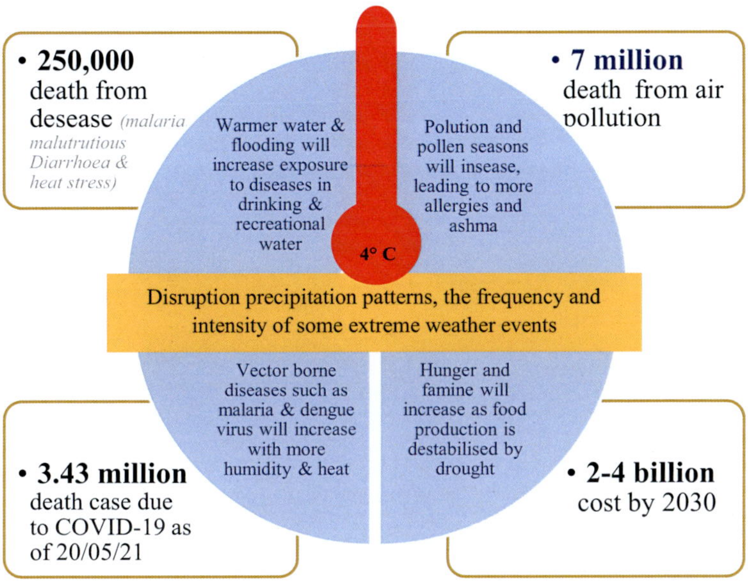

Fig. 5.14 The effect of global climate change on human. (Image source: Developed by authors after http://www.kanthalaraghu.com/2017/08/how-climate-change-impacts-health.html)

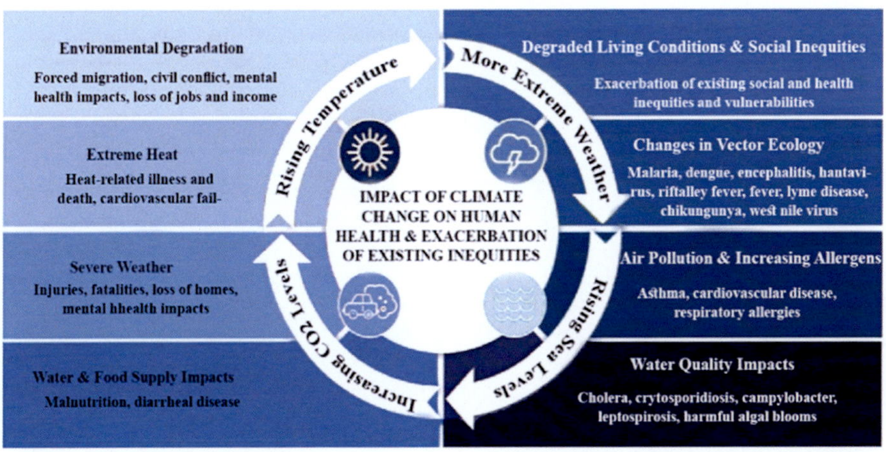

Fig. 5.15 Potential impact of climate change on human health. (Image source: https://tn.gov/health/article/climate-change)

wildfires, and decline in crop yields, fish catches, and water shortages are among the risks from the rising temperature [6] as one of the CC impacts, as listed in Table 5.3. Moreover, Japanese scientist, David Suzuki highlighted the significance of the environment to mitigate the risk of climate change (Box 5.3).

Table 5.3 Effects of climate change

Images	Effect	Discussion
 Image source: https://climate.nasa.gov/effects/	Continuous change	Worldwide atmosphere is anticipated to keep on changing over this century. The extent of environmental change beyond the following couple of decades depends essentially on the measure of warmth-catching gases transmitted universally
 Image source: https://climate.nasa.gov/effects/	Continuous rising temperature	Since human-instigated warming is superimposed on a normally shifting atmosphere, the temperature rise has not been, and will not be, uniform or smooth over time
 Image source: https://climate.nasa.gov/effects/	Prolonged frost-free season (and growing season)	In a future in which warm-catching gas emissions keep on growing, increments of a month or more in the lengths of the ice-free and developing seasons are anticipated crosswise over a large portion of the USA before the century is over, with marginally smaller increments in the northern Great Plains
 Image source: https://climate.nasa.gov/effects/	Change in precipitation modules	Precipitation average of the USA has expanded since 1900; however, a few regions have had increments more prominent than the national average, and a few territories have had diminishes. More winter and spring precipitation is anticipated for the northern USA and less for the southwest
 Image source: https://climate.nasa.gov/effects/	Drought and heat waves	Summer temperatures are anticipated to keep rising, and a decrease in soil dampness, which compounds warm waves, is anticipated for a significant part of the western and focal USA in summer. Before this century is over, what have been once-in-20-year extraordinary heat days (one-day occasions) are anticipated to happen every few years
 Image source: https://climate.nasa.gov/effects/	Stronger and more intense hurricanes	The power, recurrence, and duration of North Atlantic hurricanes, and also the recurrence of the most grounded (categories 4 and 5) tropical storms, have all expanded since the mid-1980s. The relative commitments of humans and characteristic causes to these increments are yet dubious

(continued)

Table 5.3 (continued)

Images	Effect	Discussion
 Image source: https://climate.nasa.gov/effects/	Sea level rise by 1–4 feet by 2021	Worldwide Ocean level has ascended by around 8 inches since solid record keeping started in 1880. It is anticipated to rise another 1–4 feet by 2100. This is the aftereffect of water from melting land ice and the extension of seawater as it warms. Ocean level ascent will proceed past 2100 in light of the fact that the seas take a long time to respond to warmer conditions at the Earth's surface
 Image source: https://climate.nasa.gov/effects/	Probability of the Arctic to become ice free	The Arctic Ocean is expected to become essentially ice free in summer before the mid-century

Source: Developed by authors

5.5 The Role of Biomimicry in Climate Change Mitigation and Adaptation

Biomimicry can play a significant role in climate change mitigation and adaptation. There are many ways to reduce the effects of CC by 2 °C compared to its rate before the industrial revolution. The requirements needed to reduce the harmful impacts include the decrease of the amount of released emissions and almost zero carbon emissions, and the other is the decrease of the heat-trapping gases in the next two decades [6]. This definitely means that the world will face many challenges on the economical, technological, social, and industrial as well as the environmental levels. It could witness delay in implementing such procedures due to these challenges in addition to the absence of the required advanced technologies to cope accordingly.

The assessment and predication of global climate change were recently studied, and the results from the "Integrated Research Program for Advancing Climate Models," as part of Japan's contribution to IPCC AR6 future climate projection, showed the changes of sea surface temperature and sea surface height (Fig. 5.16). This is an assessment and predication of global climate change which was presented by Professor Hiroyasu Hasumi, Atmosphere and Ocean Research Institute at the "SDGs Symposium 2021: Interdisciplinary science solutions for food, water, climate and ecosystems Sustainable Development Goals," held at the University of Tokyo on March 26, 2021 [25].

5.5.1 Climate Change Mitigation

Mitigation – reducing the causes of climate change – can be done in two ways: first, prevent or decrease the usage of GHG-generating sources such as combustion of fossil fuels for energy; and second, develop and work on the storages (sinks) that accumulate the GHGs such as soil, seas, and the green areas. The main purpose of the mitigation process is trying to stabilize and control the levels of GHG concentrations in the atmosphere over a long period to give the environment time to adapt to the changes. This will also decrease the level of harmful human contribution in CC; it will guarantee food production continuity and economic progress.

5.5.2 Climate Change Adaptation

Adaptation – adjusting to life in a changing climate – includes present or futuristic climates. The main objective is to decrease the weakness to the destructive impacts of environmental change such as SLR, exceptional climate events, or food scarcity. Similarly, it includes taking advantage of any potential gainful opportunity related to environmental change, for instance, longer farming seasons in a few areas [6]. Adaptation also embraces different procedures and policies to prepare societies for the changing process. These activities can make people adapt to the change and learn how to flourish and develop in this climatic condition. The informative expeditions concerning teaching the people how to get by in warm weather or how to act in the heat waves are considered costly and effective actions. The actions that can cost a very huge amount of money known as "gray adaptation" encompass (a) relocating homes to adapt flooding, (b) building dams, and (c) building huge storing sinks to retain water to adjust to the drought.

(a) (b)

Fig. 5.16 Assessment and predication of global climate change – changes of sea surface temperature and sea surface height. (**a**) Change of sea surface temperature (**b**) Change of sea surface height. (Image credit: Courtesy of Professor Hiroyasu Hasumi presentation "SDGs Symposium 2021: Interdisciplinary science solutions for food, water, climate and ecosystems Sustainable Development Goals" – March 26, 2021), [25]

The green adaptation is another form of adaptation methods, which comprise the use of natural solutions to increase an area's ability to adapt to CC, for example, building trees alongside a river to decrease the effect of flooding [6].

Human response to CC is taking two paths, one is *mitigation* and the other is *adaptation*; it seeks to reduce GHG emissions and eliminate the impacts of CC as shown in Fig. 5.17. Such a cyclic flow chart describes how people interact with the climate system in the following modes:

- "Climate change" leads to "impacts of climate change"
- "Impacts of climate change" lead to "responses to climate change"
- "Responses to climate change"
- "Impacts of climate change" ("adaptation")
- "Causes of climate change" ("mitigation")
- "Causes of climate change" lead back to climate change [6].

Many discussions have been pursued concerning whether the solutions are cost-effective or strain the economies of nations. The solution categories of CC are depicted in Table 5.4 which shows the solution [6, 21]. NASA, as the main source in the studies of CC and Earth's science in general, is concerned with gathering the data around climate change and making them public to the nations around the globe. This will enable the international community, the decision-makers, and the scientists to come up with strategies and policies needed. Nonetheless, the Center for Climate Strategies (CCS) in the USA published a recent study in April 2021 about the projected 2050 net GHG emissions for São Paulo's (Brazil) decarbonization pathway project [26]. Details about São Paulo's decarbonization pathway project is illustrated in Fig. 5.18. It is clear from Fig. 5.18 that the emissions from transportation and industry are expected to be the dominant sectors by 2050 in regard to net GHG emissions. Thus, a broad bundle of mitigation actions across all economic sectors is urgently needed to reach the target of net zero by 2050 [26]. On July 14, 2021, the European Commission (EC) announced that a ban on vehicles gas-powered (Diesel and Benzene) in EU countries will be enforced by 2030 [27].

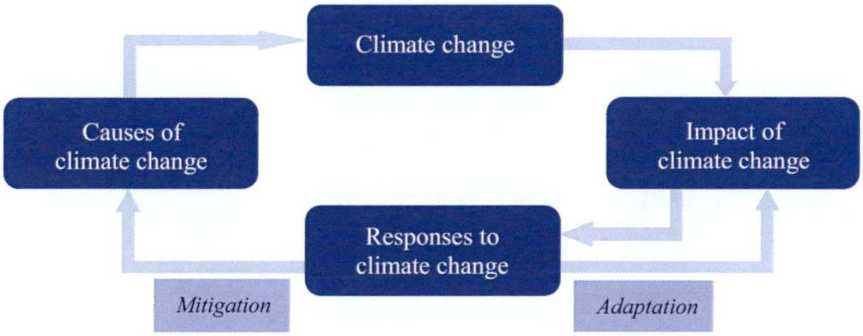

Fig. 5.17 How people interact with the climate system? (Image source: Developed by authors after https://www.e-education.psu.edu/geog438w/node/496)

By looking at the correlation between mitigation and adaptation in Fig. 5.19 which presents the Climate Change Action Plan (CCAP) 2020 in the city of Wonder, Canada [28], it shows the role of corporate and community in the mitigation and adaption of climate change (CC) including three main actions for both. In the same line, the City of Vancouver, Canada, developed a plan for the city and climate emergency which is centered on six big moves in 2020, including (a) walkable, complete neighborhoods, (b) active transportation plus transit, (c) zero-emission vehicles, (d) zero-emission space and water heating, (e) low-carbon materials plus construction planning, and (f) restored coasts and forests [28] as shown in Fig. 5.20.

The first move is part of the city plan, but the other five moves (b–e) are part of the City of Vancouver's climate emergency action plan. In addition, the City of Windsor Municipality in Canada developed objectives and supporting mechanisms in an effort to measure the degree of change in mitigating and adapting to CC as illustrated in Fig. 5.21 [28].

According to Angel Merkel, a chancellor of Germany, climate change (CC) has no boarders; it can manifest from the Pacific Islands, Europe, and Africa as well as North and Latin America [29]. The next section highlights the impact of CC in three different countries in North Africa (Egypt) and Europe (Italy and Germany) in terms of the country's position in relation to CC mitigation and adaptation activities as well as environmental action plans and adaptation strategies and implementation. In addition, it presents how these three selected countries are dealing with CC monitoring, reporting, and evaluation (sectors and actions) as well as the role of stakeholders' engagement along with highlighting the assessments based on policy and legal framework (knowledge and governance), and finally climate action plan (2050).

The following section highlights climate change in three countries (Egypt, North Africa and South Mediterranean; Italy, a European and north Mediterranean country; and Germany, an advanced North European country, leading in terms of climate adaptation policies, strategy, and climate actions for adaptation and assessment as shown in Fig. 5.22).

5.6 Climate Change in Egypt

Egypt is among the countries that ranked highly vulnerable to the CC risk. The weather in Egypt has changed drastically over time (last 1000 years) – from a wet weather to a dry or desert weather (Fig. 5.23). This resulted in the lack of interaction between humans and the environment surroundings due to the dry weather [30]. Despite the fact that Egypt has very low GHG emissions (less than 1% of the total world emissions), Egypt may not have the financial means to adapt to CC making it the most vulnerable country in the Mediterranean due to the following facts [30]:

- Rise of sea level in north coast and the delta.
- Decrease in crop productivity and hence increase in pressure on food scarcity.
- Increase in the need of water and facing the threat of drought.

Table 5.4 Solution categories of climate change

Mitigation and adaptation		Government resources	Energy innovation
Mitigation	**Adaptation**	**Climate data initiative** Data related to climate change that can help inform and prepare America's communities, businesses, and citizens	NASA's test on jet biofuels may reduce climate warming clouds
Practice energy efficiency	Preventive and precautionary measures (evacuation plans, health issues, etc.)	**US climate resilience toolkit** It provides scientific tools, information, and expertise to help people manage their climate-related risks and opportunities and improve their resilience to extreme events	A NASA study confirms that biofuels reduce a jet engine's pollution
Greater use of renewable energy	Research and development on possible catastrophes, temperature behavior, etc.	**National Oceanic and Atmospheric Administration** From supercomputers and state-of-the-art models to observations and outlooks, this site provides data, tools, and information to help people understand and prepare for climate variability and change	Electric airplanes (batteries included)
		National Climate Assessment 2014 Produced by a team of more than 300 experts and guided by a 60-member Federal Advisory Committee, this report summarizes current and future impacts of climate change on the USA	Clean, green climate change-fighting machines
Efficient means of transport implementation: Electric public transport, bicycle, and shared cars	Flexible and diverse cultivation to be prepared for natural catastrophes	**US Department of Energy** It describes strategies currently being pursued or considered to reduce carbon emissions and address global climate change	Power play: Envisioning a wind, water, and solar world
		Environmental Protection Agency Tools for learning and understanding environmental issues and recommendations for greener living	Researchers advance propulsion toward low-carbon aircraft

(continued)

Table 5.4 (continued)

Mitigation and adaptation		Government resources	Energy innovation
Electrification of industrial processes	Landscape restoration (natural landscape) and reforestation	**State of California's climate Change portal** This site provides a large number of links to state reports on climate change mitigation and adaptation options	Satellites shed light on solar system the one on your rooftop
Carbon tax and emission markets	More secure facility locations and infrastructures	**UN framework on climate Change** United Nations framework convention on climate Change online newsletter concerning issues about the UN's convention on long-term climate change	Artificial photosynthesis to make plants green

Source: Developed by authors

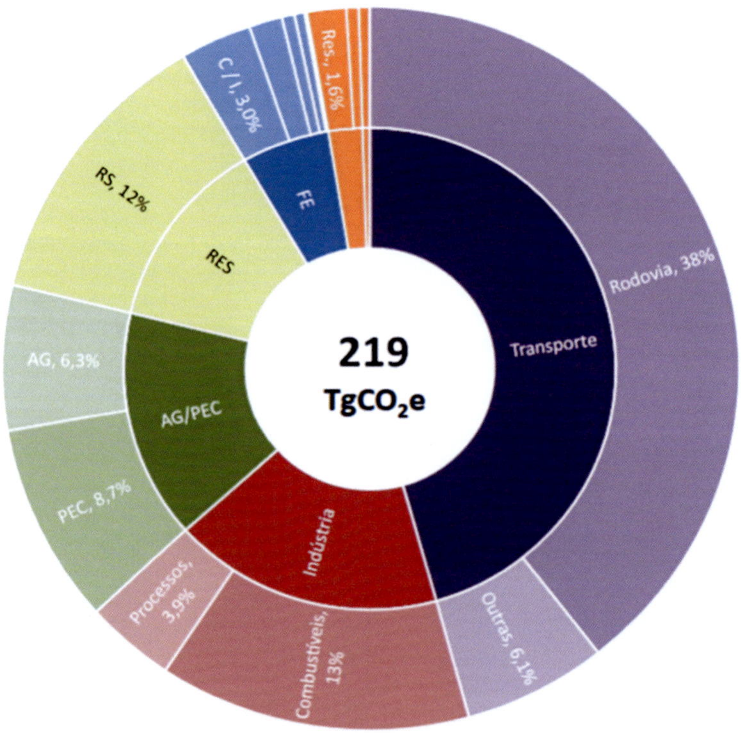

Fig. 5.18 Projected 2050 net GHG emissions for São Paulo's decarbonization pathway project. (Image source: Center for Climate Strategies (CCS), USA, https://www.climatestrategies.us/projects-all/blog-post-title-two-fx6nb)

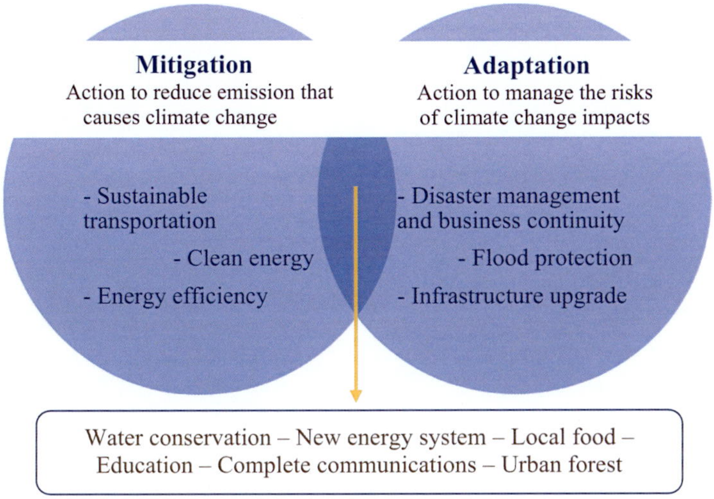

Fig. 5.19 Community role in climate change mitigation and adaptation action plan – City of Wonder, Canada. (Source: Developed by authors after the ninth edition of the Cities Forum Advancing Adaptation and Building Resilience – ICLEI Canada (2020), https://www.livablecities-forum.ca)

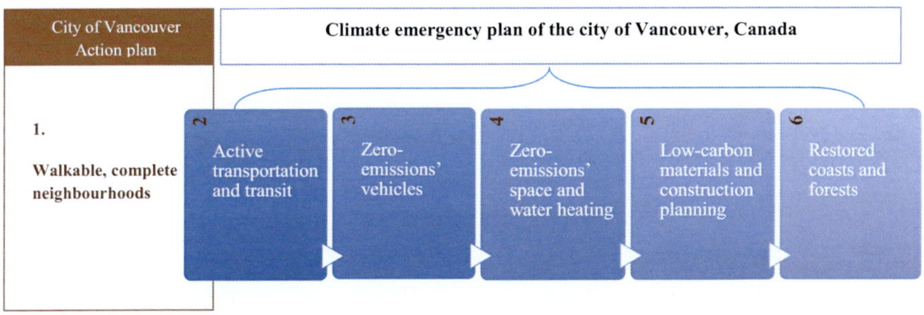

Fig. 5.20 City of Vancouver's action plan and climate emergency plan "6 Big Moves". (Source: Developed by authors after the ninth edition of the Cities Forum Advancing Adaptation and Building Resilience – ICLEI Canada (2020), https://www.livablecitiesforum.ca)

- Desertification of certain fertile lands.
- Unexpected changes in water inflow due to the change in the rain pattern.
- Deficit of water resources.
- Deterioration of biodiversity.
- Coral reef bleaching which affects tourism and touristic areas.
- Effect on public health [4, 31, 32].

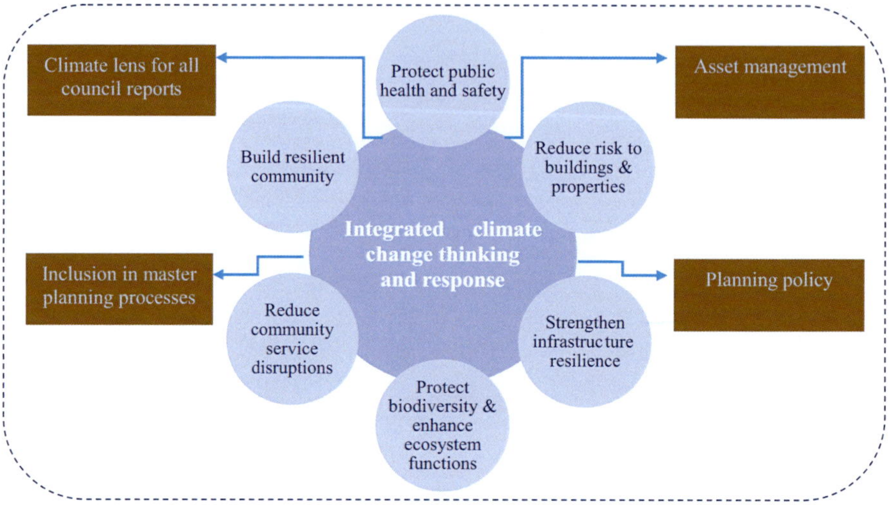

Fig. 5.21 The supporting mechanism developed by the City of Winder, Canada, for climate action plan. (Source: Developed by authors after the ninth edition of the Cities Forum Advancing Adaptation and Building Resilience – ICLEI Canada (2020), https://www.livablecitiesforum.ca)

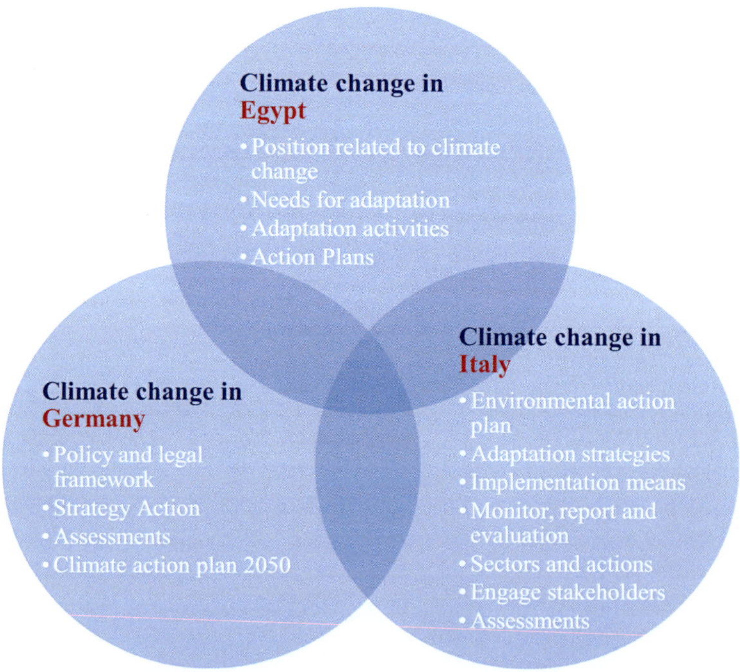

Fig. 5.22 Climate actions in Egypt, Italy, and Germany. (Image source: Developed by authors)

Fig. 5.23 Temperature rise high in Cairo, Egypt, in August 2018 due to climate change. (Image credit: Mohsen Aboulnaga)

Climate change manifestation in Egypt between 2017 and 2021 is depicted in Fig. 5.24, whereas Fig. 5.24 presents the CC solution categories. The model projection of climate hazards under a high emission scenario is illustrated in Fig. 5.26. It is clear from Fig. 5.26 that the harms in the climate resulted from high emission concentration assumption, representing high concentration path in orange [RCP] and low concentration assumption in green. The thick line describes the projection of the changes over 20 models. These values demonstrate each model solely as well as the shaded 90% of the model that measures the level of uncertainty and, if available, the annual observed records in blue. Figure 5.26 also presents days with extreme rainfall and days with rainfall and drought [31].

5.6.1 Emissions and Commitments

All countries must come together with a series of actions to reduce the increased global levels of carbon dioxides, which elevated about 80% between the years 1970 and 2010. Nevertheless, the global CO_2 levels reached a new millennial high in 2020 (414.5 ppm) despite the impact of the coronavirus (COVID-19) pandemic, which leads to a slowdown of traffic and industry [33].

Fig. 5.24 Climate change manifestation in Egypt between 2017 and 2021. (**a**) Heavy rain in the streets of Alexandria, 2017, (**b**) Floods in Alexandria streets, (**c**) November 2020, Severe storms caused havoc in Alexandria, 2020, (**d**) Torrential rain in Cairo streets, March 2020, (**e**) The Nile River after the "Dragon Storm" in mid-March 2020**,** (**f**) People drive over a bridge through fog/ very cold weather, amid the COVID-19 pandemic in Cairo, 2021. (Images' credit and source: (**a**) Al-Masry Al-Youm, https://egyptindependent.com/heavy-rain-in-alexandria-and-halayeb-on-monday; (**b**) and (**c**) Yara Sameh, https://see.nes/in-photos-bad-weather-hits-egypt; (**d**) M. Hozyen, https://commons.m.wikimendia.org/wiki/File:Ride_under_the_Rain_jpg; (**e**) https://en.wikipedia.org/wiki/Climate_of_Egypt; and (**f**) https://upload.wikimedia.org/wikipedia/commons/2/24/Morning_Fog%2C_Shubra_ElKheima_%282122704168%29.jpg)

To mitigate the effect of CC and reduce the hazard on public health, decision-makers should be provided with various data from all the sectors affecting CC such as cultivation, industry, and transportation sectors [31]. The maximum elevation in the temperature that can be tolerated compared with the previous level before the industrial revolution is 2 °C. Any increase above that level will have enormous destructive effects. To stay below this level, all countries should decrease the amount of CO_2 emitted in the air from 5.2 tons per capita to 1.6 tons per capita. Figure 5.27 presents the days with extreme rainfall and drought, while Fig. 5.28 demonstrates

Fig. 5.25 Categories of climate change solutions. (Image source: Developed by authors)

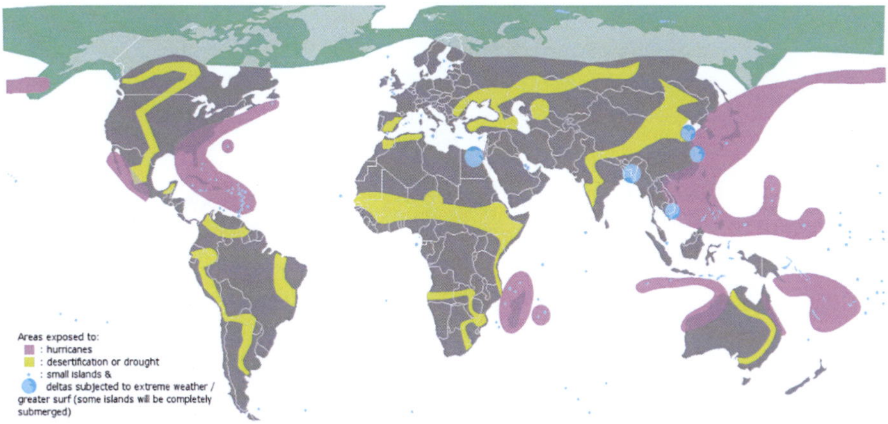

Fig. 5.26 Natural disasters caused by climate change hazards' scenarios due to high emissions – Egypt's delta among the threatened global zones. (Image credit and source: KVDP https://commons.m.wikimedia.org/wiki/File:Natural_disasters_caused_by_climate_change.jpg)

the increase of CO_2 emissions in Egypt from the year 2000 across all sectors [31]. The health community should come up with significant actions to mitigate GHG emissions and their harmful impacts. However, the predicted temperature scheme across Egypt showed that there would be a rise in all seasonal temperatures. This will require a course of actions to adapt the ecosystems and the agricultural sector to the upcoming CC [31]. The national responses have been manifested in many forms in the past years as listed in Table 5.5.

5.6.2 Egypt's Position Related to Climate Change

The impetus of CC is manifesting across the globe, causing serious devastation to cities' infrastructure and sustainable development. Egypt has developed different Intended Nationally Determined Contributions (INDCs) in response to the UNFCCC's call for governments for mitigation and adaption. The most important is presented in 2015 at COP21 – Paris Climate Agreement – which has come into effect in 2020 [34]. There are responsibilities on rich and developed nations toward

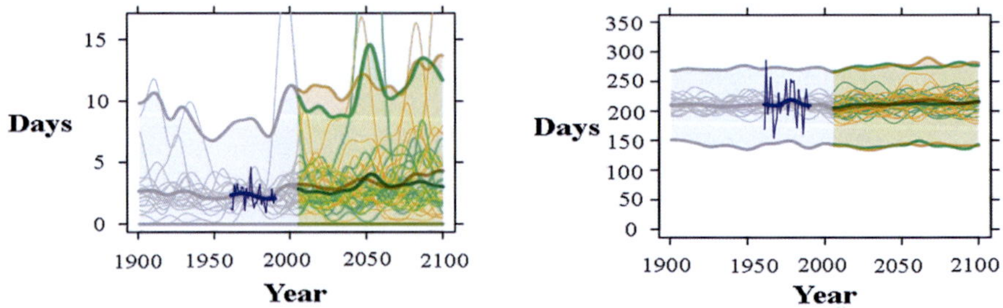

Fig. 5.27 Days with extreme rainfall and drought. (**a**) Days with extreme rainfall "flood risk," in a high emission scenario, the number of days with extreme rainfall (20 mm or more) could increase by about 2 days on an average from 1990 to 2100, increasing the risk of floods. Some models indicate larger increases, but year-to-year variability is very large. Due to variability, projected changes in flood risk are not so different under a low emission scenario (**b**) Consecutive dry days "drought," under both high and low emission scenarios, the longest dry spell is not indicated to change much from an average of about 210 days, with a continuing large year-to-year variability. (Image source: http://apps.who.int/iris/bitstream/10665/208860/1/WHO_FWC_PHE_EPE_15.06_eng.pdf)

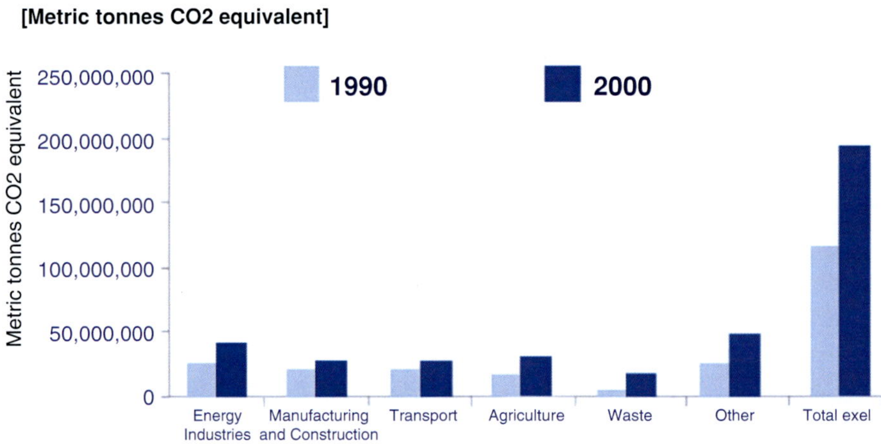

Fig. 5.28 Annual greenhouse gas (GHG) emissions – metric tons CO_2 equivalent in Egypt. (Image source: Developed by authors after UNFCCC Greenhouse Gas Data Inventory, UNFCCC – 2015)

developing nations including GHG discharge decrease, technological cooperation, adjustment fund, research and perception frameworks, and tools of measuring the response and repay of the developing nations for the atmosphere's antagonistic effects [35]. Egypt respects the continuation of arrangements with the created nations for future duties post 2012 and also believes that the issue of technology transfer is important. Hence, the implementation of national communications requires significantly increased financial support [21, 35]. In addition, climate change is presenting risks to seven main sectors in Egypt as illustrated in Fig. 5.29

Table 5.5 National responses for climate change

Year	Action
1992	Egypt signed the United Nation Framework Convention on Climate Change (UNFCCC)
1990	Egypt signed the Kyoto protocol
2010	National Environmental, economic and development study (NEEDS) for climate change
2011	National Strategy for adaptation to climate change and disaster risk reduction, The Egyptian Cabinet and Decision Support Centre (ECDSC) and UNDP
2015	National NDCs (mitigation and adaptation and MOI) in line with Egypt's SDS (sustainable development strategy) 2030
2016	National NDCs in line with Egypt's Sustainable Development Strategy (SDS) 2030
2017	National NDCs in line with Egypt's SDS 2030 – submitted to UNFCCC
2018	National NDCs in line with Egypt's SDS 2030 – submitted to UNFCCC
2019	National NDCs in line with Egypt SDS 2030 – submitted to UNFCCC
2020	National NDCs in line with Egypt's SDS 2030 – submitted to UNFCCC

Source: Developed by authors
Source: Developed by authors after UNDP – https://www.ndcs.undp.org/content/ndc-support-programme/en/home/impact-and-learning/library/indc-project-actions-and-impacts%2D%2Degypt.html

[36]. Therefore, it is vital to prioritize policies and actions for sustainable growth and inclusion of green infrastructure [37].

5.6.3 Egyptian Needs for Adaptation

There is a necessity to focus on the adaptation in Egypt [35]; it has been chosen by the UNFCCC, among ten nations, to present its National Environmental, Economic and Development Study's (NEEDS) report as an appraisal for the requirements of agriculture segment and coastal regions under adaptation, while it covers the necessities of industry under mitigation. In the meantime, there is an urgent need to find fund to provide technology, finance, and capacity building of adaptation measures in Egypt. The implementation of national communications requires significantly increased financial support. The use of mathematical models that predict risks on identifying adaptation options to cope with CC impacts that Egypt's government needs – to protect the vulnerable areas and identify needed adaptation policies – have prioritized a plan to address the CC risks and disasters' reduction and the building's capacity in the vulnerability assessment to evaluate costs of adaptation activities. Egypt believes that the issue of technology transfer is vital, especially supportive technology that increases adaptive capacity, including (a) the protection of the north coast and coastal area of the Nile Delta and (b) crops tolerant to increasing temperature, soil salinity, and water stress and livestock tolerant to high temperature [33, 35].

The Paris Climate Agreement (COP21) and COP22, COP23, COP24, and COP25 added impetus to the necessity for roadmaps to boost climate change adaptation

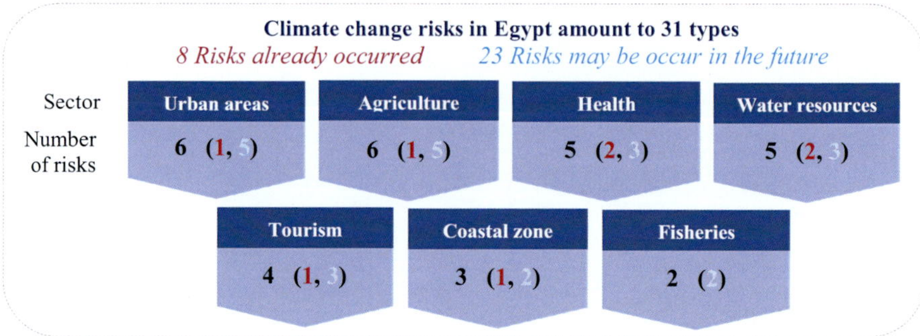

Fig. 5.29 Climate change risks in Egypt by sector. (Source: Mohsen Aboulnaga et al. [37])

(CCA) policies and actions toward strengthening climate resilience in cities [38, 39]. Global policies play a vital role in CCA, including that for energy, urban areas and infrastructure, human health as well as tourism, agriculture, and mobility and transportation [36]. Nonetheless, barriers to CCA are development, financial, information, technological, and most importantly sociocultural [36].

5.6.4 Adaptation Activities

For the adaptation activities, adjustment endeavors incorporate the readiness for the CC that are probably going to occur regardless of whether or not a 2 °C ascend in average temperatures is accomplished, among them building dams, moving populaces, and making arrangements for more productive water and power utilities [35]; this encompasses the following:

- The development of the National Strategy for Adaptation to Climate Change (NSACC).
- Conducted projects for NEEDs and financial assessment to address CC.
- The Climate Change Risk Management Program (CCRMP), a national planned program to implement the standard of adaptation and mitigation to CC harmful effects and to aid Egypt in the CC issues.
- The projects of CC that target the studying of various aspects of mitigation and the effects of CC on the Middle East.
- Estimations and procedures to adapt the elevated sea level.

Moreover, the World Bank recently approved US$ 200 million to support CC projects and climate change adaptation (CCA) policies targeting air pollution in greater Cairo – the 6-year "Greater Cairo Air Pollution Management and CC Project" [40]. This project is intended to combat the annual cost of air pollution on health which forms 1.4% of Egypt's economic output [41].

5.6.4.1 Climate Change Adaptation Policies

Climate change adaptation policies (CCAPs) are classified into two categories: vulnerability reduction and resilience enhancement [42]. These CCAPs were developed in five sectors [36]. The CCAPs and urban development are also measured by sector as listed in Table 5.6 and Fig. 5.30. However, examples of the national CCA measures are highlighted as follows:

(a) Urban and coastal areas.

- The use of renewable energy is one of the adjustment measures, but RE is 3% of the 2018 total energy mix but to reach 20% in 2022 and 42% in 235.
- Adoption of energy code.
- Use of environmentally friendly construction materials.
- Reuse of waste water and recycle of solid waste.
- Replacement or renovation of vulnerable buildings and infrastructure in urban and rural areas to be more resistant against climate extreme events.
- Use of materials that are resistant to high temperature and depend on eco-transportation.

(b) Ecosystem and tourism.

- Implementation of integrated environmental management systems in touristic sites and marine and wildlife protectorates.

(c) Water.

- Water recycling.
- Improvement of distribution network.
- Reuse of gray water in irrigation.

Table 5.6 CCA and urban development measures in Egypt by sector

Sector	Number of measures/actions
Building	7
Costal zones	8
Roads	6
Tourism	5
Agriculture	6
Livestock production	3 (a new measure enacted on December 5, 2020)
Fish wealth	3 (3 new measures were enacted in 2017, 2019, and 2020)
Health	4 (increased in the time of COVID-19 pandemic)
Tourism and ecosystem	3
Water resources and irrigation	6

Source: Aboulnaga et al. [35]
*Increased in the time of COVID-19 pandemic

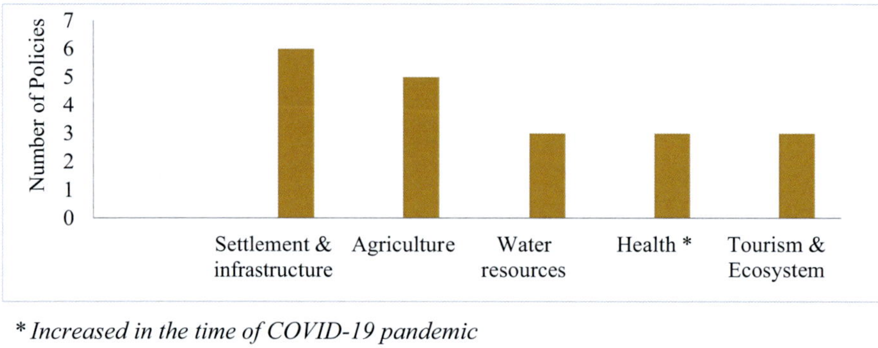

Fig. 5.30 Climate change adaptation policies in Egypt – 2017. (Image source: Developed by authors)

5.6.4.2 Renewable Energy for Green City Development: CCA

Between 2008 and 2014, the Egyptian government was subsiding electricity in Egypt, and the tariff was at a very low cost to the extent that citizens were not concerned with high energy use. Nonetheless, the government enacted policies including electricity and renewable energy laws in 2014 to liberate the energy prices and gradually increase the tariff over a span of 5 years [43]. In order to encourage the private sector to produce clean electricity from renewable energy (RE) sources, Egypt has issued many laws and green bonds to promote the production of RE [44].

5.6.4.3 CCA Policies: Climate Actions Implemented in City of Hurghada and City of Luxor

A policy design project on sustainable energy and climate action "Cleaner Energy Saving Mediterranean Cities – CES-MED" – was realized in 2018 to support South Mediterranean countries in establishing Sustainable Energy and Climate Action Plans (SECAPs) [45, 46]. The EU-funded project was coordinated with the Ministry of Foreign Affairs and Ministry of Local Development in line with the CCA national strategy and sectoral policies.

5.6.4.4 Climate Change Adaptation: City of Hurghada and City of Luxor, Egypt

The City of Hurghada and the City of Luxor realized a self-assessment of its adaptation status, putting a grade from A to D, in line with its progress regarding the Adaptation Cycle Steps (ASSs) developed by the Joint Research Center (JRC) of the European Commission [47] as well as meetings and interviews with Governorate of Red Sea officials in 2017. Tables 5.7 and 5.8 list the results of ASSs in the City of Hurghada and the City of Luxor, Egypt. For the Municipality's Score in the CCA Cycle Specific Steps in the City of Hurghada and the City of Luxor (2017), the grades A–D correspond to completion level in the city:

- "A" 75–100%
- "B" 50–75%
- "C" 25–50%
- "D" 0–25%.

The CCA Citizen Awareness Raising campaign proved to be effective as part of the CCA policy as the CCA action score of Hurghada is higher than that of Luxor, but these figures reflect the action up to the year 2017.

5.6.4.5 Implemented Public CCA Policies' Effectiveness

The policies' analysis for CCA, specifically those concerning the energy sector, indicates the effectiveness of these policies due to the reduction of the shortage in supply and the increase of the electricity surplus; yet it led to a vast saving due to less consumption in the residential buildings. The enacted policies and laws also led to advancing the renewable energy production capacity, mainly solar PV and wind power generation of a total of 5849.5 MW (Fig. 5.31). By the end of 2019 (Fig. 5.31a), it reached a total of 5955.5 MW – equal to 20,780 million kWh of total energy produced and saving of fuel 4.3 mtoe (million tons of oil equivalent) [44, 48]. For mitigating CO_2 emissions and air pollution, such policies and laws yield 11.4 million tons of CO_2 reduction in addition to creating 292,840 new jobs by 2018 (Fig. 5.31b). Consequently, the surplus of electricity is exporting to various countries since 2020.

5.6.5 Mitigation Activities

The mitigation activities in Egypt include the reduction of the amount of GHG emissions in the atmosphere by switching to other low-carbon generating power sources in sectors such as energy, industrial, agricultural, and structural as well as transportation and land-utilizing sectors. The mitigation activities have also been extended to the following:

- Developing green areas and planting forests.
- Conducting the National Strategy for Clean Development Mechanism (CDM) in 2002 followed by the Capacity Development Project in CDM (2003–2006).
- Establishing the Egyptian DNA for CDM in 2005.
- Approving 94 projects in REEE (Renewable Energy and Energy Efficiency), industrial processes, fuel switching, and solid waste.
- Total investment cost \$5.04 billion and total GHG reduction 11 mmtCO_2/year.

Table 5.7 City of Hurghada, Municipality's score in the CCA Cycle Specific Steps – 2017

Adaptation cycle steps	Actions	Grades A–D
Step 1: Preparing the ground for adaptation	1. Adaptation of defined/integrated commitments into the local climate policy	A[a]
	2. Identification of the human, technical, and financial resources	B[a]
	3. Adaptation team appointed within the municipal administration and assignment of clear responsibilities	D[a]
	4. Horizontal (e.g., across departments) coordination of/between mechanisms in place	A
	5. Vertical (e.g., across governance levels) coordination of/between mechanisms in place	A–B
	6. Setup of consultative and participatory mechanisms, fostering the multi-stakeholder engagement in the adaptation process	B
	7. Prolongation of in-place communication process	A
Step 2: Assessing risks and vulnerabilities to climate change	8. Mapping of the possible methods and data sources for carrying out a conducted risk and vulnerability assessment	A
	9. Assessment of climate risks and undertaken vulnerabilities	D
	10. Identification and prioritization of possible sectors of actions	C[a]
	11. Review of available knowledge periodically and new integrated findings	B
Steps 3 and 4: Identifying, assessing, and selecting adaptation options	12. Setup of a full portfolio of adaptation actions compiled, documented, and assessed	D
	13. Examination of possibilities of mainstreaming adaptation in existing policies and assessed plans, possible synergies, and identified conflicts	B
	14. Development of adoption of adaptation actions	C–B
Step 5: Implementing	15. Implementation of a framework set with clear milestones	C
	16. Implementation and mainstreaming of adaptation actions as defined in the SECAP document	N/A
	17. Coordination of actions between adaptation and mitigation set of measures	C
Step 6: Monitoring and evaluation	18. Monitoring framework in place for adaptation actions	C–D
	19. Appropriate monitoring and evaluation of identified indicators	C–B
	20. Regular monitoring of the progress and reporting to the decision-makers	A
	21. Adaptation strategy and/or action plan updated, revised, and readjusted according to the findings of the monitoring and evaluation procedure	D

"A" 75–100%, "B" 50–75%, "C" 25–50%, and "D" 0–25%
[a]Grades A–D correspond to completion level in the city

Table 5.8 City of Luxor, Municipality's score in the CCA Cycle Specific Steps – 2017

Adaptation cycle steps	Actions	Grades A–D
Step 1: Preparing the ground for adaptation	1. Adaptation of defined/integrated commitments into the local climate policy	D[a]
	2. Identification of the human, technical, and financial resources	C[a]
	3. Adaptation team appointed within the municipal administration and assignment of clear responsibilities	D
	4. Horizontal (e.g., across departments) coordination of/between mechanisms in place	D
	5. Vertical (e.g., across governance levels) coordination of/between mechanisms in place	C
	6. Setup of consultative and participatory mechanisms, fostering the multi-stakeholder engagement in the adaptation process	D
	7. Prolongation of in-place communication process	B[a]
Step 2: Assessing risks and vulnerabilities to climate change	8. Mapping of the possible methods and data sources for carrying out a conducted risk and vulnerability assessment	C
	9. Assessment of climate risks and undertaken vulnerabilities	C
	10. Identification and prioritization of possible sectors of actions	D
	11. Review of available knowledge periodically and new integrated findings	C
Steps 3 and 4: Identifying, assessing and selecting adaptation options	12. Setup of a full portfolio of adaptation actions compiled, documented, and assessed	D
	13. Examination of possibilities of mainstreaming adaptation in existing policies and assessed plans, possible synergies, and identified conflicts	D
	14. Development of adoption of adaptation actions	C
Step 5: Implementing	15. Implementation of a framework set with clear milestones	D
	16. Implementation and mainstreaming of adaptation actions as defined in the SECAP document	D
	17. Coordination of actions between adaptation and mitigation set of measures	C
Step 6: Monitoring and evaluation	18. Monitoring framework in place for adaptation actions	D
	19. Appropriate monitoring and evaluation of identified indicators	D
	20. Regular monitoring of the progress and reporting to the decision-makers	D
	21. Adaptation strategy and/or action plan updated, revised, and readjusted according to the findings of the monitoring and evaluation procedure	D

"A" 75–100%, "B" 50–75%, "C" 25–50%, and "D" 0–25%

[a]Grades A to D correspond to completion level in the city

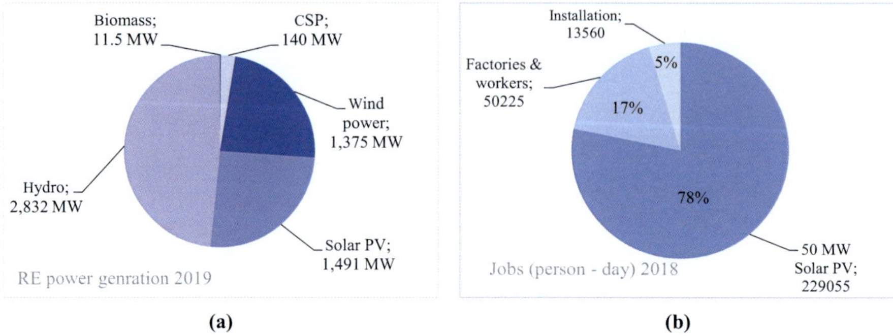

Fig. 5.31 The power generation forms renewable energy and related job opportunities. (**a**) Renewable energy power generation (**b**) Jobs in renewable energy sector by 2014–2018. (Images' source: Developed by authors after https://iclg.com/practice-areas/renewable-energy-laws-and-regulations/egypt)

- Implementing programs for fuel switching to natural gas in energy, industrial, and transportation sectors.
- Establishing the Energy Supreme Council [35].

Nevertheless, COVID-19 impacts will shrink the global economy by 3% due to the lack of climate resilience actions by 2050 [49]. The International Monetary Fund (IMF) confirmed that COVID-19 would drag the world into the same 3% GDP contraction in 2020 [50]. Moreover, COVID-19 has changed urban transportation patterns; for example, in China, the use of private cars and car-sharing has doubled from 34% to 67%, while public buses and metros have experienced a sharp drop from 57% to 4% as shown in Fig. 5.32. Such a decrease in public transportation usage may have adverse consequences on energy use, CO_2 emissions, and air pollution.

Despite the economic slowdown globally, the Minister of Planning and Economic Development (MoPED) announced on July 5, 2020, that Egypt's government has approved 691 green projects (worth EGP 36.7 billion) as part of the COVID-19 resilience plan of the fiscal year 2020–2021 [51], where both the Minister of Planning and Economic Development and Minister of Environment agreed that 30% of Egypt's investments will be implemented as per the environmental sustainability green economy concept. However, the government is planning to bring this percentage to be 100% in the next 3 years [52]. The 691 green projects account for 14% of a total public investment allocation of EGP 447.3 billion for Egypt's green agenda [53].

5.6.5.1 Climate Change Issue: An Egyptian View

The Egyptian view on climate change issue can be summarized in the following points:

- Adaptation is a first priority.
- Urgent needs to establish the Multilateral Fund with the aim to supply the needed advanced technologies, investment, and gage buildings for the adaptation and mitigation processes.
- Egypt accepts the national undertaking of appropriate mitigation actions by developing countries in the context of the National Suppliers Development Programme (NSDP) conditioned with providing technology transfer, finance, and capacity building.
- The necessity to focus on adaptation in Egypt as the most vulnerable country in the Mediterranean region [35].

5.6.5.2 Cooperation with Line Ministries

The past 5 years have witnessed partnership and cooperation between all concerned stakeholders, especially ministries for climate change adaptation in many areas such as energy, infrastructure, agriculture, transpiration, and affordable housing. These can be summarized as follows:

- Many projects and developments in the field of renewable energy resources (solar, wind, bio, hydro) and its applications are established by the Ministry of Electricity and Renewable Energy.
- Projects and procedures to insure the protection of shores and the establishment of research centers were the main focus on the ministry of agriculture and irrigation.
- Agricultural research centers study the effect of climate change on agriculture and the availability of crops in addition to the study of heat-resilient plants [35].

In closing, the role of public policies in fostering climate action and SDGs' attainment in Egypt exhibited a relatively good shift in terms of CCA between 2016

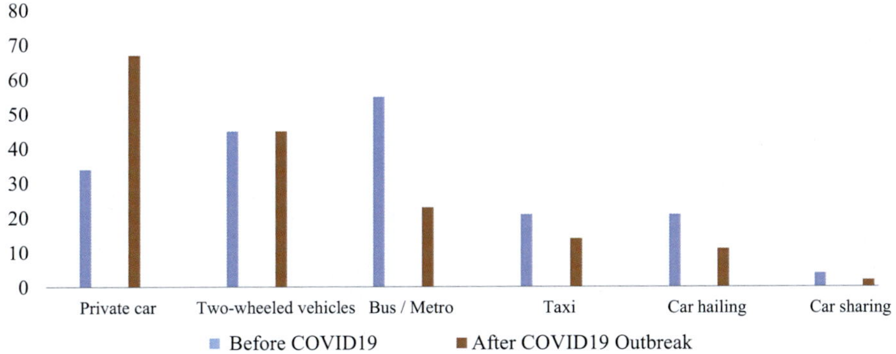

Fig. 5.32 Comparison of transportation (%) before and after the COVID-19 outbreak in China. (Image source: Developed by authors after 13 IPSOS, "Impact of Coronavirus, and Survey of 1620 consumers in China" – March 2019)

and 2020 (Fig. 5.33) and in accordance to the European Bank for Reconstruction and Development's (EBRD) green city policy menu [54, 55]. In addition, the 2 GW solar PV plant in Bin Ban, Aswan, is part of the renewable energy national target of 53% clean energy by 2030 [56] and the New Administrative Capital (a smart city) in the east of Cairo as well as El Almeen City (a green city) and City of Hurghada (green roofs, solar/wind power plants, CCA) and City of Luxor (solar PV, CCA), Tables 5.7 and 5.8, all added to the national climate actions and CCA in energy, water, and building sectors.

It is also imperative to present similar climate change adaptation in European countries and observe similar actions taken by stakeholders including governments, civil society, and citizens. In the next section, climate change in Italy and Germany will be highlighted and discussed.

5.7 Climate Change in Italy

Climate change (CC) is also manifested in Italy like the rest of the world. Climate observations already confirm an increase in the average temperature as well as an upward trend in extreme temperatures. Projections also show an increase in heavy precipitations and extreme temperatures. Italy is prone to natural hazards and CC severe events that are expected to increase its vulnerability to climate-related hazards over the next decades [57]. Italy, a member of the G7, is located in an area identified as particularly vulnerable to CC. It is located in the southern part of Europe and includes the Italian peninsula that stretches from the Alps into the Mediterranean Sea and several islands including Sicily and Sardinia. The Italian peninsula lies between latitudes 47° and 35° N and longitudes 6° and 18°E. Italy's national area is 301,340 km^2, of which 40% is mountainous. Approximately, 10.5% of the national surface land is included in protected areas [57]. Italy's population was 59.4 million in 2011. Compared to 2001, the population grew by 4.3%, mainly due to the migratory movements that counterbalanced the negative natural trend.

Fig. 5.33 Impact of policies, laws, and projects in main sectors based on EBRD's menu of green city policy. (Image source: Developed by authors after EBRD)

Currently, the population in 2020 census is 60,461,826 people (nearly 60.5 million) with 23.2% that are elderly citizens most vulnerable to CC severe events such as heat waves and floods [58]. Between July 31, and August 3, 2017, Italy was hit by a severe heat wave which caused droughts of green areas in Rome and power cut in Florence for 8 h. Figure 5.34 illustrates the impact of CC severe events such as heat wave and droughts in Rome, while Fig. 5.35 shows the heat wave in the City of Florence, Italy, at 4.00 pm, where no one is in streets due to extreme heat (43 °C).

Italy submitted its Nationally Determined Contribution (NDC) as an EU Member State in 2015 and its Seventh National Communication in 2018 [57]. Climate change adaptation (CCA) was also the focus of the Third European Climate Change Adaptation (CCA) conference "Our Climate Ready Future," which was held in Glasgow in June 2017 [57].

Cities are hotspots for measures to mitigate global warming and CC impacts since 74% of the European population lives in urban areas. In addition, cities do not only contribute to a large portion of GHG emission but also are vulnerable to the adverse impacts of CC in Italian cities as shown in Fig. 5.36. Thus, it is imperative that cities implement low-carbon policies and CCA strategies in close collaboration with all stakeholders and citizens [57].

5.7.1 Climate Change Mitigation

Italian cities focus more on mitigation than adaptation due to a national law enacted in 1991 (no. 10/1991) which obliges large Italian cities (> 50.000 citizens) to have an energy plan, mainly aimed at reducing energy consumption rather than reducing the emission of GHGs [59]. As a result, a large number of Italian cities have developed mitigation plans and carried out decarbonization strategies.

(a) (b)

Fig. 5.34 Impacts of climate change in Rome, Italy, on July 30, 2017. (**a**) Affected green areas near Rome Fiumicino Airport due to drought and heat wave, (**b**) Aerial view of the affected areas due to severe heat wave in Rome. (Images' credit and source: Mohsen Aboulnaga)

Fig. 5.35 Impacts of climate change severe event (heat wave) in Florence, Italy, on July 31, 2017. (Image credit and source: Mohsen Aboulnaga)

Currently, under the guidance of the Covenant of Mayors (CoM), nearly 3000 Italian towns and cities have submitted their Sustainable Energy Action Plans (SEAPs) in pledge to mitigate GHG emissions in these towns/cities by at least 20% in 2020 [59]. The Covenant of Mayors (CoM) initiative aims at engaging and supporting cities and towns to commit to reaching the European Union (EU) climate mitigation and adaptation targets. COM is the world's largest movement for local climate and energy action [60].

5.7.2 Climate Change Adaptation

On the climate change adaptation (CCA) side, the situation is completely different. An assessment was carried out by Pietrapertosa et al. [60], which concludes that only two large cities such as Bologna (Fig. 5.37) and Ancona have adaptation plans and address CC impacts on extreme rain events, landslides, coastal erosion, infrastructure, cultural heritage, heat waves, and droughts. These two cities are considered active cities in Italy [60]. Other cities have civil protection plans that focus on these issues as well but do not have an adaptation plan that integrates all impacts of CC [60]. Nevertheless, Italian cities have a reactive approach to making the urban environment more resilient, like taking action after disasters. The essence of CCA, however, is a proactive approach [59].

According to Filomena Pietrapertosa, Italian cities seem to be lagging behind on other European countries "Italian cities have started late due to the lack of a national

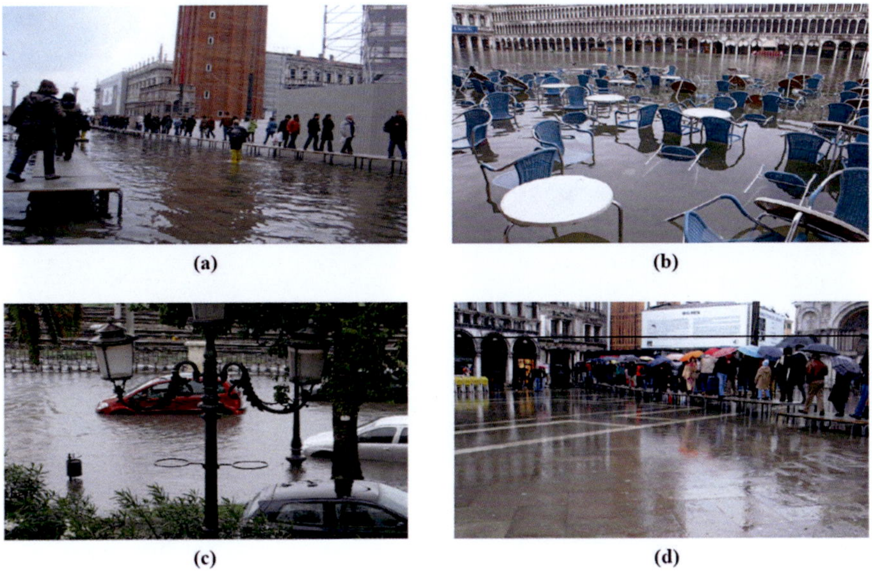

Fig. 5.36 Impacts of climate change on severe events in Italian cities. (**a**) Venice flooding near unprecedented level in 53 years – a climate change impact, (**b**) Empty café in a plaza in Venice amidst severe flooding due to high tides of climate change, (**c**) Severe floods in Calabria due to heavy rain of 450 mm of rain in two days, (**d**) People while crossing St. Mark Square amid severe flood. (Images' source: (**a**) https://commons.wikimedia.org/wiki/File:Flooding_in_Venice.jpg (**b**) https://www.flickr.com/photos/massimo_riserbo/10869670583/in/photolist-hyvUnv-dqc267-7smVgg-8Ye9Rr-7saZH3-dD65Jp-8gUfGh-8F2rwp-7rW9p5-7uWnM5-5MorkJ-5Mjcip-7wB7Gz-doznqk-y3kH-4Mqm8F-doznfB-NFiJK-dsKXkt-8eYPiq-2jsAz56-r8tuT-2kVfbk6-8oqAbv-2hQimzw-2hR9FEA-aEogLx-9eEf8M-b117jV-qAQMV4-9eHmff-fnhbP-zquLj-fc1uKm-2hQkSoU-o72Yj8-7tuyvm-KuTsBe-2hRQtoP-2hQmV9h-8NrSTu-2hKdMSo-rD84Ae-r8t6L-8ajvet-drqaQT-jM5fWM-nAB1ik-cvNTbm-5EB7R5 (**c**) https://upload.wikimedia.org/wikipedia/commons/5/5c/Flood_-_Via_Marina%2C_Reggio_Calabria%2C_Italy_-_13_October_2010.jpg (**d**) https://www.flickr.com/photos/mrcrash/284369877/in/photolist-r8tfV-8Ye9Rr-dD65Jp-8WC2WV-2m2VmLD-d6zUV-hmYN2e-4Mqm8F-q9PAXz-2m2TLVD-7uWj2d-2i5pfSi-2351bbx-4EC5DK-fnhbP-dLABTg-2351cdH-4EgeEP-2351c6D-jM5fWM-2351bJr-NyMorG-amjK1P-2351bUM-8W2Ny6-8PuvuJ-7ubHxs-up8yYt-drteiS-2jdbKkZ-dto9F5-6E7ydk-8YmBrN-2sJwWP-4BEyNK-r8t6L-bqZ5g1-29iZvnt-bqZ5as-2ix6S7D-bqZ7hJ-NwvbCd-87d8VD-bDU3sz-5kXWzr-bDU69g-2RMBP6-PwbUAR-8FEePh-bDU2sK)

adaptation strategy, which was only approved in June 2015." Pietrapertosa explains that related to this are the lack of awareness of the possible impacts of CC and the lack of money [61]. When it comes to CCA, it is not mandatory to draw up a plan for CCA, but *it depends on voluntary initiatives to what extent cities take action.* Climate change will probably hit the south of Europe hardest. As a result, cities in the south of Italy seem to be more vulnerable to CC than those in the north [59]. In Italy, the protection of the environment, ecosystem, and cultural resources is under the exclusive competency of the central government. The Ministry for the

(a) (b)

Fig. 5.37 City of Bologna seeks climate change adaptation plans due to consequent heavy rainfall events and heat waves. (**a**) San Petronio Piazza Maggiore viewed from Asinelli Tower (**b**) View of Piazza Maggiore, Bologna Vista. (Image credit and source: (**a**) Goldmund100 (Luca Volpi) https://commons.m.wikimdia.org/wiki/File:Bologna_seen_from_Asinelli_tower.jpg (**b**) Gaspa, https://pt.m.wikipedia.org/wiki/Ficherio:Bologna-vista02.jpg)

Environment, Land and Sea (IMELS) carries out activities on CC at the national and international levels [62].

5.7.3 The Environmental Action Plan and Adaptation Strategies in Italy

The growth rate of the world's population indicates an increasing trend although the pace is slowing down, while some countries, such as Italy, have come to a standstill. The most industrialized countries constantly deplete resources, such as water and energy from the environment. Such countries also spread pollutants and waste and thus irreversibly affect the quality of land, air, and water. However, while economic systems need an ever-increasing growth, the environment demands balance and stability. In order to attain sustainable development (SD), the self-subsistence and self-organization of ecosystems must coexist with anthropic processes; otherwise, their imbalance will in turn lead to self-induced entropy [62].

According to the European path for adaptation, the IMELS adopted in 2015 the National Adaptation Strategy (NAS) to CC. The NAS is a tool for encouraging adaptation actions in planning activities at national, regional, and local levels. After approval from the "State-Regions Unified Conference" on October 30, 2014, the NAS was finally adopted in June 2015 by a Directorial Decree of the Climate and Energy General Directorate, establishing specific objectives to be reached by December 31, 2016 [59].

The Directorate General for Climate and Energy of IMELS has been working for the implementation of the NAS through the Italian National Adaptation Plan – NAP – for CC. The NAP provides institutional guidance to national and local authorities for the elaboration of regional strategies or plans and for the integration of climate change adaptation within spatial and sectoral planning [59]. The NAP has been developed with the support of the Euro-Mediterranean Center on Climate Change (CMCC). During 2017 and the first half of 2018, IMELS carried out a check on the Draft NAP by a consultative process involving national, regional, and local authorities, universities, research centers, and other stakeholders. The approval of the NAP is occurring through an agreement with the State-Regions Conference [59].

5.7.3.1 Sustainable Development and Environment

The population of industrialized countries exploits ten times more natural resources per capita than those living in developing countries. The economic and demographic growth of developing countries, entailing new consumptions, can only resort to natural resources. On the other hand, nowadays and in the years to come, wealth standard requires unpolluted air, water, and food, unspoiled landscapes, enticing seawaters and shores, and towns wisely striking a balance between their huge historical heritage preservation needs and their ever-changing operational and organizational requirements [61].

At the end of the second millennium, our civilization conceived sustainable development (SD), and designed to meet the present needs while preserving the expectations and requirements of future generations, pursuing at the same time different aims, such as quality of life, peace, an equitable wealth, and a clean and healthy environment. This is not a brand-new concept, but many cultures have thoroughly accepted the need to strike a balance among different economic, social, and environmental requirements. Today, this standpoint has been restated in industrialized and developing countries eventually aware of the exhaustibility of global natural resources. However, SD urges a changeover of growth patterns and socioeconomic relations [61].

5.7.3.2 Environmental Action

New sustainability-oriented projects are most needed in order to reset ecological balances, change consumption and production patterns, promote ecological efficiency, and restore social equity conditions. The environmental action, as part of this overall planning, aims at easing frictions between the anthropic sphere and the environmental action strategy for the SD in Italy.

The environmental action plan in Italy includes many positive measures: (a) eliminating the exploitation of nonrenewable natural resources, (b) phasing out pollutants and enhancing the value of waste through its reuse, (c) recycling and recovery of both energy and secondary raw materials, (d) changing the balances of generation and absorption of GHG emissions, (e) interrupting biodiversity erosion, and (f) desertification processes and safeguarding landscapes and habitats [63]. Although few positive results have been achieved, Earth warming, biodiversity shrinkage, disposal of household and hazardous waste, and poor quality of urban areas, as well as increasing exploitation of natural resources and the spreading of environmental criminality, have become crucial issues, thus jeopardizing the environment [63].

Sustainable development (SD) cannot be based on a mere environmental action nor can it promote only bans, rules, and limits. A sustainable economy must be focused on a long-lasting stable development, including high employment rates, low inflation rates, and international trade stability. Economic growth is generally measured by the GDP rate, which does not register any environmental damages and thus their social costs [61].

The time has come, especially in the age of COVID-19, to measure the economic growth and recovery initiatives by a wide range of physical and monetary parameters integrating environmental and economic factors so as to show the achieved results in protecting the environment and the quality of life as well as the stock decrease or increase of natural resources. Social sustainability should also be integrated and should include distributive equity; human and civil rights; social conditions of children, teenagers, women, elderly, and disabled people; and immigration and cooperation between countries.

The SD targeted actions and commitments are closely linked to the implementation of policies to eliminate social exclusion and poverty. Such objective, as expected by the 2001–2003 NAP of interventions and social services, can be achieved through a fair distribution of resources, a decrease of the unemployment rate, and an accomplishment of economic measures by means of investments in the national health system, in education, and in social programs that guarantee the access to services and social cohesion [61]. The inspiring principles of the environmental action strategy for SD in Italy include ten principles: (a) integration of environmental issues into other policy making processes, (b) the preference for an aware economic and environmental lifestyle, (c) an increase in the global efficiency of resource usage, (d) refusal of the "end cycle" intervention approach and promotion of prevention policies, (e) general waste reduction, (f) stretching the lifetime of goods, (g) ending of material cycles of production consumption, (h) development of local markets and local productions, (i) promotion of typical products and traditional cultures, and (j) involvement of social parties in setting goals and commitments and sharing responsibilities [61] as summarized in Fig. 5.38.

5.7.4 Implementation Means

The Italian National Adaptation Plan (NAP) provides an indication of roles for implementing integrated adaptation actions. It also defines the possible sources of funding and assessing the necessary financial and human resources. A possible financial resource for adaptation can be obtained from the application of the Directive 2003/87/CE – Art.3 – concerning the setting up of "carbon market" for trading CO_2 allowances. The revenues of the auctions can be used, up to the extent of 50% of the total, to support adaptation actions [59].

5.7.5 Monitoring, Reporting and Evaluation

A monitoring system, in order to evaluate progress in implementing the NAS, is not yet in place, neither at the national nor regional level. However, the ongoing process toward the NAP, which started in 2016, is currently taking into consideration the need for a monitoring, reporting, and evaluation (MRE) system. Guidelines and indicators are therefore defined in order to monitor the state of implementation and the effectiveness of adaptation actions [59]. In 2016, the National System for Environmental Protection (NSEP), which is composed of the National Institute for Environmental Protection and the Regional Environmental Protection Agencies (Law No. 132/2016), set up a national working group on "impacts, vulnerability, and adaptation to climate change." The main objective is to define a set of CC impact indicators, according to the knowledge gaps currently existing in Italy. Furthermore, a first portfolio of 150 potential CC impact indicators has been identified and a prioritization phase, and it is currently being implemented. The resulting indicators will build therefore a quantitative knowledge based on the impacts of CC as a reference framework for the MRE [59].

5.7.6 Sectors and Actions

The areas for actions in the Italian National Adaptation Strategy (NAS) were selected according to a sectorial approach, based on socioeconomic relevance and vulnerability to the impacts of CC. The sectors considered in the NAP are coherent with the NAS [59].

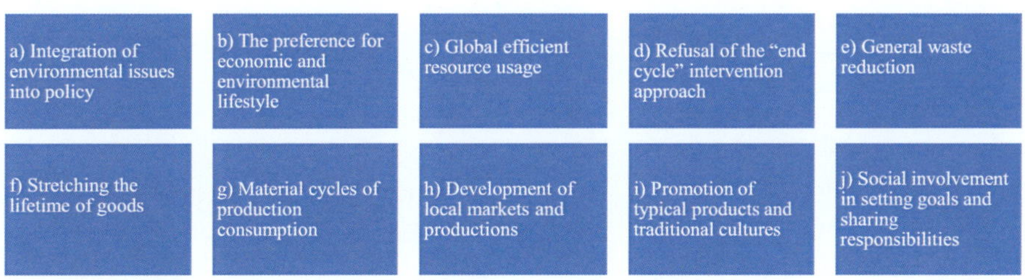

Fig. 5.38 Inspiring principles of the environmental action strategy for SD in Italy. (Image source: Developed by authors after [62])

5.7.6.1 Sectors Addressed in NAS/NAP

There are 18 sectors considered in Italy's Draft NAP as shown in Fig. 5.39. The Draft NAP provides a set of adaptation actions for the different sectors taken into consideration. The actions were evaluated through a set of selected criteria, and a subset of relevant actions were identified. These actions are focused on biodiversity, coastal zones, hydrogeological instability, and urban settlements, even though they are connected to all sectors of the NAP [59]. The MELS supports adaptation actions through the following measures:

- Notice on small landslides: hydrogeological risk reduction due to mountain municipality landslides in order to promote climate change adaptation.
- National program on resilience increase of natural and seminatural forestry systems, through the recovery and the structural and functional restoration of ecosystems and their service functionality and by consistent actions with the biodiversity protection and conservation (flora, fauna, vegetation, natural, rural landscape) in the protected areas crossed by the fire.
- Measures for energy efficiency, sustainable mobility, and CCA in the small islands [59].

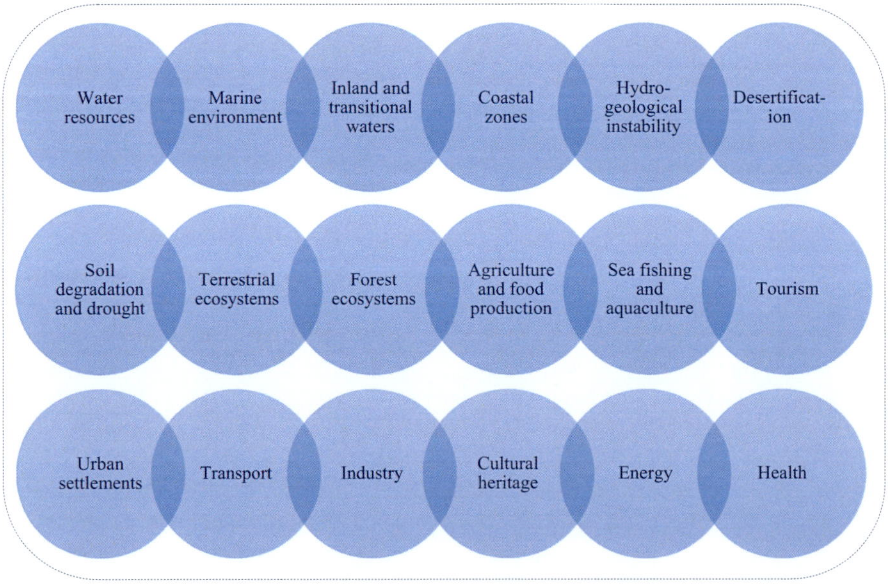

Fig. 5.39 Sectors addressed in the NAP, Italy. (Image source: Developed by authors after [60])

5.7.6.2 Mainstreaming of Adaptation

Under the European cooperation for regional development programs (ERDF), the IMELS coordinated a project to support the regional environmental authorities of four Italian regions (convergence objective regions: Calabria, Sicily, Campania, and Apulia) to implement the principles of the NAS in future regional adaptation plans. A report on the state of climate change adaptation for each region has been released in October 2016 [59]. Moreover, the framework of the Italian project *Competenze e Reti per I'ntegrazione Ambientale e per il Miglioramento delle Organizzaziono della* (CReIAMO) (skills and networks for the environmental integration and for the improvement of the organization of the CReIAMO) aims to better organize the public administration. Through these activities, the IMELS started mapping the the current state of adaptation in the Italian regions [66]. Moreover, Lombardy, Emilia-Romagna, and Sardinia approved their regional adaptation strategies [59].

5.7.7 Engaging Stakeholders

Engaging stakeholders is a vital process for CC adaptation strategy. The idea of involving stakeholders, mainly institutions (governmental and regional entities), defines the adaptation strategy and the plans that contributed to improving the information exchange between central government and local government (municipality offices) and raising awareness about the need of adaptation planning to curb CC impacts [59].

5.7.7.1 Governance

For implementing the Italian NAP, governance is a vital tool to ensure adaptation compliance. The crucial elements for the elaboration of the Italian NAS were the following:

- The establishment of a scientific panel.
- The establishment of an Institutional Panel of Ministries (IPoMs), national and regional authorities, and local authorities.
- A national public consultation [65].

A similar approach was followed in the Draft NAP definition. Precisely, the IMELS established an interregional panel with the representative designed by the Environment and Energy Commission of the Conference of Regions (EECCoRs) and an interministerial panel. The first goal of these panels was to (a) build a ground base for adaptation and (b) evaluate the state of the implementation of adaptation actions at national, regional, and local levels. To strengthen the Draft NAP, it was subject to a scientific review, and two consultations were carried out [59].

5.7.7.2 Knowledge

In order to provide access to climate change adaptation (CCA), relevant information, and data and documents to support local decision-making, a national platform is currently under preparation. In 2018, the IMELS commenced the activities of action L5: strengthening of administrative capacity for climate change adaptation, with the aim to support regions and municipalities with guidelines for developing local adaptation strategies or plans and for training administrators on CCA. The aforementioned activities allow for awareness raising and knowledge exchange on CCA between national, regional, and local institutions [59, 65].

5.7.7.3 International Dimensions

Since 2015, the bilateral cooperation activities promoted by IMELS have been expanded to new areas of the world. The strategy underlying the priority countries is defined in agreement with the Ministry of Foreign Affairs and International Cooperation [67]. In particular, the objective supports the countries that are most vulnerable to CC impacts and risks: the Small Island Developing States and the African Countries (TSIDSAC). In this regard, the IMELS signed bilateral agreements with developing countries including regional groups to support mitigation and adaptation actions, facilitate access to climate finance, and provide capacity building and technology transfer.

In the field of research, Italy fully contributes to the main international projects and programs in climate prediction and simulation activities. Italy contributes significantly to international programs involving land-, ocean-, and space-based measurements through very long instrumental temperature record in addition to actively contributing to long-term observations of polar regions [67, 68].

5.7.8 Assessments

A comprehensive observational network has been established covering the Italian territory with the intent of monitoring climate data, weather forecasting, and an early warning system. The Italian observational network is managed by several institutions and agencies such as the Agrarian Research Council, the Italian Air Force Meteorological Service, the National Agency for New Technologies (NANTs), the Energy and Sustainable Economic Development (ENEA), the National Research Council (NRC), the Institute of Environmental Protection and Research (ISPRA), several universities, and the Civil Protection and Regional Environmental Agencies (REAs) as shown in Fig. 5.40 [59, 64, 65]. However, the next section presents four main parts of the assessment: observations and projections, impacts and vulnerability assessment, research, and monitoring progress.

5.7.8.1 Observations and Projections

As far as the current and recent past climate variability and trends are concerned, time series belonging to national and regional monitoring networks are used to calculate climate indicators over Italy. However, time series have different characteristics in terms of continuity, completeness, spatial coverage, and data quality. The trends of mean temperature and accumulated precipitation, as well as their extremes, are updated regularly and disseminated through the System for Climate Data Collection and Dissemination (SCIA) and reported annually by ISPRA [59, 65].

5.7.8.2 Impacts and Vulnerability Assessment

In terms of impacts and vulnerability assessment, the NAS furnishes a state-of-the-art report on the scientific knowledge of impacts, vulnerabilities, and adaptation of climate. This illustrates in-depth, for national territory, sectors that are affected by

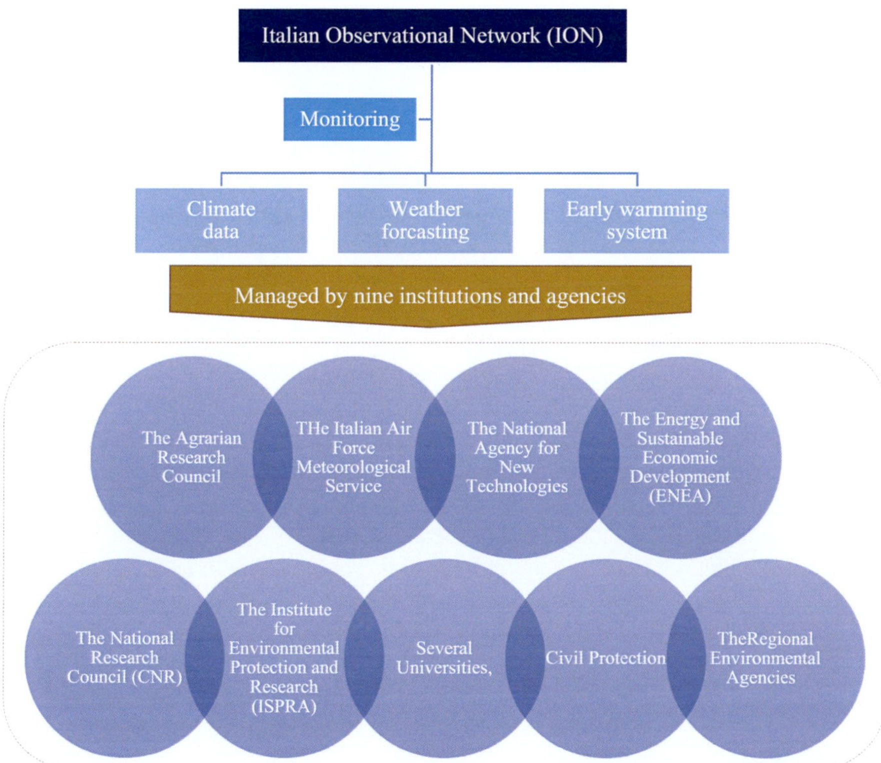

Fig. 5.40 Sectors addressed in the NAP, Italy. (Image source: Developed by authors after [60])

future CC impacts as well as the vulnerabilities for the Alps and Apennines. The Draft NAP also provides an evaluation of impacts and vulnerabilities of the main sector identified coherently with the NAS [59, 64].

5.7.8.3 Research

The National Research Program 2015–2020 (PNR – *Programma Nazionale della Ricerca*) is the main national policy on research including climate change. It sets a long-term general action to coordinate coherently several subprograms and funds with the European Horizon 2020 program and with other national and regional policy instruments. The Ministry for Economic Development (MISE) and the Ministry of Education, University and Research (MIUR) have developed the National Operative Program (NOP) "Research and Innovation" 2014–2020. The program is financed through the European Regional Development Fund (ERDF) and Social European Fund (SEF). The NOP also focuses on the southern regions, with the aim of increasing the research infrastructure of the less developed areas of Italy. A large number of national universities, public and private institutes, and other organizations are involved in climate research or climate-related research and systematic observation [59, 69].

5.7.8.4 Monitoring Progress

A monitoring system that is capable of evaluating progress in the NAS implementation at the national or regional level has been developed. However, the ongoing process toward the NAP, started in 2016, is currently taking into consideration the need for a monitoring, reporting, and evaluation (MRE) system. Guidelines and indicators are therefore defined in order to monitor the state of implementation and the effectiveness of adaptation actions [59].

Furthermore, the National System for Environmental Protection (NSEP), which encompasses the National Institute for Environmental Protection (NIEP) and the Regional Environmental Protection Agencies (REPA) (Law No. 132/2016), set up a national working group on the impacts, vulnerability, and adaptation to CC. The main objective is to define a set of CC impact indicators according to the knowledge gaps existing in Italy. The first portfolio of 150 potential CC has been identified, and a prioritization phase has also been implemented. The resulting indicators are creating a quantitative knowledge base on CC impacts as a reference framework for the MRE initiative on adaptation [59, 60].

5.8 Climate Change in Germany

Germany, with a total area of 357,409 km^2, is located in Central Europe bordering the Baltic Sea and the North Sea, between the Netherlands and Poland, south of Denmark. Agricultural land, forests, settlements, and water are covering approximately 52%, 32%, 13.7%, and 2%, respectively.

The German economy is the fifth largest economy in the world in public-private partnership (PPP) terms and Europe is the largest [70]. Gross domestic product – GDP – (adjusted for inflation) is 3468.9 billion US dollars and increased by 1.9% in 2016, the seventh year in succession. However, the GDP in 2020 was 3803.01 billion US dollars in 2020, about 9.6% increase in 4 years despite the severe impact of COVID-19 [71]. According to the Federal Statistical Office (Destatis), the country's population was 83.2 million in 2020, of which more than 77% live in urban areas [72]. Germany is part of the temperate, rainy climate zone of the midlatitudes. The maritime influence generally keeps winters mild and ensures that summers are not too hot.

Based on the Seventh National Communication – NDC 2017 – the mean air temperature has risen by 1.4 °C between 1881 and 2015 and the mean annual precipitation increased by over 10% for the same period. Germany submitted its NDC as an EU Member State in 2015 [70]. In addition, the German government's Climate Action Programme 2030, adopted in December 2019, does not entail enough policy measures to meet its emission mitigation targets in 2020 and 2030; Climate Action Tracker says [73]. The GHG emission in 2020 is expected to be 9–11% lower than in 2019 due to restricted mobility resulting from the COVID-19 pandemic [73]. Although efforts are made for climate action and adaptation, Germany faced many natural hazards and resulted in fatalities and economic impacts. Figure 5.41 shows the recent divested severe floods in Berlin and other cities in Germany, a severe impact which costs more than five billion Euros, while Fig. 5.42 presents different natural hazards and resulted fatalities and economic impacts between 1950 and 2013 in Germany [74]. Germany's climate action plan (CAP) 2050 outlines the process for achieving the national climate targets for all sectors in line with the Paris Climate Agreement 2015, including energy, the built environment, transportation, industry and business, and agriculture and forestry. The CAP also defines emissions' mitigation targets for individual sectors by 2030, yet it provides guidance on how to achieve those targets [75].

In addition, the CAP takes into account the processes for monitoring progress and updating the plan, if necessary, in line with the five-yearly stock take of the NDCs specified in the Paris Agreement. The main goal of the plan is to achieve carbon neutrality by 2050. The interim goal of the plan is to reduce Germany's total emissions by 55% by 2030, compared to 1990 [76].

Nonetheless, concrete measures to achieve these targets laid out by the CAP include (a) establishing a governmental commission to addressing the question of how to combat climate change while maintaining growth and development, (b) developing more robust energy standards for new buildings and those undergoing

(a)

(b)

(c)

Fig. 5.41 Recent destructive floods in some towns of Germany – July 2021. (**a**) Severe and destructive floods in Altenahr, Altenburg, Germany, on July 15, 2021, (**b**) Flood damage at the corner of Rathausstrasse in Hagen, Germany, July 2021, (**c**) Flooded streets in Clausen municipality, Rhineland Palatinate, Western Germany, July 2021. (Image credit and source: (**a**) CnndrBrber, https://commons.m.wikimedia.org/wiki/File:Hochwasser_in_Altenahr_Altenburg.jpg; (**b**) Bärwinkel, Klaus, https://commons.m.wikimedia.org/wiki/File:Rathausstraβe_Hochwasser_(2).jpg#; and (**c**) Tristan Schumurr, https://commons.m.wikimedia.org/wiki/File:Flooded_street_in_Clausen_(51313737196).jpg#)

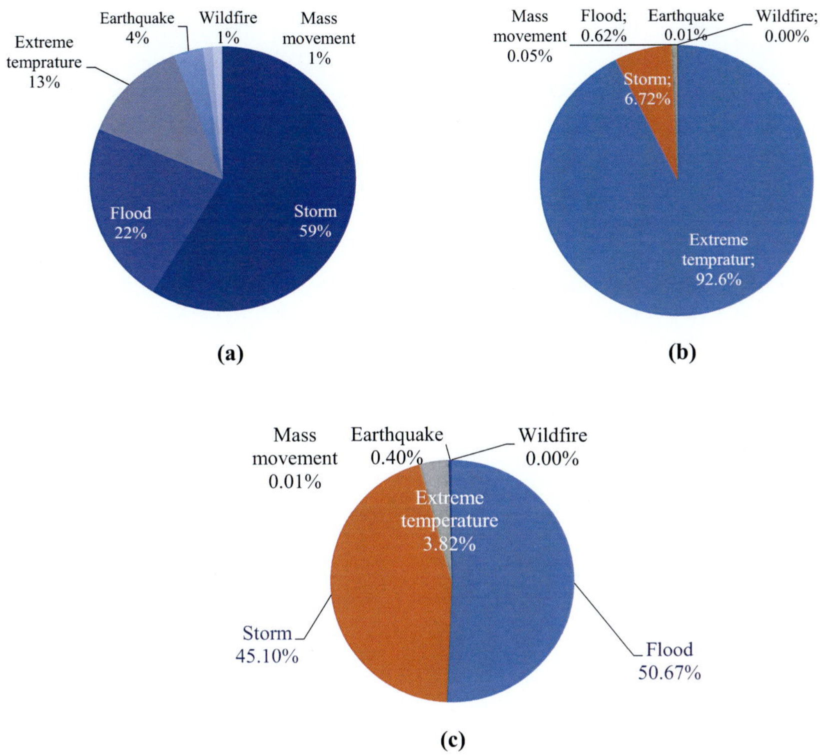

Fig. 5.42 Distribution of different natural hazards and resulted fatalities and economic impacts from 1950 to 2013 in Germany. (**a**) Natural severe events in Germany (85 in nos.) (**b**) Fatalities from natural severe events' type (10,173 in nos.) (**c**) Economic damage due to natural severe events in Germany (29 billion €). (Images' source: Developed by authors after EM-DAT 2014 [63])

substantial renovation, (c) formulating a complete strategy for reducing emissions in the transportation sector, (d) launching a research and development program aimed at reducing greenhouse gas emissions from industrial processes, (e) fully implementing existing provisions governing fertilizer use, (f) improving carbon sequestration through afforestation, and (g) strengthening economic incentives, including taxes, to encourage polluters to reduce emissions [75] as summarized in Fig. 5.43.

5.8.1 Policy and Legal Framework

The German Adaptation Strategy (GAS) aims at reducing the vulnerability of climate change (CC) impacts, sustaining the adaptive capacity of natural, societal, and economic systems. It is clear that the adaptation to CC in Germany is a permanent task established along an agreed and politically adopted institutional and

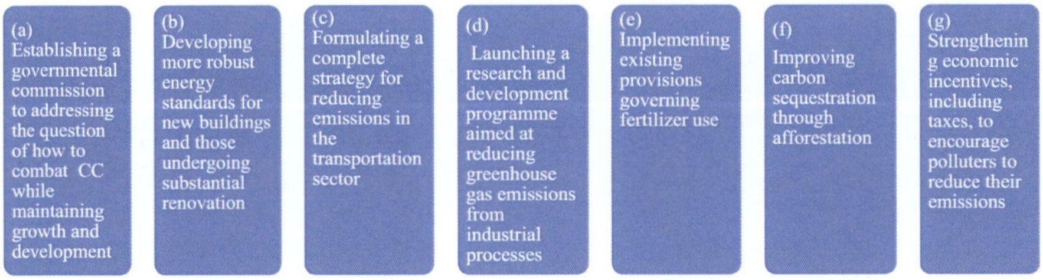

Fig. 5.43 Measures to achieve targets of Germany's climate action plan 2050. (Image source: Developed by authors)

methodological framework. Scientific research programs, participation and consultation processes, and the establishment of ongoing reporting systems are set up. On the national level, nearly all federal ministries are represented in the "interministerial working group on adaptation to climate change" (IWG adaptation), led by the German Federal Ministry for the Environment, Nature Conservation and Nuclear Safety. To coordinate adaptation activities with the federal states, the Conference of Environmental Ministers has been established in June 2009 as standing committee for the adaptation to CC [77].

5.8.1.1 Adaptation Strategies

In 2015, the Federal Government of Germany adopted the first progress report of the DAS. This report furnishes an overview of the primary federal activities since the adoption of the DAS in 2008 and the Adaptation Action Plan (APA I) (2011). The action plans of the GAS are updated every 5 years together with "progress reports" and adopted by the cabinet [77]. The progress reports contain an updated action plan with concrete steps for the further development and implementation of the GAS. The reports draw on findings and results from four main sources: monitoring report, vulnerability assessments, adaptation action plan, and evaluation of the adaptation strategy (Fig. 5.44). All form a reporting system for the planning process for adaptation to CC in Germany [77]. The process can be divided into four phases based on the adaptation policy cycle (APC), which includes the KLiVO – Climate Preparedness Portal (September 2018) – and the German Climate Preparedness Portal (GVPP). KLiVO is the one governmental meta-information platform that guides users to relevant and verified climate services. Users can self-define their individual needs in terms of steps in the adaptation cycle, sector, region, and climate hazard and user group [77].

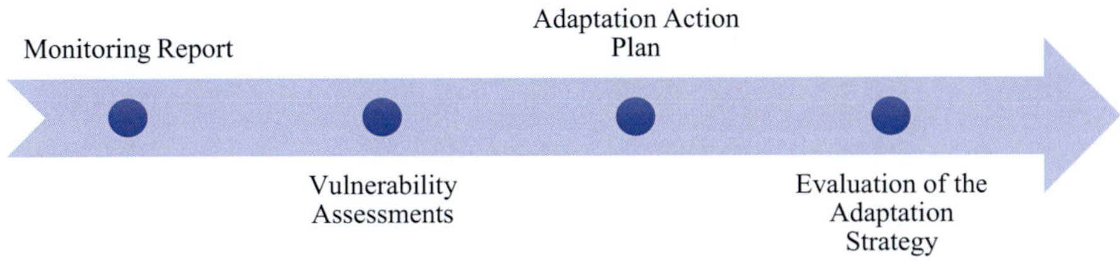

Fig. 5.44 Progress report of action plans of the German adaptation strategy GAS. (Image source: Developed by authors after [77])

5.8.1.2 Implementation Means

The adaptation action plans (APAs) specify the current and future measures on the federal level to adapt to CC. These measures are based on the scientific findings and results of the climate impact and vulnerability analysis (KWVA). The APAs thus describe the implementation of the GAS through specific measures that demonstrate links with other national processes. The implementation of these measures described in the APAs lies within the responsibility of the relevant ministries [77].

5.8.1.3 Monitoring, Reporting, and Evaluation

The monitoring report provides an overview of the observed impacts of CC and the adaptation measures already introduced in Germany. This provides a compact overview of the changes that can already be observed as a result of CC using measured data. The monitoring report will be updated every 4 years.

The KWVA, developed for the first time in 2015 and updated every 6 years, identifies the fields of action, climate impacts, and regions, which are particularly affected by CC and where there are strong needs for taking preventive actions. The strategy process and its implementation of the GAS are evaluated every 4 years and carried out according to a methodology adopted by the interministerial working group on adaptation [77]. For the sectors addressed in NAS/NAP, the strategy is subdivided into 15 fields of action as shown in Fig. 5.45.

5.8.2 Assessments

The first vulnerability assessment of climate change (CC) impacts for selected fields of actions in Germany was conducted in 2005. Many studies have investigated this topic in recent years commissioned by different agencies [77]. For advancing the development of the DAS and the prioritization of climate risks and fostering actions, an up-to-date, consistent cross-sectoral vulnerability assessment, covering the whole of Germany, was prepared in 2011 until 2015 [77].

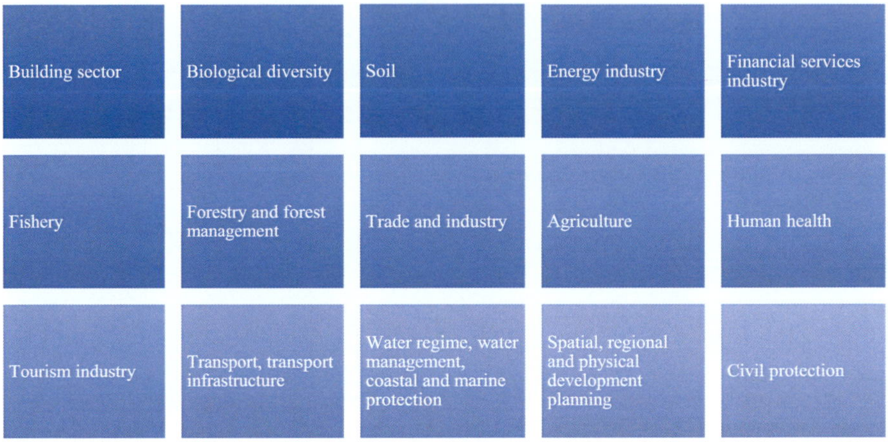

Fig. 5.45 German adaptation strategy's fields. (Image source: Developed by authors after [77])

In a screening procedure, using a developed and consistent method, the regions and topics are identified at particular risk due to CC for each of the examined sectors. The results include the mapping of CC impact with a spatial resolution of municipalities. The vulnerability assessment (VA) was published by the end of 2015 [77]. However, the VA has been elaborated through the years.

5.8.3 Climate Action Plan (CAP) 2050

The German government adopted the climate action plan (CAP) 2050 in November 2016, making Germany one of the first countries to submit the long-term low greenhouse gas (GHG) emission development strategy to the UN as required in accordance to the Paris Climate Agreement. The CAP 2050 confirms and defines in more detail the German government's ambitious climate targets. Germany's long-term goal is to become largely GHG neutral by 2050. This is based on the goal stipulated in the Paris Agreement to achieve global carbon neutrality in the second half of the century. Such target also reflects the country's particular responsibility as a leading industrialized nation and the EU's strongest economy [75].

The medium-term target is to cut GHG emissions in Germany by at least 55% by 2030 compared to 1990's levels. As part of the CAP 2050, the German government also lays down 2030 targets for the individual sectors, describing the necessary development pathways, listing initial measures for implementation, and establishing a process for monitoring and updating policies and measures. With this plan, Germany is pursuing its part to achieve the international target set out in the Paris Agreement to keep global warming to significantly below 2 °C or even limit it to no more than 1.5 °C [75]. The CAP 2050 maps out the process for achieving Germany's climate targets for all areas of action (energy, buildings, transport, trade and industry, agriculture and forestry) in line with the Paris Agreement. The plan also lays down the first emission reduction targets for individual sectors for 2030, which provides guidance for strategic decisions over the coming years. In addition, the plan envisages monitoring and public participation. The first program of measures under the CAP 2050, including measures to ensure that the 2030 target is achieved, has been submitted at end of 2018 [75].

5.8.3.1 Development of Climate Action Plan 2050

In the development of CAP 2050, the 2013 coalition agreement stipulated, "In Germany, we intend to codify the further reduction steps in light of European targets and the results of the 2015 Paris Climate Change Conference up to the target value of 80–95% in 2050, and to back this up with measures (climate action plan) as part of a broad dialogue process," and the dialogue process began in summer 2015 and was concluded in March 2016 [75, 78]. Nevertheless, the German government drew up the CAP 2050 on the basis of scientific studies and scenarios and in light of the Paris Agreement, taking into account the proposals from the broadly based dialogue. Previously, federal states, municipalities, associations, and citizens had jointly developed proposals for strategic climate measures up to 2030. However, the aforementioned entities presented the German Environment Minister with the resulting list of 97 proposed measures.

To limit global warming to a maximum of two degrees centigrade, industrial countries must reduce their GHG emissions by 80–95% in 2050 as compared to 1990. Therefore, the German Federal Government has adopted a national CAP 2050 starting 2016, and it is ongoing to reach such target [75, 78].

5.8.3.2 Goals and Contents of CAP 2050

In reaching a low-carbon economy in Germany, the CAP 2050 summarizes a modernization strategy for the necessary transformation to reach such ambition. This has been developed in three levels as follows [78]:

- Contains specific guiding principles for the individual areas of action for 2050, leaving scope for innovations and striving to maximize sustainability.
- Outlines robust transformation pathways for all areas of action, highlights critical path dependencies, and describes interdependencies.
- Underpins goals, particularly the interim GHG target for 2030 of at least 55% reduction in GHG emissions compared to 1990, with emission targets for all sectors, concrete milestones, and strategic measures, taking impact and cost analyses into account.

For shaping the transition toward a GHG-neutral economy and society, the German government has established the goals and management rules of the national sustainability strategy which need to be taken into account [78]. In this regard, to attain the UN Sustainable Development Goals (SDGs), the focus should be geared toward mitigating GHG emission and increasing energy efficiency, but social and economic requirements must be considered when structuring the transformation [75, 78]. According to recent UN 2021 report, COVID-19 has delayed the governments' efforts worldwide in achieving SDGs between 2020 and 2021 [79, 80].

5.8.3.3 Sectoral Targets in the Climate Action Plan 2050

By looking at sectoral targets in the CAP 2050, it depicts action areas such as energy, transport industry, buildings, agriculture, land use, and forestry. It also positions overarching targets and measures to reach such sectoral targets of the CAP 2050 (Fig. 5.46).

It is a vital aspect to develop a restructuring plan for the energy sector. The energy transition has laid an essential foundation in this field. With the new expansion of renewable energy (RE) and the phaseout of electricity from fossil fuels, the energy sector's emissions will be reduced from 61% to 62% by 2030 compared to the 1990's level. One action for the energy sector is concerned with the establishment of a commission by the German government for growth, structural change, and regional development.

In June 2018, the government, with the participation of other ministries, established the commission as part of the Federal Ministry for Economic Affairs and Energy. Such commission consisted of representatives of federal states, municipalities, trade unions, affected businesses or sectors, and local stakeholders to drive this plan [75, 81].

To guarantee successful transformation, it was necessary to make the transformation process work effectively and hence establish a realistic outlook for the industry. To secure the affected areas, agree on the implementation strategy, and bring about the necessary financial position, the commission published its final report in January 2019. It recommended the phaseout of coal-fired power plants in Germany no later than 2035 and no later than 2038. The commission also made recommendations on specific prospects for the coal regions [75, 81].

The German government also clarified greenhouse gas (GHG) emission targets for all sectors of the climate action plan 2030–2050. The 2016 mitigation target for the industrial sector was set at 49–51% by 2030 compared to the 1990's level. The German government has worked with the industry to initiate research and development (R&D) programs based on the goal of GHG neutrality with the goal of reducing GHG emissions in industrial processes. The industrial CO_2 recycling options (carbon capture and storage – CCU) are considered in this context.

To contribute toward the mitigation of GHG emissions through energy efficiency, measures such as utilizing the potential of existing waste heat are implemented. Fuel accounts for about 70% of the final energy demand of the current industry, which reflects the large amount of generated heat, including waste heat. The future goal is to use this waste heat consistently and strategically in both industry and home. All options are contemplated, such as the development and use in district

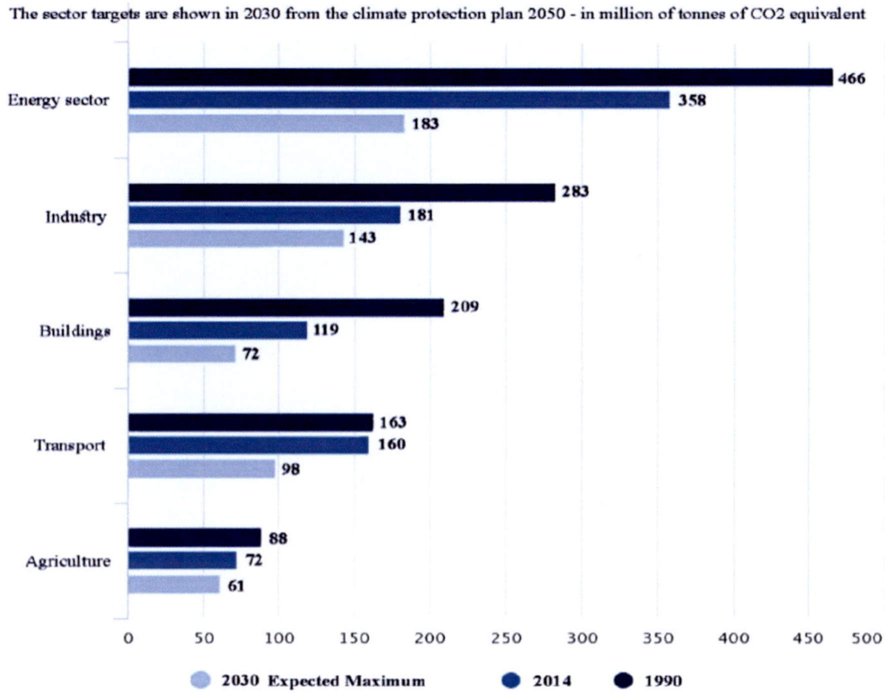

Fig. 5.46 Sectoral targets in the climate action plan 2050. (Image source: Developed by authors after https://www.bmu.de/en/topics/climate-energy/climate/national-climate-policy/greenhouse-gas-neutral-germany-2050/)

heating systems and district heating networks, used based on existing programs and measures [75, 81].

For the building sector, a roadmap has been created to achieve a virtually climate-neutral building inventory. The life of the building is very long, so lay the foundation for 2050 as soon as possible. The aim is mainly to reduce emissions from 66% to 67% by 2030 compared to the 1990's level. By 2050, achieving a de facto climate-neutral building inventory will phase out ambitious standards for new buildings, long-term strategies for upgrading existing buildings, and fossil heating systems. It depends on what you do.

As a result, the new zero-energy building (ZEB) standards, which will be enforced from 2021 to achieve the virtually climate-neutral new medium-term standards. This makes the installation of a new heating system – that is, efficiently using renewable energy – a much more attractive option than operating such system on fossil fuels. Appropriate incentives for using and constructing buildings that produce more energy than they use to support this goal are reviewed. Existing buildings will also be upgraded by 2050 to improve energy efficiency and expand the use of renewable energy, effectively meeting the criteria for a climate-neutral building. Therefore, the energy requirements of existing buildings will be developed gradually by 2030, taking economic aspects into account [75, 81].

A reduction from 42–40% – compared to 1990 – to the 2030 climate target will be achieved by the transport sector (Table 5.9). It also requires measures' identifications for achieving this target through climate strategies. Greenhouse gas (GHG) emissions can be reduced by 2030 by a climate strategy for road transport. Emissions from cars, light and heavy commercial vehicles and issues related to GHG-free energy supply, the requisite infrastructure, and the interlinking of sectors, through electric mobility, are addressed by the strategy. The key role in the transport sector is by using alternative drives, local public transport, rail transport, cycling, walking, and a digitalization strategy [75, 81]. In addition, mitigation from 34–31% (compared to 1990) to the 2030 target will be pursued in the agricultural sector as shown in Table 5.9. This includes a significant reduction in nitrous oxide emissions due to overfertilization. Mitigation is generally limited in the agricultural sector. The German government, in cooperation with the federal state, has made the fertilizer bill sustainable in Germany, especially by ensuring full enforcement and consistent enforcement of fertilizer ordinances and planning ordinances for the practice of good nutritional management of farms. The goals of Germany's Sustainable Development Strategy (SDS) are achieved through the implantation of a 70 kg/N/ha in agriculture between 2028 and 2032 [75, 81].

Based on the above overview and discussion of climate change and actions in Egypt, Italy, and Germany, Table 5.10 lists the summary of the outcomes and main findings.

5.9 Biomimicry and Climate Change

It is imperative, therefore, to highlight an important question regarding "the ability of biomimicry to assist in developing design solutions for mitigating CC and related adaptation actions." The appropriate response is yes – it can help in driving biomimetic solutions and it can grow more reasonable arrangements as far as mitigation and adaptation to CC impacts [4]. The commitment to CC actions necessitates the development of solutions for the urban areas and cities, which is divided into two

Table 5.9 Emissions in the areas of action contributing to the target – Germany

Area of action	1990 (in million tons of CO2e)	2014 (in million tons of CO2e)	2030 (in million tons of CO2e)	2030 (reduction in % compared to 1990)
Energy sector	466	358	175–183	62–61
Buildings	209	119	70–72	67–66
Transport	163	160	95–98	42–40
Industry	283	181	140–143	51–49
Agriculture	88	72	58–61	34–31
Subtotal	1209	890	538–557	56–54
Others	39	12	5	87
Total	**1248**	**902**	**543–562**	**56–55**

Source: Federal Ministry for the Environment, Nature Conservation & Nuclear Safety (2016), CAP 2050

Table 5.10 Summary of the comparison of climate actions in Egypt, Italy, and Germany

Country	Mitigation	Adaptation	Actions
Egypt	Development of green areas and planting forests Conduction of national strategy for clean development mechanism (CDM) in 2002 followed by capacity development project in CDM (2003–2006) Establishment of Egyptian DNA for CDM in 2005 Approval of 94 projects in REEE, industrial processes, fuel switching, and solid waste Total investment cost \$5.04 billion and total GHG reduction 11mmtCO_2/year Implementation of programs for fuel switching to natural gas in energy, industrial, and transport sectors Established of the Energy Supreme Council	The use of RE is one of the adjustment measures, but RE is 3% of the 2018 total energy mix Adoption of energy code Use of environmentally friendly construction materials. Reuse of waste water and recycle of solid waste Replacement or renovation of vulnerable buildings and infrastructure in urban and rural areas to be more resistant against climate extreme events Use of materials that are resistant to high temperature and depend on eco-transportation Implementation of integrated environmental management systems (IEMS) in touristic sites and marine and wildlife protectorates Water recycling Improvement of distribution network Reuse of gray water in irrigation Adopting of IMELS in 2015 the Italian National Adaptation Strategy to CC The Directorate General for Climate and Energy of IMELS is currently working for the implementation of the NAS through the Italian National Adaptation Plan – NAP – for CC	Egypt signed the United Nations framework convention on climate change (UNFCCC). Egypt signed the Kyoto Protocol The National Strategy for Adaptation to Climate Change and Disaster Risk Reduction, Egyptian Cabinet and Decision Support Center (ECDSC), and UNDP National NDCs (mitigation and adaptation and MOI) in line with Egypt SDS 2030. National NDCs in line with Egypt's Sustainable Development Strategy (SDS) 2030 National NDCs in line with Egypt's SDS 2030 – submitted to UNFCCC ((remove extra space here) from 2016 to 2020)

(continued)

Country	Mitigation	Adaptation	Actions
Italy	Nearly 3000 Italian towns and cities have submitted their Sustainable Energy Action Plans (SEAPs) in pledge to mitigate GHG emissions Reducing energy consumption Developed mitigation plans and carried out decarbonization strategies Submitted their Sustainable Energy Action Plans (SEAPs) in pledge to mitigate GHG emissions in these towns/cities by at least 20% in 2020 Engaged and supported cities and towns to commit to reaching the European Union (EU) climate mitigation and adaptation targets	Addressed CC impacts on extreme rain events, landslides, coastal erosion, infrastructure, cultural heritage, heat waves, and droughts Taking action after disasters Adopting IMELS in 2015 the Italian National Adaptation Strategy (NAS) to CC. The NAS is a tool for encouraging adaptation actions in planning activities at national, regional, and local levels The NAP provides institutional guidance to national and local authorities for the elaboration of regional strategies or plans and for the integration of climate change adaptation within spatial and sectoral planning	Eliminating the exploitation of nonrenewable natural resources Phasing out pollutants and enhancing the value of waste through its reuse Recycling and recovery of both energy and secondary raw materials Changing the balances of generation and absorption of GHG emissions Interrupting biodiversity erosion and desertification processes and safeguarding landscapes and habitats A fair distribution of resources A decrease of the unemployment rate and accomplishment of economic measures by means of investments in the national health system, education, and social programs that assure the access to services and social cohesion Integrating environmental issues into other policy processes The preference for an aware economic and environmental lifestyle An increase in the global efficiency of resource usage Refusal of the "end cycle" intervention approach and promotion of prevention policies General waste reduction Stretching goods' lifetime The end of material cycles of production consumption Development of local markets and local productions Promotion of typical products and traditional cultures Involvement of social parties in setting goals, commitments, and sharing responsibilities Guidelines and indicators are defined to monitor the state of implementation and the effectiveness of climate adaptation actions (CAA) In 2016, the National System for Environmental Protection (NSEP), which is composed of the National Institute for Environmental Protection and the Regional Environmental Protection Agencies (Law No. 132/2016), set up a national working group on "impacts, vulnerability, and adaptation to climate change." The main objective is to define a set of CC impact indicators, according to the knowledge gaps currently existing in Italy A first, a portfolio of 150 potential CC impact indicators has been identified and prioritized, and it is currently being implemented Notice on small landslides: hydrogeological risk reduction due to mountain municipality landslides in order to promote climate change adaptation National program on resilience increase of natural and seminatural forestry systems, through the recovery and the structural and functional restoration of ecosystems and their service functionality and by consistent actions with the biodiversity protection and conservation (flora, fauna, vegetation, natural, rural landscape) in the protected areas crossed by the fire Measures for energy efficiency, sustainable mobility, and CCA in the small islands Strengthening of administrative capacity for CCA, with the aim to support regions and municipalities with guidelines for developing local adaptation strategies or plans and for training administrators on CCA The IMELS started the mapping of the current state of adaptation in the Italian regions Involvement of stakeholders, especially institutions (governmental and regional entities) in defining the adaptation strategy and the plans that contributed to improving the information exchange between central government and local government (municipality offices) and raising awareness on the need of adaptation planning to curb CC impacts The establishment of a scientific panel The establishment of an Institutional Panel of Ministries, national and regional authorities, and local authorities A national public consultation Build a ground base for adaptation Evaluate the state of the implementation of adaptation actions at national, regional, and local levels Support regions and municipalities with guidelines for developing local adaptation strategies and/or plans for training administrators on CCA. The aforementioned activities allow for awareness raising and knowledge exchange on CCA between national, regional, and local institutions The Italian Ministry for the Environment signed bilateral agreements with developing countries including regional groups to support mitigation and adaptation actions, to facilitate access to climate finance, and to provide capacity building and technology transfer In the field of research, Italy fully contributes to the main international projects and programs in climate prediction and simulation activities Italy contributes significantly to international programs involving land-, ocean-, and space-based measurements through very long instrumental temperature record and to contributing to long-term observations of polar regions A comprehensive observational network has been established covering the Italian territory with the intent of monitoring climate data, weather forecasting, and an early warning system

(continued)

Country	Mitigation	Adaptation	Actions
Germany	Use of renewable energies for electricity production Use of renewable energies for heating	In 2015, the Federal Government of Germany adopted the first progress report of the DAS Adaptation action plan 2011 The action plans of the GAS are updated every 5 years together with "progress reports" and adopted by the cabinet The reports draw on findings and results from four main sources: monitoring report Vulnerability assessment Adaptation action plan and evaluation of the adaptation strategy The adaptation action plans (APAs) specify the current and future measures on the federal level to adapt to CC Finalizing negotiations in the Environment Council for an EU climate law, designed to achieve a climate-neutral EU by 2050 Implement the EU biodiversity strategy for a healthy natural environment Harness digitalization for environmental protection and innovation	Establishing a governmental commission to address the question of how to combat CC while maintaining growth and development Developing more robust energy standards for new buildings and those undergoing substantial renovation Formulating a complete strategy for reducing emissions in the transportation sector Launching a research and development program aimed at reducing greenhouse gas emissions from industrial processes Implementing of existing provisions and governing fertilizer use Improving carbon sequestration through afforestation Strengthening economic incentives, including taxes, to encourage polluters to reduce their emissions For the building sector, a roadmap has been created to achieve a virtually climate-neutral building inventory In 2016, the German government clarified GHG emission targets for all sectors of the CAP 2030–2050

main courses of actions. The first is decreasing the amount of GHG emissions into the atmosphere, whereas the second is the adaptation process of the urban environment to the current and futuristic climate changes as illustrated in Fig. 5.47. The prevention of using any GHG emitting substance will not solve the problem completely as we will still face the effects of the previous harm done to the atmosphere [4].

It is observed that the rising in the levels of GHG emissions despite all the international protocols and policies is still manifested. The building segment alone contributed in increasing GHG emissions by 2% annually since the year 1970, while generating the needed energy. The environment specialists and professionals (architects, engineers, urban designers, and planners as well as city leaders) must constantly and rapidly work on mitigating CC and drive climate action plans for

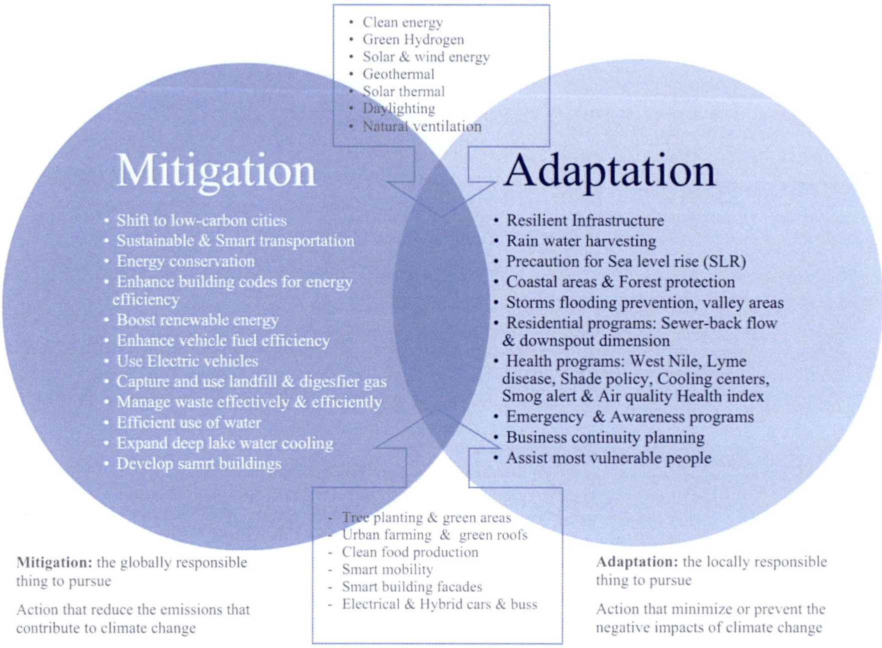

Fig. 5.47 Mitigation and adaptation to climate change. (Image source: Developed by authors after http://www.adorngeo.com/3-responding-to-climate-change.html https://www.adaptationclearinghouse.org/resources/taking-action-on-green-resilience-climate-change-adaptation-and-mitigation-synergies.html)

adaptation. The main threat currently facing the globe and humanity is the risks of CC and its severe impacts on all sectors.

Little has been manifested regarding CC mitigation and adaptation despite the efforts of the United Nations and all the policies enacted, mainly between COP21 in Paris and COP25 in Spain. Hope is high for COP26 held between November 1 and 12, 2021 in Glasgow, Scotland, UK. Thus, biomimicry is a science that studies the nature forms, processes, and behaviors along with implementing them to reach suitable solutions that can face the CC impacts [4].

Biomimicry plays a major role in CC mitigation and adaptation. In addition, biomimicry has the ability to find other alternative methods of human behaviors and activities in order to decrease the percentage of air, water, and land pollution in the environment. Moreover, biomimicry can create a different and sustainable urban environment. Looking to the natural organisms, one can find that their survival depended on their ability to adapt to the change. The utilization of such science in the designing process remains unnoticed and not spread enough [4]. The proper utilization of biomimetic methods will result to not only reaching sustainable solutions for the urban environment but also regenerating and restoring the natural ecosystem [4].

5.9.1 How Biomimetic Design Responds to Climate Change?

Biomimicry has had a particularly significant impact in the architectural field, where it has led to innovations in climate control, ventilation, and structural integrity, among other features. The Melbourne's Council House 2 was inspired by gas and heat exchange in termite mounds. Also, London 30 St. Mary Axe building, affectionately called "The Gherkin," was inspired by structural reinforcement in sea sponges. For many centuries, we were sightless to the possibility that animals and plants, seen as lower life forms, had anything to teach us, but these examples show how much we have to learn, inspire, and innovate [4].

The building sector is one of the sectors that can mitigate CC impacts with the least amount of expenses unlike other sectors [82]. In this regard, biomimicry case studies demonstrate the approach's ability to guide eco-integration. As such, architects are uniquely positioned to embrace biomimicry and become leaders of a sustainable transition. Biomimicry has proven to be not only ecologically viable and economically viable but also socially viable, which was highlighted in the age of COVID-19, in the face of critics who hang on to a careless view of humans. Criticism leveled against biomimicry as an approach to design, includes the following:

- Biomimicry diminishes the role of the designer.
- Design inspired by biological models is too tightly bound by evolutionary instrumentalism.
- Eco-conscious approaches like biomimicry urge humans to repress their impulse to build; some critics even go so far as to argue that.
- Biomimetic architecture reflects an impoverishment of human learning [82].

In light of CC, their impacts are expanding; strategies, policies, and activities to lessen GHG emissions must be manifested and extend quickly. Implementing these actions is a necessity worldwide, especially in the industrialized nations such as the USA, China, Japan, and Europe. The civilized environment contributed to increasing the levels of GHG emissions by 30%. Moreover, it will face the process of adjusting to CC impacts as the core of cultural and social life. The architectural design of buildings to sustain CC has an aim to reach zero effect. Also, the urban environment also should exercise a beneficial effect on the climate and move away from affecting the environment negatively. The definition of this kind of architecture is "regenerative architecture" [83]. The good understanding and applying of analogies or metaphors in ecological and biological systems would assist in the architectural constructions. The mimicking process of an organism or systems and applying that into architecture will have great impacts on creating flexible and more adaptable built environment and also regenerative ecosystems [28].

According to the growing body of international research on biomimicry, researches stated that there are many problems facing us while implementing biomimicry into the designing process. The first barrier that architectural designers face is the absence of clear model or procedure to biomimicry implementation, especially if the goal is more flexible built regions. These matters have been discussed in several occasions [28]. It is essential to realize that not all of the untraditional and revolutionized processes and solutions will be appropriate or reasonable for human usage. As an expert of sustainability and the built environment, more actions should be in place to ensure that more critical and troublesome issues that identify CC impacts are outlined and manifested. It is also suitable to look at cases of how similar issues have been dealt with by other living life forms or biological systems in similar climatic conditions. If not giving full and effortlessly transferable arrangements, it might at any rate give new regions of investigation [28]. Be that as it may, biomimicry can be exploited to confront environmental change as well as CC and achieve climate neutrality. Nonetheless, biomimicry illustrates that there are chances to move past narrow prospective to join a general comprehension of the implementation of ecosystems into structural designs. The ability of well-understood design techniques in mitigating CC impacts and risks is noticed, depicted, and evidenced; however, there isn't an expansive work study that talks about adjusting new or existing structures to CC in addition to recycling materials. Recent methods and technologies which can add to mitigation and adjustment to CC with advantages in the economy, society, and environment sectors can be uncovered by intense studies of the behaviors of certain organisms and the biological systems [84].

5.9.2 Biomimicry to Mitigate Greenhouse Gas Emissions in the Built Environment

In mitigating greenhouse gas (GHG) emissions in the built environment, adopting biomimicry in architecture can be one of the rapid and productive solutions. Numerous cases of new advancements on moderating GHG outflows can be categorized into three methodologies as follows:

(a) Mimicking the energy utilization of living biological systems and organisms where the main target is reducing the amount of fossil fuels hence mitigates GHG levels.
(b) Reaching better methods for generating another form of energy with non-GHG emissions and decreasing our dependence on fossil fuel [85]. These methods are most widely recognized to decrease environmental change related to the urban regions.
(c) Dealing with mitigating GHG discharges through examining living biological systems to discover cases and behaviors that enabled them to isolate and store carbon.

In mimicking these approaches, the aim is to reach zero GHG emissions from human activities [85]. Basic moderation methodologies in an urban environment status encompass the following:

- Energy utilization and discharges from vehicles.
- Creation and keeping up of urban woods and green areas.
- Design ways for energy conservation.
- Maintenance of renewable energy sources.
- Carbon stockpiling or sequestration.

5.9.2.1 Biomimicry for Energy Efficiency

There are various cases of living organisms and ecosystems that utilize energy efficiency. This results in a comprehension of how people implement their method without the reliance on GHG emitting resources of energy; even notable biomimicry models fall under such category [85]. For example, the Eastgate Building was intended to have a generally thermally stable inside condition that utilizes insignificant mechanical cooling and hence generates less GHG emissions. The constructed outline was developed, to a certain extent, with respect to generating thermal energy as depicted in the termite hills of South Africa (Fig. 5.48). This creates a huge decrease in the utilization of energy by 17–52%, unlike other similar buildings in Harare [85]. In this building, air-conditioning is performed by passing underground water and cleaning it through sewers and then reusing it to cool the air. This is an example of how certain organisms use underground water as cooling-vaporizing method, which results in decreasing the water temperature, especially at nighttime. The water traveling through the shower towers additionally cools the encompassing air, which is then used to ventilate the business premises on the ground floor [85]. The water proceeds to the cellar where it goes through a framework that keeps up the cool temperature by utilizing phase change materials' characteristics. The CH2 building is accounted for saving 85% less power and 70% less water than regular similar structures. Enhancing the usage of energy effectively and efficiently to mitigate CC is the main goal; however, it ought to be viewed as a middle-of-the-road step still [85].

(a) (b)

Fig. 5.48 A mimicry example of energy efficiency in South Africa. (**a**) Termite mounds in Africa (**b**) Eastgate building in Harare. (Images' credit and source: (**a**) J Brew, https://commons.m.wikimedia.org/wiki/File:Catherderal_Termite_Mound_Brewbooks.jpg, (**b**) Gary Bembridge, https://commons.m.wikimedia.org/wiki/File:Zimbabwai_Harare_Eastgate_Shopping_Mall.jpg)

Biomimicry used in generating energy mitigates the causes of CC. Technologies and systems based on biomimicry are intended to substitute the utilization of fossil fuels as the essential energy generating source, trying to relieve GHG emissions. Cases that are being developed incorporate mimicking the procedure of synthetic [85].

Photosynthesis in solar panel innovations and the advancement of sea energy technologies are mimicking the mobility of kelp (algae) in the water. In this regard, the Australian organization BioPower has created submerged power generators that vibrate in sea waves and streams instead of swirl-like turbines [86]. Using lasting magnet engines, the low speed and high torque of those vibrations are changed into fast low-torque-high-speed rotation. Bio-wave is planned for use at depths of 40–45 m. It is secured at the ocean depths by the utilization of a little rootlike gadget to prevent intense, complicated damages to drillers and installation tools. The large parts are drifts which enable the structure to move forward and backward with the current. As the waves influence the surface to move vertically, the current under the surface makes a horizontal movement, which the buoys take after [86]. Discovering techniques to supplant fossil energy with renewable power sources is a way to mitigate CC, resulting in a long-term arrangement. Time and assets are required to develop and execute such innovative technologies, while displacing the use of petroleum derivatives. This approach may prevent extra GHG emissions from being developed or discharged. Likewise, biomimicry offers methods to effectively deal with abundant carbon dioxide present in the environment, to mitigate high GHG levels [85]. The bull kelp is great at withstanding strong and severe water waves. However, a couple of fascinating systems that the bull kelp utilizes to affect the bio-wave's design are as follows:

- Strong rootlike framework called holdfast. The holdfasts are not called roots since they do not dive deep into the ground but instead remain nearby to the surface making numerous centralizing points that relocate vertical and horizontal forces.
- Buoyancy: The kelp has gas-filled bladders that buoy and keep the kelp upright. From the bladders, they develop edges that should be near the surface to assimilate daylight for photosynthesis.
- Kelp moves: They follow the streams opposite to them. That way it can withstand extremely solid streams for photosynthesis [85].

The three strategies stated above were reused in bio-wave design as follows [85]:

- Holdfast-like grapple systems enable it to not penetrate the ground, limiting expense and harm to the ocean bottom.
- Buoys permit buoyancy with the under surface streams.
- Adaptability of the kelp duplicated by utilizing a swivel model enables the structure to pivot toward the current. It additionally can set down near the ocean bottom in the presence of heavy current, limiting the power needed for the structure to withstand harm and dislocation.

5.9.2.2 Biomimetic Segregation and Storing Carbon

Looking at biomimetic segregation and carbon storage, carbon in nature can be stored, isolated, and reused by a few life forms and procedures. This fact is researched in the advancement of innovations and technologies implemented in the industrial sector and in the urban environment – one of the largest examples is in the USA by Occidental. In Quebec, Canada, CO_2 arrangements are creating carbon imprisoning containers that mirror mammals' behaviors, which empower them to control carbon dioxide (CO_2) amid breath. The CO_2 arrangement innovations copy the compound carbonic anhydrate, which can change over CO_2 into bicarbonates. The procedure works at barometric pressure and surrounding temperature and subsequently does not require a lot of energy [87].

Such process generates bicarbonate, which can be utilized in industrial waste management or can be changed once more into CO_2 and can be changed into carbonate compounds, for example, limestone which is used in some industries like cement production. The watery arrangement, where the change of CO_2 to bicarbonate occurs, is reutilized in a closed cycle. The innovative technology can be adjusted to existing applications, for example, control plants, cement production, aluminum smelters, and power generating facilities. Introductory testing on an Alcoa aluminum smelter in Deschambault – Quebec – showed an evacuation of 80% of CO_2 supposed to be discharged in the air. Also, the use of waste is a critical piece of the way toward cycling supplements and is a principal part of keeping up the soundness of biological systems. In utilizing biomimicry to address the abundance of carbon in the climate, it might be conceivable to utilize it as an asset instead of a generator of contamination or waste [85]. A notable case from nature is the means by which plants use CO_2 amid the photosynthesis procedure, transforming it into the elements required for plant development and improvement, for example, cellulose. For plants' survival, CO_2 is an important asset at the right amounts, as opposed to contamination. There are a few extra strategic, financial, innovative, and ecological issues with current endeavors at carbon sequestration [85]. The advantage of this method is that such procedures and techniques may modify and adjust existing building foundations while reducing GHG emissions in the short term. The current urban environment should adopt and implement long-term procedures to tackle CC risks and impacts in light of the generally long existence of structures and the moderate rate of renewal compared with consumable products like clothes and electronic tools [85]. In this line, the urban environment consumes around 40% of the materials used by the worldwide economy. Building materials that store carbon over a long period, or those that are produced using CO_2 and don't discharge a similar rate of CO_2 on decaying, if opt, sustainable, and safe, could have an enormous influence on decreasing the causes of CC or even bringing down the concentration of barometrical carbon over the long period, in case it is joined with different activities [85].

(a)

(b)

Fig. 5.49 CH2 building in Australia. (**a**) Exterior of the building (**b**) Interior of the building. (Images' credit and source: (**a**) Roryrory, https://commons.m.wikimedia.org/wiki/File:CouncilHouse2.jpg#mw-jump-to-license, (**b**) https://www.melbourne.vic.gov.au/building-and-development/sustainable-building/council-house-2/Pages/council-house-2.aspx)

5.9.3 CH2 Building: Best Biomimicry in Architecture Case Study

The best example of biomimicry in architecture is the case of CH2 building in Australia. This case study of $11.3 million structure was exploited into maintenance characteristics, which are relied upon to be redeemed and compensated in a decade. The CH2 building, Fig. 5.49, illustrates sound and productive working environments to its occupants, while diminishing the building's effect on nature through brilliant thinking in both design and development. More beneficial structures enhance users' wellness and livability indoor. This building contributes positively by lowering GHG emissions and having less impact on the environment. By and large, the CH2 building is considered a superb model of biomimicry approach in design as it accomplished supportability, exceptionally in ventilation as biomimicking termite hill [85].

5.10 Benefits of a Biomimetic Approach to Architecture

The responses to climate change (CC) by exploiting biomimicry approaches and levels from various examples of nature (short, medium, and long terms) are illustrated in Table 5.11. Some of these examples are more effective than others; nevertheless, in the short and medium terms, technologies and innovations are crucial. For the long term, solutions addressing CC impacts could be inspired and obtained from biomimicry. This is particularly vital in the advancement of technologies to supersede the utilization of nonrenewable energy sources, in both the advancement of technologies to control CC impacts and the developments of urban environment systems utilizing biological biomimicry [86]. In fact, it is considerably profound to empower a comprehension of natural processes in mimicking architecture.

Table 5.11 Timeline for biomimicry approaches to address climate change

Time Mitigation and Adaptation	1–5 years Short term (5 years)	6–40 years Medium term (40 years)	41–80+ years Long term (80+ years)
Biomimetic mitigation of climate change causes in the built environment	Biomimicry to increase energy efficiency		
	Biomimicry for carbon confinement/ storage		
		Biomimicry to replace the use of fossil fuels	
Biomimetic adaptation to climate change in the built environment	Biomimetic technology to address direct impacts		
		Systemic improvement-ecosystem mimicry	

Source: Developed by authors after https://www.tandfonline.com/doi/abs/10.3763/asre.2008.0065

It is progressively discussed in new studies, if the empowering and incorporating of biological biomimicry in designs enough to change how buildings affect the environment. This is imperatively ought to be discussed. Nonetheless, various researchers clearly uttered the connection between bio-affected designs and enhanced mental, physical well-being and livability.

Regarding financial advantages, there is proof of economical sustainability for biomimicry examples unlike the traditional ones [86]. Since GHG emissions are progressively being managed, structures that do not meet lawful or execution expectations might be harder to rent, market, and secure resulting in high monetary life cycle expenses and revenues. In such cases, structures can provide biological community benefits. Hence, this may result in an extra income for proprietors or occupants.

Connections with expanded human efficiency in business structures and educational organizations that show parts of bio-affected designs have additionally been archived, giving more monetary advantages of biomimetic planning [87].

5.11 Conclusion

What has been illustrated expresses the amazing capability of biomimicry to mitigate human-made GHG emissions, yet to adjust to environmental changes and effects by looking at living organisms and ecosystems. There is more to discover in the applications of biomimicry in the built environment. The expansion in exploiting biomimicry in the urban built environment can be extended and comprehended by mimicking nature further beyond metaphor. It has likewise been shown that biological system biomimicry is a technique for managing the plentiful methods of building sustainable structures. The change required will not happen through new advancements and technologies solely, but also by applying new ideologies, innovations, and targets regarding the functionality of urban environments.

Climate change causes and impacts have been discussed. Climate change (CC) in countries such as Egypt, Italy, and Germany is presented and illustrated in terms of biomimicry and how it could mitigate CC. Also, how biomimetic design can respond more efficiently to CC is highlighted. In addition, biomimicry‘s role in mitigating GHG emissions from the built environment through energy efficiency and energy generation to offset the causes for CC is discussed. Moreover, biomimetic sequestering and storing carbon are presented and discussed for more comprehension of the relation between biomimicry and CC. The optimum case study of CH2 building in Australia has been portrayed as the best example of biomimicry when it comes to climate change mitigation. Furthermore, the benefits of the biomimetic approach to architecture over time have been highlighted and discussed. However, the main question remains if biomimicry is authentically the optimum approach to be followed to achieve sustainability and climate change mitigation and adaptation. It is questionable whether it is applicable and could be applied through different levels of biomimicry approaches and concept levels and could be effective in achieving sustainability and reducing GHG emissions. Perhaps, further research and studies are needed to answer these questions.

Did biomimicry face criticism? In the next chapter, the criticism toward biomimicry will be discussed to find out if it is really a solution to protect the environment and realize a more sustainable architecture.

References

1. Newburger, E. (2021). *Disasters caused $210 billion in damage in 2020, showing growing cost of climate change*. www.cnbc.com Retrieved 30 April 2021, from https://www.cnbc.com/2021/01/07/climate-change-disasters-cause-210-billion-in-damage-in-2020.html
2. *The costs of climate change – Mehr Dürren – Mehr Strüme – Mehr Ernteausfälle kosten mehr Geld – KlimaWandel koschtet* – Generation Nachhaltigkeit. Generation-Nachhaltigkeit.de. (2021). Retrieved 30 April 2021, from https://generation-nachhaltigkeit.de/the-costs-of-climate-change-mehr-duerren-mehr-stuerme-mehrernteausfaelle-kosten-mehr-geld-klimawandel-kostet-geld-und-leben/
3. Watkiss, P., Downing, T., Handley, C., & Butterfield, R. (2005). *The impact and cost of climate Change* (pp. i–vii). AEA Technology Environment and Stockholm Environment Institute – Oxford Office, Commissioned by European Commission – DG Environment. Retrieved from, https://www.ec.europe.eu/clima/sites/default/files/strategies/2020/docs/final_report2_en.pdf
4. Zari, M. (2010). Biomimetic design for climate change adaptation and mitigation. *Architectural Science Review, 53*(2), 172–183. Retrieved 1 April 2018 from https://doi.org/10.3763/asre.2008.0065
5. Climate Change Evidence: How Do We Know? (n.d.). *Global climate change: Vital signs of the planet*. Retrieved 20 December 2017 from https://climate.nasa.gov/evidence/
6. Jackson, R. (2021). *The effects of climate change. Climate change: Vital signs of the planet*. Retrieved 30 June 2021, from https://climate.nasa.gov/effects
7. Greene, T., & Jacobs, P. (2021). *2020 tied for warmest year on record, NASA analysis shows*. NASA. Retrieved 27 June 2021, from https://www.nasa.gov/press-release/2020-tied-for-warmest-year-on-record-nasa-analysis-shows
8. Matawal, D., & Maton, D. (2013). Climate Change and global warming: Signs, impact and solutions. *International Journal of Environmental Science and Development, 4*(1), 62–66. Retrieved 27 June 2021, from https://doi.org/10.7763/ijesd.2013.v4.305
9. The Paris Agreement | United Nations. (n.d.). United Nations. Retrieved 9 July 2021, from https://www.un.org/en/climatechange/paris-agreement
10. COP1 to COP25 – Find all the information & News about COP From COP1- COP25. (n.d.). Down To Earth | Latest news, opinion, analysis on environment & science issues | India, South Asia. https://www.downtoearth.org.in/climate-change/coplist
11. COP22 Marrakech Morocco - Sustainable Innovation Forum in partnership with United Nations Environment Programme (UNEP) | COP22 Marrakech. (n.d.). COP22 Marrakech

Morocco – Sustainable Innovation Forum in partnership with United Nations Environment Programme (UNEP) | COP22 Marrakech. https://www.cop22.org/
12. COP-1 to COP-25 Summary for UPSC – PSCprep-The biggest community of UPSC/PCS Aspirants. (n.d.). PSCprep-The biggest community of UPSC/PCS Aspirants Retrieved 9 July 2021, from https://pscprep.com/cop-1-to-cop-25-summary/
13. United Nations. (2016). Retrieved 9 July 2021, from https://unfccc.int/process-and-meetings/conferences/past-conferences/marrakech-climate-change-conference-november-2016/marrakech-climate-change-conference-november-2016
14. Cho, R. (2020). *COVID-19's long-term effects on climate change—For better or worse*. State of the Planet. Retrieved 9 July 2021, from https://news.climate.columbia.edu/2020/06/25/covid-19-impacts-climate-change/
15. *2020 Tied for Warmest Year on Record, NASA Analysis Shows*. NASA (2021). Retrieved 28 October 2021, from https://www.nasa.gov/press-release/2020-tied-for-warmest-year-on-record-nasa-analysis-shows
16. IPCC. (2014). In Core Writing Team, R. K. Pachauri, & L. A. Meyer (Eds.), *Climate change 2014: Synthesis report. Contribution of working groups I, II and III to the fifth assessment report of the intergovernmental panel on climate change* (p. 151). IPCC. Retrieved 27 June 2021, from https://www.ipcc.ch/site/assets/uploads/2018/05/SYR_AR5_FINAL_full_wcover.pdf
17. Sodangi, I. A., Izge, A. U., & Maina, Y. T. (2011). Climate change: Causes and effects on African agriculture. *Journal of Environmental Issues and Agriculture in Developing Countries, 3*(3), 23–25. https://www.oalib.com/paper/2789287#.YXr_GZpByUk
18. Climate Change Indicators: Atmospheric Concentrations of Greenhouse Gases | US EPA. US Environmental Protection Agency EPA. (2021). Retrieved 27 June 2021, from https://www.epa.gov/climate-indicators/climate-change-indicators-atmospheric-concentrations-greenhouse-gases
19. Change, N. (2021). *Carbon dioxide concentration | NASA global climate change. Climate change: vital signs of the planet*. Retrieved 20 December 2017, from https://climate.nasa.gov/vital-signs/carbon-dioxide/.
20. Kaddo, J. R. (2016). *Climate change: Causes, effects and solutions*. A with Honors Projects. 164. Retrieved 27 June 2021, from https://spark.parkland.edu/cgi/viewcontent.cgi?referer=&httpsredir=1&article=1163&context=ah
21. Harrabin, R. (2021). *Ban on new petrol and diesel cars in UK from 2030 under PM's green plan*. BBC News. Retrieved 27 June 2021, from https://www.bbc.com/news/science-environment-54981425.
22. LYNX. (2019). *Electric vehicles in Egypt opportunities and challenges*. Retrieved 27 June 2021, from http://lynxegypt.com/assets/pdfs/IndustryNotes1-2019.pdf
23. Heatwaves: Adapting to climate change. (2018). UK. Retrieved from https://publications.parliament.uk/pa/cm201719/cmselect/cmenvaud/826/826.pdf
24. Coronavirus disease 2019 (COVID-19) Situation Report – 51. (2020). Retrieved from https://www.who.int/docs/default-source/coronaviruse/situation-reports/20200311-sitrep-51-covid-19.pdf?sfvrsn=1ba62e57_6
25. |Campaigns | Springer Nature. Springernatue.com. (2021). Retrieved 3 May 2021, from https://www.springernature.com/jp/campaign/20210326-E/
26. *CIOESTE, São Paulo, Brazil – Climate Strategies*. Climate Strategies. (2021). Centre for Climate Strategies, USA. Retrieved 15 July 2021, from https://www.climatestrategies.us/projects-all/blog-post-title-two-fx6nb
27. Hetzner, C. (2021). *Automakers blast Europe's plan to ban gas-powered cars*. Fortune. Retrieved 15 July, 2021 from https://fortune.com/2021/07/14/automakers-europe-ban-of-new-combussion-engine-cars-by-2030
28. *Livable Cities Forum 2020 | ICLEI Canada*. Icleicanada.org. (2021). Retrieved 15 July 2021, from https://icleicanada.org/project/livable-cities-forum-2020
29. Mostafa, A., Wheida, A., El Nazer, M., Adel, M., El Leithy, L., Siour, G., et al. (2019). Past (1950–2017) and future (−2100) temperature and precipitation trends in Egypt. *Weather and Climate Extremes, 26*, 100225. https://doi.org/10.1016/j.wace.2019.100225
30. Fahmy, K. (2016). *Under the United Nations framework convention on climate change*. Egyptian Environmental Affairs Agency (EEAA). Retrieved 2 July 2021, from https://unfccc.int/sites/default/files/resource/TNC%20report.pdf

31. Egypt State of Environment Report 2008. (2009). *Arab Republic of Egypt Ministry of State for Environmental Affairs Egyptian Environmental Affairs Agency*. Retrieved 2 July 2021, from: http://www.eeaa.gov.eg/portals/0/eeaaReports/SoE2009en/Egypt%20State%20of%20Environment%20Report.pdf
32. Flis, A. (2021). *Global CO2 levels reach a new record high in 2020 despite the industry and Traffic slowdown during Covid-19 pandemic*. Severe Weather Europe. Retrieved 6 May 2021, from https://www.severe-weather.eu/news/global-co2-high-covid-19-fa/
33. *Egypt Submits its Climate Action Plan Ahead of 2015 Paris Agreement*. Unfccc.int. (2021). Retrieved 6 May 2021, from https://unfccc.int/news/egypt-submits-its-climate-action-plan-ahead-of-2015-paris-agreement/
34. Climate and Health Country Profile 2015 Egypt. UN & WHO. Retrieved 2 July 2021., from http://apps.who.int/iris/bitstream/10665/208860/1/WHO_FWC_PHE_EPE_15.06_eng.pdf
35. Aboulnaga, M., Elwan, A., & Elsharouny, M. (2019). *Urban climate Change adaptation in developing countries* (1st ed.). Springer. https://www.springer.com/gp/book/9783030054045
36. United Nations Office for Disaster Risk Reduction – UNODRR. (2020). Retrieved 6 May 2021, from https://www.undrr.org/news/covid-19-crisis-opportunity-cities-resilience
37. What do adaptations to climate change and climate resilience mean? (2020). *United Nations framework convention on climate change*. Unfccc.int. Retrieved 7 May 2021, from https://unfccc.int/topics/adapatations-and-resilience/the-big-picture/what-do-adaptation-to-climate-change-and-climate-resilience-mean
38. *What is the Paris Agreement?* (2020). *United Nations Framework Convention on Climate Change*. Unfccc.int. Retrieved 7 May 2021, from https://unfccc.int/process-and-meetings/the-paris-agreement/what-is-the-paris-agreement
39. *New Project to Support the Improvement of Air Quality and the fight Against Climate Change in Greater Cairo*. World Bank (2021). Retrieved 6 May 2021, from https://worldbank.org/en/news/press-release/2020/09/03/new-project-to-support-the-improvement-of-air-quality-and-the-fight-against-climate-change-in-greater-cairo
40. The National News – Environment. (2020). *The World Bank Approves 200m for Egypt's Air pollution and climate change project,* Retrieved from https://www.thenationalnews.com/business/economy/world-bank-approves-200m-for-egypt-s-air-pollution-and-climte-change-project-1.1086633
41. El-Batran, M., & Aboulnaga, M. (2014). Climate Change adaptation: An overview on challenges and risks in cities, regions affected costs and benefits of adaptation, and finance mechanisms. In W. Leal (Ed.), *Handbook of climate Change adaptation* (1st ed.). Springer. Retrieved 7 May 2021, from https://doi.org/10.1007/978-3-642-40455-9_124-1
42. *'Electricity Tariff'*. (2021). Egyptian Electricity and Consumer's Protection Institution, Retrieved 7 May 2021, from http://egyptera.org/ar/t3reefa.aspx
43. Ministry of Electricity and Renewable Energy – MoERE. (2020). https://www.moere.gov.eg
44. The Cleaner Energy Saving Mediterranean cities – CES-MED. (2011). Cesmed.eu. European Union. Retrieved 7 May 2021, from https://www.ces-med.eu
45. Guidebook – 'How to Develop a Sustainable Energy Action Plan (SEAP) in South Mediterranean Cities'. (2014). European Commission, EC– Joint Research Centre – Institute for Energy and Transport. EUR 27016 EN, https://doi.org/10.2790/392701. Retrieved from http://publications.jrc.ec.europa.eu/repository/handle/JRC93697
46. Bertoldi P., Cayuela, D. B, Monni, S., & Raveschoot, R. P. (2010). *Existing Methodologies and Tools for the Development and Implementation of Sustainable Energy Action Plans – SEAP*. European Commission Joint Research Centre (EC-JRC). Italy. Retrieved 7 May 2021, from https://publications.jrc.ec.europa.eu/repository/bitstream/JRC56513/reqno_jrc56513_existing_methodologies_and_tools_for_the_development_and_implementation_of_sustainable_energy_action.pdf
47. *'Egypt's Renewable Energy Laws and Regulations'*. (2020). Iclg.com. Retrieved 7 May 2021, from https://iclg.com/practice-areas/renewable-energy-laws-and-regulations/Egypt/
48. The Economist Intelligence Unit. (2021). *Global economy will be 3 percent smaller by 2050 due to lack of climate resilience*. Retrieved from https://www.eiu.com/m/global-economy-will-be-3-percent-smaller-by-2050-due-to-lack-of-climate-resilience

49. World Economic Forum Agenda. (2020). *International Monetary Monitory Fund, IMF-economy-Coronavirus-covid-19-recession*. Retrieved 14 January 2021, from https://www.weforum.org/agenda/2020/04/imf-economy-coronavirus-covid-19-recession
50. Ministry of Planning and Economic Development. (2021). Retrieved 6 May 2021, from https://www.mop.gov.eg
51. *Egypt approves 691 green projects for the FY 2020/21*. Internationalfinanace.com. (2020). Retrieved 6 May 2021, from https://intentionalfinance.com/egypt-approves-691-green-projects-fy-2020-21/
52. Moniem, D. (2020). EGP 36.7 billion in Egypt green projects to be implemented in FY2020/2021. *Alahram Online*. Retrieved 6 May 2021, from https://english.ahram.org.eg/NewsContent/3/12/373650/Business/Economy/EGP--bn-in-Egypt-greenprojects-to-be-implemented-.aspx
53. *'Green Cities Framework: Green City Action Plan'*. (2019). Ebrd.com. European Bank for Reconstruction and Development Retrieved 9 May 2021, from https://www.ebrd.com/cs/Satellite?c=Content&cid=1395287068226&pagename=EBRD%2FContent%2FContentLayout&rendermode=live%3Fsrch-pg
54. *'Green Cities Programmes- Methodology'*. (2016). European Bank for Re-construction and Development, OECD, and ICLEI for local Governments. Retrieved 9 May 2021, from https://www.ebrd.com/documents/technical-cooperation/green-city-action-plan-in-triana.pdf
55. IRENA – International Renewable Energy Agency. (2018). *Renewable energy outlook EGYPT*. IRENA. Retrieved 9 May 2021 from https://www.irena.org/publications/2018/oct/renewable-energy-outlook-egypt
56. From Appendix to the Metro Vancouver Climate 2050 Discussion Paper. (2018). Retrieved from http://www.metrovancouver.org/services/air-quality/AirQualityPublications/GlobalclimagechangetrendsevidenceAppendix2.pdf
57. Varrella, S. (2021). *Italy: share of elderly population in Italy 2009–2020 | Statista*. Statista Retrieved 9 May 2021, from https://www.statista.com/statistics/785104/elderly-population-in-italy/
58. Country Italy. Climate Change Knowledge Portal. (2021). Climateknowledgeportal. Retrieved 9 May 2021, from https://climateknowledgeportal.worldbank.org/country/italy#
59. *Covenant of Mayors for Climate and Energy Europe*. Climate Adapt. (2021). Retrieved 9 May 2021, from https://climate-adapt.eea.europa.eu/eu-adaptation-policy/convenant-of-mayors
60. Pietrapertosa, F., Salvia, M., De Gregorio Hurtado, S., D'Alonzo, V., Church, J., Geneletti, D., et al. (2019). Urban climate change mitigation and adaptation planning: Are Italian cities ready? *Cities, 91*, 93–105. https://doi.org/10.1016/j.cities.2018.11.009
61. M S. Planning for climate change in Italian cities: barriers, opportunities and future perspectives [Internet]. *ClimateChangePost*. 2017, Retrieved 9 April 2021]. Available from: https://www.climatechangepost.com/news/2017/6/27/planning-climate-change-italian-cities-barriers-op/.
62. Ministry for the Environment and Territory. (2021). *Environmental Action Strategy for Sustainable Development in Italy*. Available at: Retrieved 2 July 2021. Available from: https://www.un.org/esa/agenda21/natlinfo/countr/italy/Italian%20NSDS.pdf
63. European Commission. (2019). *Country Report – Italy*. Brussels. Retrieved from https://ec.europa.eu/environment/eir/pdf/report_it_en.pdf
64. Giuffrida, A. (2021). *Italian climate activists sue government over inaction*. The Guardian. Retrieved 6 July 2021, from https://www.theguardian.com/world/2021/jun/05/italian-climate-activists-sue-government-over-inaction
65. Lombardini, M. (2021). *Italy's energy and climate policies in the post COVID-19 recovery*. IFRI Center For Energy & Climate. Retrieved from https://www.ifri.org/sites/default/files/atoms/files/memo_lombardini_italy_necp_in_an_european_context_fev_2021.pdf
66. Italy – Adaptation Fund. Adaptation Fund. (2021). Retrieved 6 July 2021, from https://www.adaptation-fund.org/about/partners-supporters/italy/.
67. Ministry for the environment land and sea. (2019). Fourth Biennial Report, Italy. Retrieved from https://unfccc.int/sites/default/files/resource/BR4_2019%20Italy.pdf
68. Lazaro, O. (2017). *Analysis of National Initiatives for Digitising Industry: Italy Piano Nazionale Industria 4.0*. Retrieved from https://ec.europa.eu/futurium/en/system/files/ged/it_country_analysis.pdf

69. *Sharing adaptation information across Europe*. Climate Adapt. (2019). Retrieved 9 April 2021, from https://climate-adapt.eea.europa.eu/countries-regions/countries/italy
70. *Germany – Gross domestic product (GDP) 2026 | Statista*. Statista. (2021). Retrieved 23 June 2021, from https://www.statista.com/statistics/375206/gross-domestic-product-gdp-in-germany/
71. *Current population*. Federal Statistical Office. (2021). Retrieved 22 June 2021, from https://www.destatis.de/EN/Themes/Society-Enviornment/Population/Current-Population/_node.html
72. *Germany | Climate Action Tracker*. Climateactiontracker.org. (2021). Retrieved 22 June 2021, from https://climateactiontracker.org/countries/germany/
73. Kreibich, H., Bubeck, P., Kunz, M., Mahlke, H., Parolai, S., Khazai, B., et al. (2014). A review of multiple natural hazards and risks in Germany. *Natural Hazards, 74*(3), 2279–2304. https://doi.org/10.1007/s11069-014-1265-6
74. Climate Action Plan 2050. (2016). *Grantham research Institute on Climate Change and the Environment*. Retrieved 9 April 2021, from https://climate-laws.org/geographies/germany/policies/climate-action-plan-2050
75. Climate change: EU to cut CO2 emissions by 55% by 2030. BBC News. (2021). Retrieved 6 July 2021, from https://www.bbc.com/news/world-europe-56828383
76. World Bank Climate Change Knowledge Portal. Climateknowledgeportal.worldbank.org. (2021). Retrieved 9 April 2021, from https://climateknowledgeportal.worldbank.org/country/germany
77. *Climate Action Plan 2050 – Germany's long-term low greenhouse gas emission development strategy*. Bundesministerium für Umwelt, Naturschutz und nukleare Sicherheit. (2021). Retrieved 23 June 2021, from https://www.bmu.de/en/topics/climate-energy/climate/national-climate-policy/greenhouse-gas-neutral-germany-2050/
78. Responding to the COVID-19 Pandemic Leaving No Country Behind. (2021). United Nations (ESCAP), the Asian Development Bank (ADB) and the United Nations Development Programme (UNDP). Retrieved 7 July 2021, from https://www.undp.org/sites/g/files/zskgke326/files/publications/UNDP-RBAP-SDG-Responding-to-COVID-19-Pandemic-Leaving-No-Country-Behind-2021.pdf
79. Shulla, K., Voigt, F., Cibian, S., et al. (2021). Effects of COVID-19 on the Sustainable Development Goals (SDGs). *Discover Sustainability, 2*, 15. Retrieved 7 July 2021, from https://doi.org/10.1007/s43621-021-00026-x
80. Climate adapts sharing adaptation information across Europe. (2019). Retrieved 9 April 2021, from https://climate-adapt.eea.europa.eu/countries-regions/countries/germany
81. Leordean, L. (2016). *Biomimicry in architecture – Mitigation and adaptation to climate change (Master)*. Politecnico di Milano. Retrieved 7 July 2021, from https://issuu.com/ioanaleordean/docs/thesis_book_online-compressed
82. Intergovernmental Panel on Climate Change. (2007). *Climate change 2007: Impacts, adaptation and vulnerability*. Retrieved 7 July 2021, from https://www.ipcc.ch/site/assets/uploads/2018/03/ar4-wg2-intro.pdfs
83. Zari, M., & Storey, J. (2007). An ecosystem based biomimetic theory for a regenerative built environment. In *Biomimetic architecture and cities*. Retrieved 15 April 2018, from https://www.irbnet.de/daten/iconda/CIB11734.pdf
84. Abd Elrahim, S. (2021). *Ministry of Environment – EEAA > Topics > Air > climate change*. Eeaa.gov.eg. Retrieved 20 December 2018, from http://www.eeaa.gov.eg/en-us/topics/air/climatechange.aspx
85. Steemrs, K. (2010). Towards a research agenda for adapting to climate change. In *Building research and information* (pp. 291–301). University of Cambridge, Online. https://www.tandfonline.com/doi/abs/10.1080/0961321032000097692
86. Sctebn, M., & Leary, D. (2009). Climate Change and renewable energy from the ocean and tides: Calming the sea of regulatory uncertainty. *The international Journal of Marine and Coastal Law, 24*(4), 617–651. Retrieved 15 April 2018, from https://brill.com/view/journals/estu/24/4/article-p617_3.xml
87. Jamei, E., & Vrcelj, Z. (2021). Biomimicry and the built environment, learning from Nature's solutions. *Applied Sciences, 11*, 7514. https://doi.org/10.3390/app11167514

Chapter 6
Biomimicry in Criticism: Argument, Defense, and the Direction Toward Sustainability

> Biomimicry is basically taking a design challenge and then finding an ecosystem that's already solved that challenge, and literally trying to emulate what you learn. (Janine Benyos)
> I think the biggest innovations of the 21st century will be at the intersection of biology and technology. A new era is beginning. (Late Steve Jobs)

6.1 Introduction

Biomimicry is a fair approach in the global efforts to save our planet and preserve our environment amidst climate change impacts and the coronavirus (COVID-19) pandemic (Figs. 6.1 and 6.2). As stated by the late Steve Jobs, former founder and CEO of Apple, the biggest innovations of the twenty-first century will be through the intersection of biology and technology. The combination of elements of architecture with nature and technology would be the best way to solve the current problems of cities' resilience. Figure 6.3 shows the cactus shell surviving with the minimum water consumption. Such behavior of the cactus plants can be emulated in buildings to save water consumption through technology and make these buildings sustainable and resilient.

Although biomimicry, as an idea, is not discovered today, it was already practiced from the start of the creation, as mentioned earlier. Considering biomimicry as an approach and applying it in different fields have been successful. It should take into contemplation the profound meaning of biomimicry, not only mimicking the forms of what exists in nature (Figs. 6.3 and 6.4), which will assist in developing a better environment and sustainable life. However, the real application of biomimicry should address the behavior level and ecosystem level of the organisms and yet translate their ways of adaptation to problems in nature as presented in Fig. 6.4. The creation of these organisms reflects a high percentage of perfection in a sense that is inspired from nature (Fig. 6.5). In addition, it is a tool and a language that could be applied on a human scale and fulfilling to our needs. In this chapter, arguments and criticisms of biomimicry will be reviewed and discussed.

M. Aboulnaga, S. E. Helmy, *Biomimetic Architecture and Its Role in Developing Sustainable, Regenerative, and Livable Cities*,
https://doi.org/10.1007/978-3-031-08292-4_6

Fig. 6.1 Save our planet – forests in Helsinki, Finland, October 2019. (Image credit and source: Mohsen Aboulnaga)

6.2 Arguments on Biomimicry

There is an argument about biomimicry; however, it is critical to be practical about where nature's systems and strategies will and will not help. There are certainly a few disadvantages to the way life is designed, which most likely would prefer not to emulate [1]. In some literature such as Steven Vogels' book, *Cats' Paws and Catapults* is a piece of writing that shows the negative arguments toward biomimicry. Vogel refers to this as "naive biomimicry." The word "naïve" referred to the scientists and inventors who were using the same and exact technology found in nature without any adjustments, turning it into something at human scale. By implementing such architectural designs, the majority of these trials and examples were unsuccessful [1].

In an appraisal by the Yale Environment Review (YER), it raises an important question: Does nature hold the answer to sustainability [2]?

> Biomimicry is the science of studying nature's models and taking inspiration from it to solve human problems. A recent study examines this concept and its philosophical origins – and whether it can help humankind tackle the challenges of sustainability. (Timila Dhakhwa, Yale School of the Environment)

Biomimicry as ecological innovation is vital due to the growing concern of unsustainable utilization of natural resources. According to a recent European Union report – Manifesto for a Resource-Efficient Europe – that stressed the mounting pressure on resources and the environment, the EU has no option expected to

Fig. 6.2 Preservation of our environment is crucial amidst climate change risks and the coronavirus (COVID-19) pandemic. (Image credit and source: Mohsen Aboulnaga)

shift toward sustainability and thus the rest of the world [2, 3]. It was stated that this transition has crafted interest in what is called "biomimetic approaches to technology and innovation." Is this new approach humanity's ultimate answer? Timila Dhakhwa says [2]. In fact, this question has been answered by a group of

(a) (b)

Fig. 6.3 Cactus plants' shell in hot areas inspiring biomimicry in architecture (**a**) A Cactus plant (**b**) A group of Cactus shells shading each other. (Images' credit and source: Mohsen Aboulnaga)

researchers at Wageningen University in the Netherlands. They attempted to critically assess the theoretical and philosophical foundation behind biomimicry, which is by and large known as biomimicry [2].

Biomimicry is one of the biological sustainable approaches that assist in having environmentally friendly architecture and cities. In the field of architecture, there are nine possible levels in which biomimicry can occur, taking into account that nature works in systems. Architects can mimic a natural "concept," natural "process," "morphology," "form," "structure," "skin," and "material"; also "expressions" and "symbols" can be borrowed from nature to complete the architectural experience.

Seemingly, it is imperative to mention that architecture similarly works in a system. Figure 6.6 illustrates the nine levels of biomimicry in the fled of architecture. All these levels can be borrowed from nature to complete the architectural experience. Nevertheless, there are many thoughts and arguments toward imitating nature's ideas, concepts, and behaviors from its organisms; on the other hand, there are many defenses toward this approach and many proofs that it is an excellent solution in different aspects.

However, there is a question that needs to be thought of. Do all biomimicry concept applications, through their different levels, contribute to attaining sustainable and environmentally friendly architecture and achieving the sustainable development goals (SDGs)? Nonetheless, this question could be answered in a thesis.

Two concepts were studied and used by a Dutch research team: the first is centered on the conceptualization of nature; and the second is based on the conceptualization of mimesis or imitation to describe biomimicry [2]. Figure 6.7 illustrates

Fig. 6.4 Geometric pattern shell inspiring biomimetic architecture. (Image credit and source: Mohsen Aboulnaga)

these explanations in detail with the two concepts of biomimicry [2], while Fig. 6.8 highlights the difference between strong biomimicry and weak biomimicry.

By considering these two concepts as a core, the research team establishes what is described as strong and weak biomimicry. In this regard, strong biomimicry assumes perfection of nature and takes mimicry as reproduction of nature, whereas weak biomimicry acknowledges deficiencies in nature as well as the role of mimicry to supplement nature [2]. Nevertheless, strong biomimicry bears the risk of crafting the argument that biomimicry is the definitive solution due to being natural, while weak biomimicry avoids this mistake. In conclusion, the team reached a deduction that strong biomimicry could be more challenging conceptually than weak biomimicry.

An Egyptian pioneer architect Professor Ali Raafat also mentioned in his book *Architecture of the future* that sometimes metamorphic designs look primitive, shallow, and naïve as they imitate elements that look like a human face or an elephant, for example. This occurred throughout the history of architecture, and it is not taken in a serious matter, unless it was designed for a political or ideological matter, such

Fig. 6.5 Vertical cactus tower – inspiration from nature. (Image credit and source: Mohsen Aboulnaga)

as the Sphinx and the status of goddess and kings as illustrated in Figs. 6.9 and 6.10, respectively.

Designs from nature ought to be an elegant and sentimental assurance in which natural design is somehow better than synthetic innovation. The clearest counterarguments Steven Vogel realizes are motion via air and water [4]. In the early beginning, inspiring the idea of inventing the flight from the flying creatures put the designers in the wrong direction. Feathered creatures are smaller than individuals;

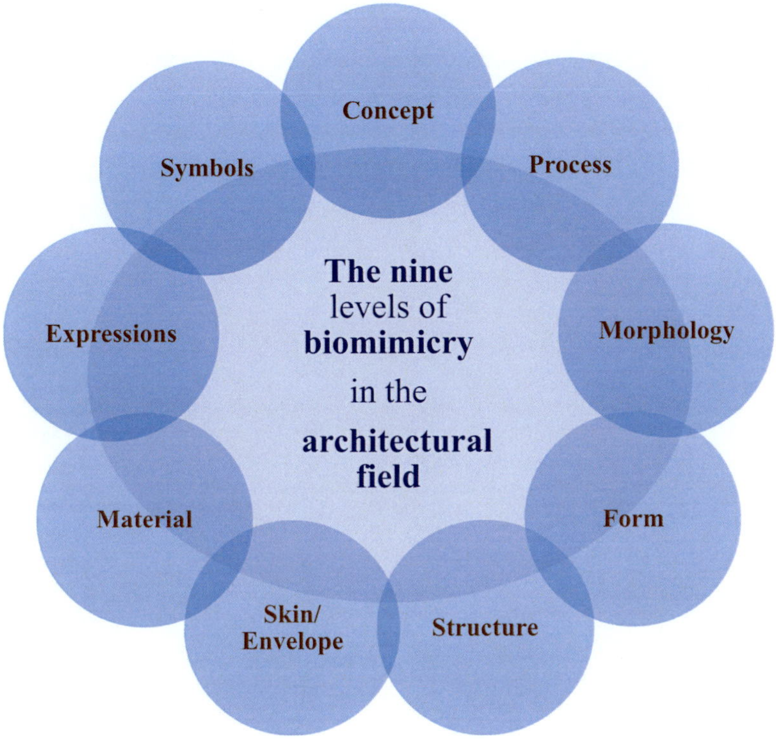

Fig. 6.6 The nine possible levels of biomimicry in the field of architecture. (Image source: Developed by authors)

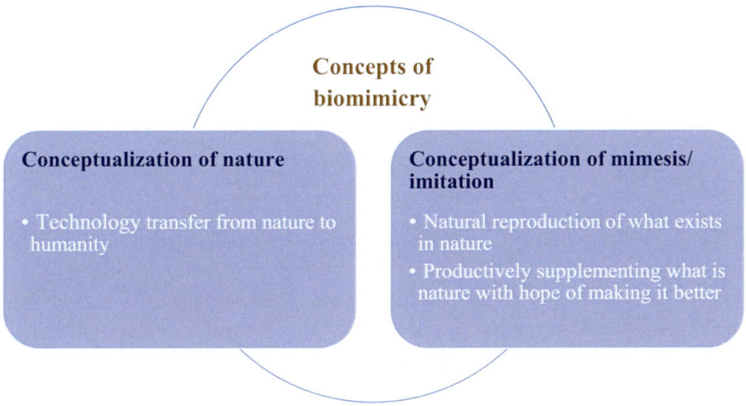

Fig. 6.7 Two concepts of biomimicry. (Image source: Developed by authors)

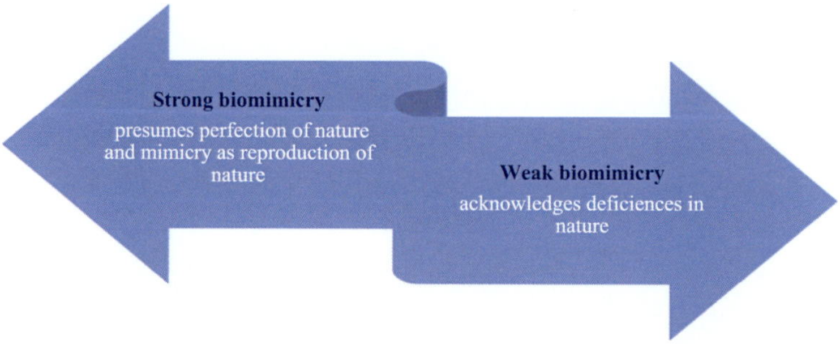

Fig. 6.8 Strong biomimicry versus weak biomimicry. (Image source: Developed by authors)

(a) **(b)**

Fig. 6.9 Sphinx, status in Giza, Egypt. (**a**) The face of pharos of Sphinx (**b**) Sphinx, a mimicked face of a pharos on a lion body. (Images' credit and source: (**a**) https://commons.wikimedia.org/wiki/Files:Sphinx.jpg, (**b**) https://commons.wikimedia.org/wiki/Files:Great_Sphinx_of_Giza_2008716a.jpg)

thus, the attributes to invent a flight should be different and innovative. How can huge and heavy elements stay up high? They need to go quicker to stay up high. The same concept was used in getting inspired from the ship form and the movement of the marine creatures in the water, but the problem was in the different two scales [4]. Therefore, getting inspiration from nature was just in the form, but it was not applicable in the function according to the big difference [4].

This book compares and contrasts human innovation with common innovation in areas such as structure, shape, and materials as well as headway, utilizing loads of cases from the universes of science and antiques. Steven Vogel clarifies the physical standards, advantages, and tradeoffs for the diverse plan approaches, utilizing words to portray the fundamental math. The material would be considerably essential with energized number of crunchers that demonstrated the changing properties of flight [4].

The debate is made with a light hand, and the bulk of the book comprises of superb correlations and complexities between different methods for tackling design issues. With regard to biomimicry, Vogel states and argues the best cases that include

Fig. 6.10 Status of goddess and kings. (Image credit and source: Marcus Cyron, https://commons.m.wikimedia.org/wiki/File:King_Menkaura,_the_goddess_Hathor,_and_the_deified_Hare_nome_1jpg)

a few parts of a characteristic design, for example, (a) a dolphin's streamlined shape rousing streamlined vehicles, (b) a scarab's jaws motivating cutting tool teeth, and (c) the cement qualities of burr inspiring Velcro [4, 5]. These models are adjusted from nature to the particular plan issue and materials required for the human necessity [4, 5].

The critique of naive biomimicry concentrates on a great extent around the working attributes of the innovations, mainly how they take care of the displaying plan issues of structure and movement. In pursuing this, Vogel misses a couple of key focuses on how and why human architects should need to imitate nature [4]. Nonetheless, Vogel clarifies that human innovation can use considerably higher temperatures and different temperature ranges than the source of inspiration creatures [4]. It is clear that this clarification highlights the importance of human innovation in supporting mimicry architecture.

Another topic that the book does not address at all is waste since human production processes have tended to create vast quantities of frequently harmful waste and air and soil pollution, which is one of the sectors influencing climate change. In addition, the nonbiodegradable plastics, heavy metals in rivers, and fertilizer created dead zones in oceans. Nonetheless, natural processes tend to consume byproducts instead of creating waste, perhaps due to the fact that they evolved at a slower scale in the context of ecosystems. An additional example is that of birds digesting fruit pulp and excreting the seeds that help grow another plant. Moreover, animals, at the end of their life, become a source of food for vultures, larvae, and bacteria [4]; hence, waste is minimized as such.

In the same line, human mechanical advances have been colossally inefficient in terms of energy and material consumption. Our way of life needs more viable procedures, not because it sounds sentimental but rather in light of the fact that the present metaphor will not last. Steven Vogel's understanding is that characteristic models are best with adjustments not imitating [4, 6]. There are different areas where science and innovation have gone past the data accessible to Vogel when he composed the book ten years back [4].

It seems that not all the examples in the architectural field that mimic nature, especially at organism level or in form, should be environmentally friendly or sustainable. Thinking ecologically and looking at nature for inspiration should be pursued with reasons not only to imitate the form of the organisms but also to interact with nature and to apply and sustain the same concept in the building. But some examples are not reflecting these concepts rather than imitating creatures' forms of animals or plants to shape buildings. Thus, they cannot be considered biomimicry; however, it is just imitating shapes and forms.

6.3 Naïve Biomimetic Architectural Design Examples

In Steven Vogel's book, *Cats' Paws and Catapults*, a piece of literature brings to reality some of the negative arguments toward biomimicry. He refers to this as "naïve biomimicry" which means scientists and inventors who used the exact same technology found in nature and without any adjustments, turning it into something at human scale [7]. All implemented examples were unsuccessful according to Vogel. Box 6.1 shows biomimicry versus naïve biomimicry.

Examples of such naïve biomimicry are depicted in buildings around the world that were built to look like all sorts of creatures from cats to koalas. Imaginative architects have been designing buildings worldwide that are shaped like animals. Pet lovers can visit hotels, churches, and museums resembling dogs, cats, and even elephants while on their travels [8]. The awe-inspiring structures, designed by a variety of animal enthusiasts, can be found in countries such as Australia, Germany, India, New Zealand, and the United States. One of the oldest animal structures is Lucy, a 65 ft. elephant in New Jersey, USA, which was built in 1881 as a landmark for the town of Margate as shown in Fig. 6.11a [8]. The Trojan

Horse hotel in Durbuy, Belgium, which has its own lowering bridge, has each room, following a fantasy theme (Fig. 6.11b). Other buildings in animal's configuration include the Big Sheep in New Zealand, a former wool and craft shop (Fig. 6.11c), and the Cat School in Germany, which is built as a place for people to take their cats to be looked after (Fig. 6.11d). In addition, there are other naïve biomimicries such as the space-age like the Dog Bark Park Inn, created by Frances Conklin and Dennis Sullivan, in Cottonwood, Indiana, USA, as a dog-themed hotel in the shape of a beagle for dog lovers (Fig. 6.11e); the Seminole Hard Rock Hotel & Casino in Florida, USA (Fig. 6.11f); the Shrine to Anglers building in the shape of a jumping Muskie fish at the Fresh Water Fishing Hall of Fame in Hayward, Wisconsin, USA (Fig. 6.11g); and the fish building which serves as an office for the National Fisheries Development Board in Hyderabad, India (Fig. 6.11h). Although the Seminole Hard Rock Hotel & Casino in Florida is considered a naïve example of architecture, it can be related to expressionist naïve architecture rather than a naïve biomimicry.

Moreover, the Universe Science Center is a clam-shaped building in Bremen, Germany, where the building is a science museum built in the shape of a clam that originally opened in September 2000 [8] as presented in Fig. 6.11i; and the Kakadu Crocodile Hotel in Jabiru, Australia, is in the shape of the reptile and has even been built using green materials as shown in Fig. 6.11j. Furthermore, the Big Duck – a duck shop (Fig. 6.11k) – and the Giant Koala tourist complex in Australia are shown in Fig. 6.11l. Contemporary examples of naïve biomimicry exist in Thailand and the UAE as illustrated in Fig. 6.12 [8].

By looking at these examples, it is clear that these buildings, which are created by imaginative architects across the world, are ridiculously imitated shapes of the animals with no sustainable or environmental impact on the architecture.

In contrast, the lotus flower is known to be associated with purity and faithfulness. The color of the lotus flower indicates many meanings, for example, the blue

Box 6.1: Biomimicry Versus Naïve Biomimicry

Biomimicry does literally mean the 'imitation of nature', but a literal imitation of nature is not what biomimicry is. (CHY)

lotus flower is associated with the spirit of wisdom and knowledge, while the white lotus flower represents a state of mental purity as shown in Fig. 6.13. In contrast, the red or fuchsia lotus flower is associated with the compassion and passion for every Buddhist as presented in Fig. 6.14. Such fuchsia lotus flowers are associated with purity, spiritual awakening, and sincerity for all Buddhist saints. Nonetheless, the lotus leaf and flower have nano-sized hydrophobic surface [9] that allows the raindrops roll off carrying dirt particles, leaving the surface clean. Lotus leaves and flowers have also a wax-like surface for each epidermal cell [9]. The properties of cleaning surfaces in plant leaves have opened the possibility of manufacturing various ultra-hydrophobic products, such as paints, glass, and windows. An example is

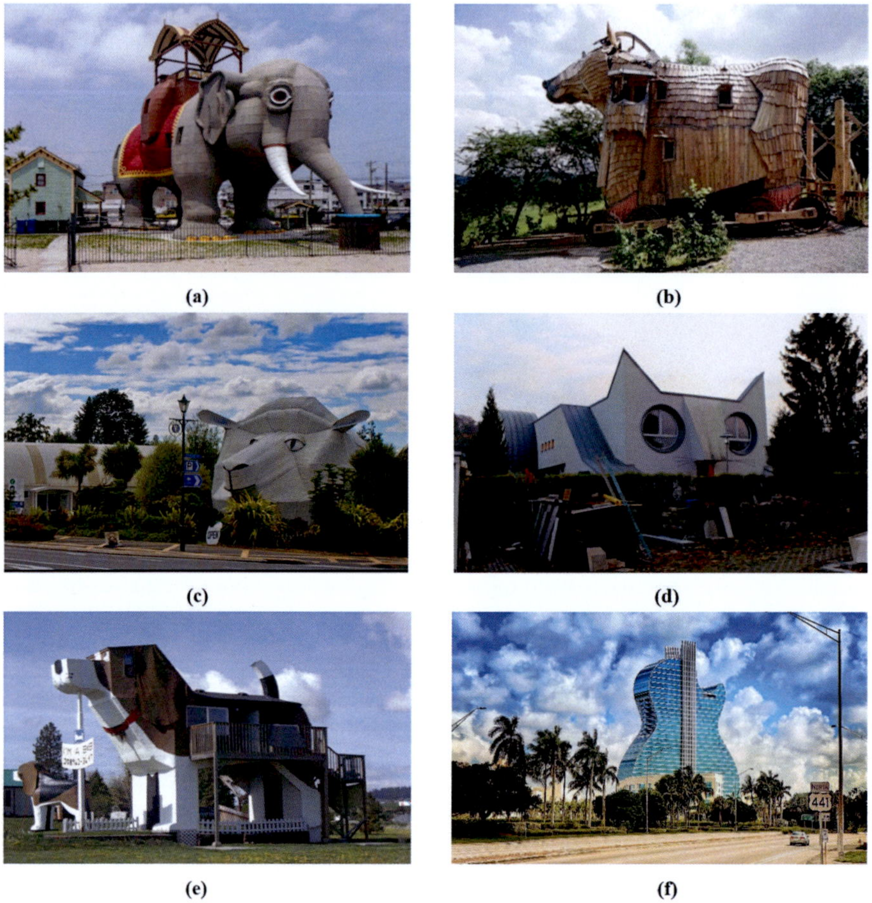

Fig. 6.11 Naïve biomimicry examples of architectural buildings worldwide, (**a**) Lucy elephant structure in New Jersey, USA (**b**) Trojan horse-shaped hotel in Durbuy, Belgium (**c**) Big sheep in New Zealand (**d**) A cat-shaped building in Australia (**e**) The Dog Bark Park Inn in New York, USA (**f**) Seminole Hard Rock Hotel & Casino, Florida (**g**) Jumping Muskie Fish, Hayward, Wisconsin, USA (**h**) Fish Building Hyderabad, India (**i**) Universe Science Center is a clam-shaped building in Bremen, Germany, (**j**) Kakadu Crocodile Hotel in Jabiru, Australia (**k**) Big Duck in New York, USA (**l**) Giant Koala Tourist complex in Australia. (Images' credit and source: (**a**) https://commons.m.wikimedia.org/wiki/File:Lucy_the_Elephant_NJ3.jpg, (**b**) https://inhabitat.com/guests-can-sleep-in-a-fantastical-trojan-horse-shaped-hotel-in-belgium/, (**c**) https://commons.m.wikimedia.org/wiki/Category:Sculpture_of_sheep_in_New_Zealand#/media/File%3AThe_dog_and_sheep_building%2C_Tirau%2C_Waikato%2C_New_Zealand%2C_3_April_2008.jpg, (**d**) https://commons.m.wikimedia.org/wiki/File:Kindergarten_Wolffartsweier14112016_1.jpg, (**e**) https://commons.m.wikimedia.org/wiki/File:Dog_Bark_Park_Inn.jpg, (**f**) https://commons.wikimedia.org/wiki/Category:Seminole_Hard_Rock_Hotel_%26_Casino_Hollywood#/media/File:Hard_Rock_Casino_Hotel_Fort_Lauderdale_Florida,_June_2021.jpg, (**g**) https://commons.m.wikimedia.org/wiki/File:HaywardMuskie-061-050507.jpg#, (**h**) https://commons.m.wikimedia.org/wiki/File:Fish_shaped_building.jpg, (**i**) https://commons.m.wikimedia.org/wiki/File:Universum®_Science_Center,_Bremen,_Germany.jpg, (**j**) https://commons.m.wikimedia.org/wiki/File:Kakadu_2427.jpg#, (**k**) https://commons.m.wikimedia.org/wiki/File:The_Big_Duck.JPG, (**l**) https://commons.m.wikimedia.org/wiki/File:Giant_Koala.JPG)

(g) (h) (i) (j) (k) (l)

Fig. 6.11 (continued)

the large-scale imitative Lotus Conference Center building in Wujin is a government center in Wuhan District, China, which resembles the pink lotus flower. The structure is developed as a combination of a sculpture and a powerful sensuous and feminine biological form as shown in Fig. 6.15. The building, which is described as a striking design inspired by the blooming of lotus flowers, is the most famous icon in the city [9].

Another building example of the lotus flower is the ArtScience Museum, which is one of the most eye-catching buildings recently constructed in Singapore. The

(a) (b)

Fig. 6.12 Additional of naïve architectural buildings. (**a**) The Elephant Building in Thailand (**b**) New Moon Building in the UAE. (Images' credit and source: (**a**) https://upload.wikimedia.org/wikipedia/commons/e/e9/ตึกช้าง_เขตจตุจักร_กรุงเทพมหานคร_Elephant_Building%2CChatuchak%2CBangkok%2CThailand_%282%29.jpg; (**b**) http://www.dailymail.co.uk/news/article-3790822/Animal-house-Buildings-world-look-like-sorts-creatures-cats-koalas.html, m and n: https://littlegreenseed.wordpress.com/2012/12/02/this-is-not-biomimicry/)

Fig. 6.13 The white lotus flowers represent a state of mental purity – Ghent, Belgium. (**a**) Fuchsia lotus flower (**b**) Lotus flowers in leaves (**c**) White lotus flowers. (Image credit and source: Mohsen Aboulnaga)

(a) (b) (c)

Fig. 6.14 The lotus flower. (**a**) Front view of imitative Lotus Conference Center (**b**) Bird's-eye view of the Lotus Conference Center. (Image credit and source: Mohsen Aboulnaga)

(a) (b)

Fig. 6.15 Wujin Lotus Building in China. (Image credit and source: https://laughingsquid.com/the-lotus-building-a-massive-sculptural-building-in-china-that-evokes-3-stages-of-a-blooming-lotus-flower/, https://www.pinterest.com/pin/513692519792551O2/)

ArtScience Museum is a unique structure designed in the shape of an open lotus, as illustrated in Figs. 6.16 and 6.17, while the inspiration of water lilies' flowers informing the design of the ArtScience Museum's iconic building are shown in Figs. 6.18 and 6.19. The skin of the giant lotus is made of stainless steel composite material; it's a noncombustible mineral-filled core, sandwiched between sheets of stainless steel, which makes it both aesthetically beautiful and highly durable as presented in Figs. 6.16 and 6.17c. Using these composite sheets, the luxurious design and high corrosion resistance are greatly reduced due to their own weight [9].

The ArtScience Museum is Singapore's largest permanent digital art gallery, enabling imaginative play and exploration; it allows the visitors to experience exploring the crossroads of art, science, technology, and culture, yet it creates a fluid combination of artistic expression, technical ingenuity, and scientific inquiry [9]. The panoramic view of the ArtScience Museum in front of Singapore's landmark Marina Bay Sands is shown in Fig. 6.20.

As far as biomimetic architecture is concerned, the rules of standard architecture still apply the notion of "form follows function." This notion is an essential rule in nature as well – every part of a plant or animal performs a crucial role in their

Fig. 6.16 ArtScience Museum in the City of Singapore, Singapore. (Image credit and source: Basile Morin, https://commons.m.wikimedia.org/wiki/File:ArtScience_Museum,_Marina_Bay_Sands,_Singapore.jpg)

survival. Therefore, if the design of a building form mimics nature, but it does not achieve efficiency or sustainability measures, then such a building cannot be classified as truly biomimetic [9, 10]. For example, the "Sky City" project furnishes its inhabitants with an oasis in the sky and overcomes pollution and buzz of the city as shown in Fig. 6.21, using the metaphor of the lotus flower which can grow into a beautiful blossom above dirty water [10]. The lotus-shaped design with a beautiful and inspiring contrast against the square gray buildings below is purely a visual gesture. If those towers were simple unadorned poles and platforms, the function of this design would be unaltered, rendering the lotus shape and form as decorative only [9, 10].

As for the metaphor of the flowering blossom above dirty water, there are some social and environmental aspects, associated with it, but it should be addressed. Are we going to great lengths here to elevate ourselves above the problems we have created rather than remediating them? Hence, creating an exclusive oasis for a small group of people could be the answer. Architectural biomimicry is a complex and multidimensional form of design, which goes beyond putting an organic shape or natural metaphor onto an inherently unsustainable building. A biomimetic built environment should function successfully for both human and ecosystem health and happiness [10] (Box 6.2).

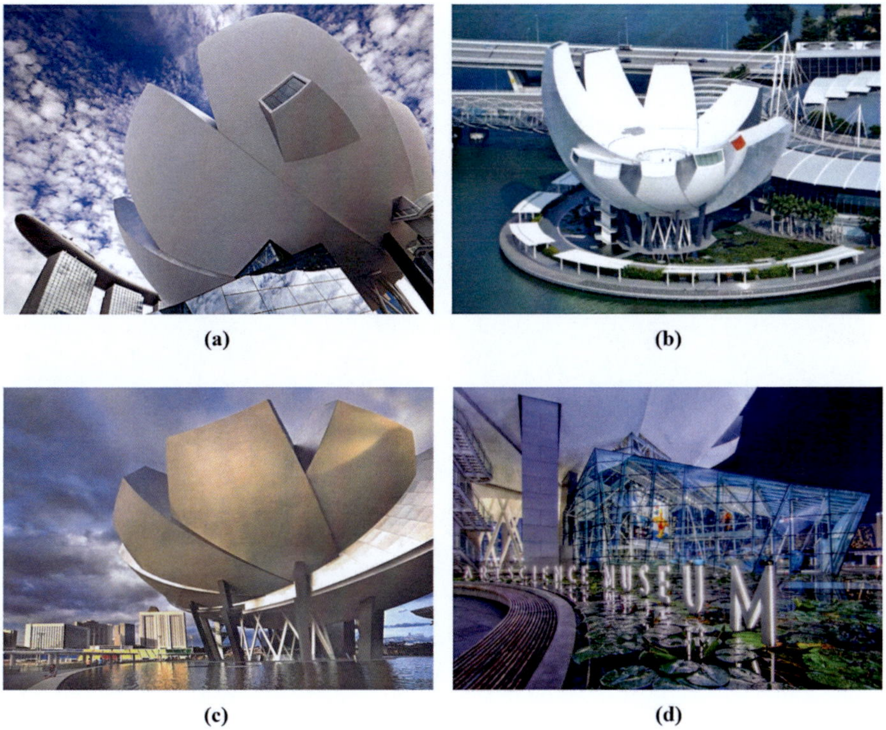

Fig. 6.17 ArtScience Museum in the City of Singapore, Singapore. (**a**) Close-up view of the lotus leaves of the ArtScience Museum – Marina Sands Bay (**b**) Waterfront façade of the ArtScience Museum (**c**) The stainless steel lotus leaves' façade of the ArtScience Museum (**d**) The entrance of the ArtScience Museum as night. (Images' credit and source: (**a**) William Cho, https://commons.m.wikimedia.org/wiki/File:ArtScience_Museum,_Marina_Bay_Sands,_Singapore_-_20110221-02.jpg#; (**b**) Zairon, https://commons.m.wikimedia.org/wiki/File:Singapore_ArtScienceMuseum_viewed_from_UOB_Plaza.jpg; (**c**) William Cho, https://en.m.wikipedia.org/wiki/File:Singapore_ArtScience_Museum_1.jpg (**d**) Erwin Soo, https://commons.m.wikipedia.org/wiki/File:ArtScience_Museum,_Marina_Bay_Sands,at_night._-_Singapore._(8416803612).jpg)

Box 6.2: Biomimicry as a Path Toward Sustainable Future

The conscious emulation of life's genius is a survival strategy for the human race, a path to a sustainable future. The more our world functions like the natural world, the more likely we are to endure on this home that it is ours, but not ours alone. (Janine Benyus)

The Sky Tower, in Wroclaw, the third tallest tower in Poland, is an example of biomimicry where the design of the tower was inspired from the vertical cactus to provide stability and verticality in a sleek way [11]. The 51-floor tower, as a part of three buildings, is a unique example of biomimetic architecture as shown in

Fig. 6.18 Water lilies' flowers inspiring the design of the ArtScience Museum in Singapore. (Image credit and source: Dietmar Rabich, https://commons.m.wikimedia.org/wiki/File:Singapore_(SG),_ArtScience_Museum,_Water_Lilies_%2D%2D_2019_%2D%2D_4538.jpg)

Fig. 6.19 The ArtScience Museum at night, Singapore. (Image credit and source: Niaz Morshed, https://commons.m.wikimedia.org/wiki/File:Singapore_Art_Science_Museum.jpg)

Fig. 6.20 The ArtScience Museum at Marina Sands Bay in the City of Singapore, Singapore. (Image credit and source: Niaz Morshed, https://commons.m.wikimedia.org/wiki/File:ArtScience_Museum,_Marina_Bay_Sands,_Singapore.jpg)

Fig. 6.21. The construction of the 212-meter Sky Tower used 25,000 tons of steel and 4 hectares of glass. Nonetheless, the footprint data of the tower is not available to evaluate its sustainability. As seen in Fig. 6.21b, the simple form and verticality of the cactus plant inform the design of the Sky Tower.

Another example of biomimicry architecture is the Lotus Tower in Colombo, Sri Lanka. The pink and yellow tower base is inspired by the lotus throne which is formed of two inverted trapezoids and comprises four floors [12, 13]. The 356-meter tower house shaped like a lotus flower is composed of 8 floors as shown in Fig. 6.22. The lotus flower opens and closes as presented in Figs. 6.23 and 6.24. The cap of the Lotus Tower is inspired from the lotus flowers when closed, as illustrated in Fig. 6.24.

The Sky Garden Skyscraper at the 20 Fenchurch Street in London, UK, exhibits a new futurism style as well as two lofty inspirations from nature. The first inspiration mimics the cactus-shaped plant as illustrated in Fig. 6.25. The second inspiration brings nature (greenery) in a three-level garden at the building's top floors. The 38-storey commercial building called "the Walkie-Talkie" has a distinctive shape like a Kyoto tree (Fig. 6.26a), and a hedgehog-shaped cactus from the top (Fig. 6.26b). The 160-m high building in London's financial district is a great model of biomimicry inspired from nature. The unique skyscraper is designed by Architect Rafael Viñoly [14].

(a) (b)

Fig. 6.21 The Sky Tower in Wroclaw, Poland. (**a**) The Sky Tower design inspired from nature (**b**) Vertical cactus plants. (Image credit and source: (**a**) Jacek Halicki, https://commons.m.wikimedia.org/wiki/File:2017_Sky_Tower_we_Wroclaiu.jpg, and (**b**) Image credit – Mohsen Aboulnaga)

The Sky Garden Skyscraper's envelope and interior are shown in Figs. 6.27, 6.28, and 6.29. It is clear from Figs. 6.26, 6.27, and 6.28 that the design of the Sky Garden Skyscraper form was inspired from nature (Fig. 6.26a, b). It also brought nature to the top of the building (Fig. 6.26c), where the public enjoy livable spaces and a panoramic view of the City of London as presented in Figs. 6.27b and 6.29. The exterior of the top floors is shown in Fig. 6.28.

6.4 In Defense of Biomimicry

In the debate and defense of biomimicry, it is imperative to highlight how biomimicry is essential for humanity and environmental sustainability. According to Janine Benyus, who perceives humility encouraged by biomimicry and not minimized self-confidence? In this context, the biomimicry philosophy promotes humility in the face of our natural limits [15]. When it comes to

Fig. 6.22 The Tulip Telecommunication Tower in Colombo, Sri Lanka – South Asia. (Image credit and source: Meshaun de Silva, https://en.m.wikipedia.org/wiki/Lotus_Tower#/media/File:Lotus_Tower.jpg)

> **Box 6.3: Strong Biomimicry**
> Humility is seen as a 'source of power, a focusing mechanism.' (Janien Benyus [22])

eco-dependencies, humans are not ranked above other species. Nonetheless, our dependence on nature cannot be denied. In fact, humility can be a source of confidence (Box 6.3).

If we admit that not all problems are solvable, we are empowered to look for alternative solutions. Architect William McDonough is an excellent example of a humble, yet confident biomimicry practitioner. McDonough has accepted that the continued prosperity of humankind is dependent on the health of the biosphere [16, 17]. McDonough rejects the traditional "cradle-to-grave" manufacturing model, in which up to 90% of materials used in production are cast off as industrial waste. Instead, McDonough advocates a "cradle-to-cradle" model. This model manufacture emulates nature's closed-loop use of materials (circular economy). In nature, an organism's waste is cycled through the ecosystem and becomes a nutrient for other living species. In this model of manufacturing, waste from one production line becomes feedstock for another. Products at the end of their lifecycle safely reenter the environment and decompose into five biological nutrients or are disassembled

Fig. 6.23 The tulip flowers opened during the day inspiring biomimicry architecture. (Image credit and source: Mohsen Aboulnaga)

Fig. 6.24 The tulip flowers closed at the sunset inspiring biomimicry architecture. (Image credit and source: Mohsen Aboulnaga)

Fig. 6.25 Sky Garden Skyscraper at 20 Fenchurch Street in London. (Image credit and source: Colin, https://commons.m.wikipedia.org/wiki/20_Fenchurch_Street#/media/File3%AWalkie-Talkie_-_Sept_2015.jpg)

into "technical nutrients" and upcycled; which means they are recycled to form higher-grade products [17] as shown in Fig. 6.30.

McDonough has developed the "cradle-to-cradle" cities for the Chinese government that could house more than a hundred million people. His blueprints are respectful of each site's hydrology, biota, wind, and solar income. In each city, all

(a) (b) (c)

Fig. 6.26 Nature inspiring the Sky Garden Skyscraper. (**a**) A special shape of a tree in Arashiyama – Kyoto, Japan (**b**) A hedgehog-shaped cactus plant, Ain Sokhna, Egypt (**c**) A garden atop a stem of a tree. (Image and source: Mohsen Aboulnaga)

(a) (b)

Fig. 6.27 The Sky Garden Skyscraper in London. (**a**) Side view of the Sky Garden Tower (**b**) The Sky Garden atop the Walkie-Talkie, London. (Image credit and source: (**a**) Keith Edkins, https://commons.m.wikimedia.org/wiki/File:20_Fenchurch_Street,_January_2014.jpg#mw-jump-to-license; (**b**) Colin, https://commons.m.wikimedia.org/wiki/File:The_Sky_Garden.jpg)

sewage is upcycled, water waste is converted into natural gas via constructive wetlands, and solid waste is used as farming compost [16]. Biomimicry is far away from encouraging diminished self-confidence, but in fact, biomimicry places great faith in designers' creative capacity, which allows them to apply biological insights suited to human application. Good ideas cannot be found in nature alone but also in how humans adapt them to their own ends (Box 6.4).

Fig. 6.28 The "Walkie-Talkie" top floors of the Sky Garden Skyscraper at the 20 Fenchurch Street in the City of London, UK. (Image credit and source: Sebastiandow5, https://en.m.wikimedia.org/wiki/File:The_Walkie_Talkie_Sky_Garden.jpg#)

Fig. 6.29 The interior of the top floors of the Sky Garden Skyscraper in the City of London, UK. (Image credit and source: Chris Down, https://commons.m.wikimedia.org/wiki/File:Sky_Garden_inside.jpg#)

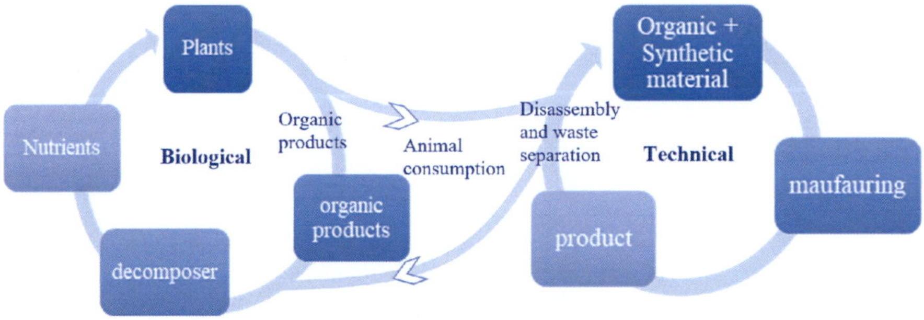

Fig. 6.30 Biological and technical nutrient cycles in "cradle-to-cradle" model. (Image source: Developed by authors after http://www.globalbuiltenvironmentreview.co.uk/Documents/10.1%20 Kennedy_Revised%20Biomimicry%20MS_Clean%20Copy_080916.pdf)

Box 6.4: Diversity of Nature and Biomimicry Laws

Despite the amazing diversity of nature, it is important for biomimicsts to take account of certain basic laws of nature when they engage in the practice of biomimicry. (Janine Benyus)

6.5 Biomimicry Is Not Constrained by Evolutionary Instrumentalism

Nature is a source of inspiration! When assessing biomimicry, whether it is not constrained by evolutionary instrumentalism, it is critical to be practical about where nature's systems will and would not offer to help. Although creatures and life systems are superior, looking into nature to solve most of humanity problems is not inclusively the solution. Nature's designs are developed through natural selection, which proceeds and advances incrementally. By and large, nature cannot choose to wipe the slate clean and redesign an organism from scratch, but nevertheless, the human designers can scrap existing design concepts and choose to start all over again with a new idea or a concept [18]. For instance, the giraffe's tall neck might be seen as a mere exemplar of a suboptimal result of nature's incremental design. However, no architect would imitate the mechanism of the nerve association between the mind and larynx of a giraffe and apply it to gain the benefits from natural systems in this creature, which can mimic it easily [18].

There is no doubt that the comparison between nature and the designer is a prior wrong, because nature is not a designer. So looking into nature for inspiration is just "design of a biological system," but the language, which can be used, is a metaphorical language so that design requires reflective awareness [19]. Rather than the comparison of the human design process with natural selection, one should think about whether this model has anything to serve the design or not. Human creativity can be stimulated in many new ways and this helps in solving several problems [20].

Moreover, the design of the giraffe's neck cannot be considered a mistake, the prolonged giraffe's neck is a source of inspiration, and the function is to avoid mortal hypertension in the head when it bows to drink. The channels in its neck naturally contract to keep blood from pooling in the direction of the gravity force. This component inspired the biomimetic "G-raffe" military pilot increasing speed suit. The texture of the suit tenses, with pneumatic force, compacts the body in vital zones to keep up even blood circulation. The anatomy of the giraffe may be different and odd, but it inspired humans to create a suit that can withstand up to nine G forces without losing tactile control, without this suit; the human loses consciousness at 4–5 G forces [21].

6.6 Biomimicry Is an Aid and Does Not Restrict Human Thinking to Build

Biomimicry is diverse and provides tools, which does not limit thoughts to building. Creating more sustainable designs is the goal of biomimicry. The chance of accumulating sustainable outcomes occurs when practitioners take into consideration the form, process, and ecosystem levels of biomimetic design [22]. Biomimicry encourages humans to keep building but to render construction consistent with life on earth. Up till now, human production and postproduction maintenance often has a negative impact on the environment. For instance, constructing a large bridge, or a tall building, or a dam would require concrete and steel; in turn, such production generates excessive greenhouse gases (GHGs), mainly CO_2 emissions [23]. Postproduction effects of a concrete dam can include soil erosion, species endangerment if dam bypass routes are not constructed for migrating species, spread of disease since water flow decreases and the river turns into a breeding ground for parasites, etc. [23].

In this context, nature nevertheless teaches lessons for life-friendly manufacture and maintenance. Consider the fact that the gross biomass of ants on earth is more noteworthy than the gross biomass of people, yet the ants do not pollute or contaminate their environment [17]. A tool called "life's principles," which has been discussed previously in Chap. 4, is embodied by the vast majority of organisms and ecosystems on earth. If we use life's principles as a benchmark for good design, the same neutrality possessed by the ants may be achieved. The socioeconomic processes will no longer create friction with the ecological processes on which we depend [23].

6.7 The Limitations in Biomimetic Architecture

Biomimetic architecture is not infinitely open, but it has limitations. A current trend in architecture has been outlining environmentally dependable buildings in light of the dangers of depleted resources, the expanded impacts of global warming in addition to climate change risks, and the public growing awareness of human effects on the habitat. The introduction of the Green Building Councils in the 1990s and beyond, the LEED (low-energy electron diffraction) affirmation patterns, and the Green Building Movement are all usage cases of this trend inside the field of architecture. Regardless of this thriving pattern, designers and analysts are swinging to biomimicry as a new move for design by imitating natural processes and characteristics and applying them in the design phase of construction [24]. Biomimicry could be called bio-inspiration. The latter term refers to the lessons learned from nature to overcome human problems, as shown in Box 6.5. The attempt to apply biomimicry to multiple disciplines such as engineering, chemistry, medicine, and computational systems may be appropriate [24].

There are existing limitations in applying biomimicry concepts, especially in the architectural field, such as the following:

- Excellent imitation of a great idea, but in the wrong context.
- Inapplicable on human scale.
- Cost a lot of money.
- Mimicking nature in an organism level or behavior level with no benefit for the environment.

This means that there is a clear difference between how nature works and how it goes in its system, human tends and needs. The idea of biomimicry tends to link how to be inspired from nature with practical functionality related to human needs.

Box 6.5: Various Terms of Biomimicry

Biomimicry, or bio inspiration as some call it, very simply means applying lessons learned from nature to solve human problems. ([24])

Biological systems are described as “systems of interrelations among living beings and their condition in a characterized space” [24]. All nature rules and systems are taken into consideration when living-being inside a biological system are subject to the procedures and frameworks of different living-being inside that same condition. A basic illustration is the oxygen-carbon dioxide cycle: creatures take in oxygen and trade it into carbon dioxide (CO_2), while plants and trees take in CO_2, trading it for oxygen as shown in Figs. 6.31 and 6.32. For however long, such two life forms continue to exist within that given condition where there will be a gas adjustment [24].

In a similar pattern, architecture could be net-zero structures, which endeavor to offset energy utilization by generating energy themselves – net zero-energy building (nZEB). Similarly, the food chain is dependent on organisms feeding on other organisms within that environment that fall lower in the food chain. When the food chain is in check, populations are maintained and the ecosystem functions as it should be. The moment the population of a predator or prey fluctuates, the entire food chain is affected, sometimes threatening the existence of the entire ecosystem

Fig. 6.31 Combining nature in the built environment. (Image credit and source: Mohsen Aboulnaga)

[24]. This is something that has been recently discussed with regard to overfishing and other human influences during the Brexit deal with the European Union (EU) and other places.

Despite the fact that a hippopotamus can excrete a combination of chemicals that act as a sun block, insect repellent, and antifungal ointment, what good does this do if the plants and food that hippos rely on become scarce? These very basic examples obviously appear that they can be classified as common knowledge; however, they are not thoroughly weighed by designers and architects who claim that biomimetic

Fig. 6.32 Flowers and plants during the oxygen-carbon dioxide cycle – City of Westminster London Borough, Central London, UK. (Image credit and source: Mohsen Aboulnaga)

architecture is the solution to future problems. When working in conjunction with other organisms and natural systems, nature operates optimally, as shown in Fig. 6.33 [24].

Even though biomimetic plans motivated by the manners – which nature works hypothetically – seem like sheltered and functional building arrangements on an individual scale, the manners in which the outlines work firmly are frequently dismissed. In her TED talk on biomimicry in real life, Janine Benyus clarifies biomimicry as "the marvel in which individuals are starting to recollect that creatures are doing accurately what we have to do keeping in mind the end goal to have the capacity to live nimbly and for an expanded time."

Benyus has provided different cases of life forms, which have been considered as "diagrams or formulas" for impersonation in man-made designs, for example, microscopic organism repellant motivated by sharks presently being utilized as a part of healing facilities. Be that as it may, the larger part of Benyus' illustrations work on an individual scale and not many of the executed cases are specifically identified with design (trees and bones as motivation for spans and the coral-propelled recipe of adding carbon dioxide (CO_2) to concrete being the most pertinent). It is not necessarily the case that these individual biomimetic cases couldn't all be connected to a building; however, applying singular nature-motivated designs to work without considering and inquiring about the outcomes of consolidating these designs could bring about further practical and perhaps deadly responses [24].

Fig. 6.33 Three types of trees in Bangkok, Thailand, where nature operates optimally. (Image credit and source: Mohsen Aboulnaga)

The reason is that it is so basic to comprehend the manner in which arranged designs cooperate inside a biological system, that when this system ends up unequal, the reactions can be deadly. Similar to how a failing biological community can endanger the lives of living beings or entire species, a failed biomimetic building design can endanger the people for whom it was built. Also, mistakes would probably happen in a man-made impersonation of nature.

One of the announcements Benyus opens her discussion with is "envision outlining spring." Benyus additionally clarifies that it's something which happens yearly around a similar time and includes the joint effort of numerous living beings and designs. With imitation and flow innovation, it would be incredibly troublesome for people to facilitate a season with insignificant mistakes.

Masterminding the softening of snow, sprouting of blossoms and plants, end of hibernation, and expanded precipitation is only a portion of the things that would need to be considered and executed at a particular time. It is not a new revelation that people have defects and cannot generally fabricate or configure consummate items, frameworks, or designs. In reproducing nature-roused frameworks, people are basically imitating nature – not making a perfect and cloned duplicate.

Therefore, human error must be taken into account with biomimicry, especially with the ecosystem; if one piece of the system is unbalanced, the completely larger system can in turn collapse. Since the biomimetic systems could be mimicked falsely from the start or simply become unbalanced through a chain effect.

Unlike biomimetic chemistry, which is tested and contained in the lab before being distributed (and even after being tested can still have serious repercussions), biomimetic architecture would serve spaces that are often inhabited by people, whether in the workplace or at home. Although architecture can be tested to a certain degree, social and health effects are usually not noticed until years later, when the architecture has actually been occupied and used by people [24].

Despite the fact that biomimicry may hold answers to future building and city challenges amidst COVID-19 and climate change, the application to engineering is not yet at large scale. Nevertheless, it could demonstrate a valuable approach.

Utilizing engineering to copy a characteristic found in nature could be effectively connected on a smaller plan scale, for example, for a dress or a device, research ought to lead the process of applying biomimicry to building designs [24]. To make a genuinely useful building biomimetic displays for the general public, the outline would not just need to impersonate particular attributes or frameworks found in nature but the larger frameworks as well. Biomimetic engineering should mimic the environments and, later, the minor frameworks that coexist within them. In this context, Benyus addresses the relationship between urban areas and biological communities, saying that she and HO + K Architects trust that urban areas should work in any event and additionally; they trust the environments they supplant, without digging significantly further. This is vital and, in some cases, skimmed over part of biomimicry. For powerful biomimitic design, urban areas should work closely and collectively with a biological community, where singular structures go about as individual life forms which have a place within a greater adjusted framework that depends on the living beings as much as the living beings depend on it.

Regardless of the fact that the current research in biomimicry provides interesting options to the eventual fate of engineering, it should first examine and observe how every design functions on a larger scale and how they are altogether interconnected before applying it to design.

6.8 Strength Versus Weakness of Biomimicry Comparison

In comparing the strength and weaknesses of biomimicry, Benyus has defined biomimicry as a copy or reproduction of natural processes, which therefore embraces the naturalistic concept of biomimicry [17, 25]. In fact, it is necessary to distinguish between strong and weak concepts of biomimicry. Hence, Janine Benyus has expressed the solid idea of biomimicry that is conceptualized naturalistically as the impersonation of nature's models in order to tackle human issues.

The primary target of biomimicry is to reflect the thoughts of nature into our own lives. In this contest, biomimitic researchers are investigating nature's gems: photosynthesis; self-gathering; regular choice; self-supporting biological systems; eyes, ears, skin, and shells; talking neurons; and normal prescriptions, and that's just the beginning, and after that, they replicate these outlines and assemble procedures to tackle our own issues [25]. Box 6.6 summarizes the target of biomimicry. Such a journey is called biomimicry and the mindful copying of life's genius and innovation inspired by nature.

It is imperative to state that although mimicry is stimulated and engulfed by nature, this motivation is principally comprehended as revelation and impersonation of nature and speaks to along these lines the naturalistic idea of mimicry that we find in Aristotle's physics. This solid idea of biomimicry is, as per Benyus, mostly persuaded by the possibility that nature is a measure to judge the moral rightness of our technological innovations [25, 26]. According to Benyus, biomimetic innovation and advancement can claim to be morally right because normal standards are thought to be standards that contribute to the environmental well-being and respectability of the biological communities of planet Earth [25, 26].

Box 6.6: Target of Biomimicry

The primary target of biomimicry is to resonate the thoughts of nature in our own lives. ([25])

In addition, it is imperative to state that by utilizing the same outline standards – as normal substances and designs – and by displaying our innovative plans on common standards, biomimicry holds fast to bio-inclusive morals that empower us to resituate our innovative plans inside the biological points, of confinement of the biosphere: "If individuals are to 'act from inside nature,' as the ethos of biomimicry infers we should, at that point the prerequisite of reclusiveness applies to us, to our office. We should permit our finishes and additionally our methods, our plans, to be formed by who and what is out there in our condition" [25, 26].

6.8.1 *Strong Concept of Biomimicry*

The strong concept of biomimicry includes nature as a normative principle of conduciveness to ecological health and integrity as a design principle, that is, as a standard of ecological ethics, contrary to conceptualizations of biomimicry that exclusively look at these principles of nature in mechanistic terms [26]. In order to adhere to natural principles, as a normative standard of technology and design, however, mimicry has to be understood in the strict sense of copying or reproduction of nature. On the one hand, by copying or echoing the design principles of nature, our technological innovations can claim to be more "natural" and to be more "adaptive" to the capacities of the biosphere. On the other hand, biomimetic designs can only claim to be ethically advantageous if they actually copy these natural principles.

Alternatively, biomimicry is understood in the strong sense as the copying or reproduction of natural processes. However, Benyus, on the other hand, thought biomimicry was both bio-inclusive and ethical [26].

In examining a strong concept of biomimicry, there are three propositions to be thoughts as presented in Fig. 6.34. The first proposition is a strict distinction between the absolute discovery of systems that exist in nature (imitated) and the invention of new products (invented). This strict distinction is questionable from an Aristotelian viewpoint, since mimesis concerns not just the representation or increase of what is already in nature, as we have seen, but the supplementary flawlessness of what nature isn't equipped to deliver itself [26]. Such trouble in recognizing revelation and development is likewise affirmed by Godin's history of the idea of advancement:

> The possibility of curiosity is without a doubt enters the idea of innovation as opposed to the idea of impersonation, yet creation can be utilized both for finding and for making. At last, if the theory of (leadenness of perception) is acknowledged, it should be admitted that every discovery just like every invention involves a construction. ([26])

A second proposition of the strong idea of biomimicry is the epistemic sufficiency to "know" the plans of nature, with a specific end goal of resounding the thoughts of nature or "getting plans from the bigger life design," as well as strength, ought to have been taken to full access of nature and yet to have the efficiency to comprehend the common procedures themselves. It is faulty, be that as it may, regardless of whether the characteristic procedures are completely open and justifiable. The many-sided quality of common wonders confined the likelihood of the finding and after that resounding nature's models to deep innovative issues [26].

The third proposition is a strict distinction between technological intervention in nature, the abuse of the first industrial revolution, and natures unadulterated receptivity, detecting and profiting from the second industrial revolution. In the event of recognizing the intricacy of nature, it likewise requires identifying the need to decipher and translate common wonders to investigate their pertinence to innovative issues. With the presentation of the idea of interpretation and translation, in any case, the moral preferred standpoint of biomimetic configuration winds up faulty, because a strict qualification among proliferation and development can never again be made [26]. Figure 6.35 shows nature that could inspire strong biomimicry.

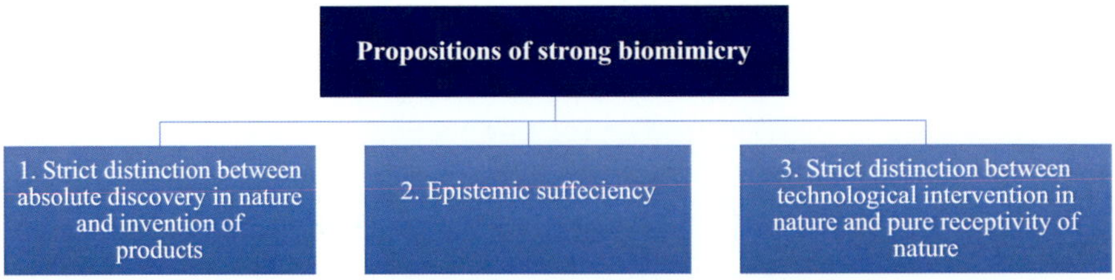

Fig. 6.34 Three propositions of strong biomimicry. (Image source: Developed by authors)

Fig. 6.35 The Sky Tower in nature to inspire strong biomimicry. (Image credit and source: Mohsen Aboulnaga)

6.8.2 Weak Concept of Biomimicry

In contrast to the strong concept of biomimicry, there is a weak concept of biomimicry. It is somehow a weaker but more sophisticated concept of biomimicry. At the Kavli Institute for Bionano Science and Technology of Harvard University, the co-director introduced biomimetics as a strategy: "To mimic high-tech solutions that nature can offer, reformulate natural materials, natural strategies and to create new materials and devices, which outperform anything that we have today" [27]. For the sophisticated concept of biomimicry, mimicry does not consist of the reproduction or duplication of natural solutions, rather these natural solutions are taken as inspiration to create new materials and devices.

As Bensaude-Vincent et al. put it, "biomimetic is more a question of inspiration in the poetic sense, that is the intention of original forms or processes starting from a natural motive that solved a similar problem either borrowing directly from living things, or mimicking nature's structures, function and processes" [28]. Contrary to the first assumption of the strong concept of biomimicry, the weaker concept perceives mimicry not as the duplication of natural solutions but primarily as a creative solution inspired by nature. Dissimilar to the second and the third presuppositions of the strong concept, the weaker concept acknowledges, "Some form or procedure of interpretation or translation from biology to technology is required." Therefore, in the weaker concept of biomimicry, the complexity of natural phenomena is acknowledged as well as the necessity of translation in order to transfer natural problem-solving to technological problem-solving [28].

Mimicry cannot be understood as an invention inspired by nature, because mimicry can only be an emulation or perfection of nature if it is in fact oriented toward nature as a model. The distinction between discovery and invention does not assist in characterizing the nature of biomimicry and seems to introduce a completely new category [28]. Nature is somehow deficient because it cannot produce everything and has to be complemented by technology. On the basis of Heraclitus' famous thesis, it states that nature has the tendency to conceal itself, but we can conceptualize this deficiency of the phases as the tendency of nature to conceal itself, and therefore, nature has to be supplemented by a technological representation in order to appear sustainably [28].

The concept of mimesis already makes it clear that such a concept is theoretically underdeveloped in the current literature on biomimicry, but its conceptual analysis could provide possible guidelines for future research in the field of biomimicry. The concept of supplementary, for example, could be useful in biomimetic practices because it can aid in reconsidering the clear distinction between discovery and invention, as well as intervention on one another [28]. Biomimicry is then not the discovery of nature's models and the copying of natural processes into technological designs, as is claimed by the strong concept of biomimicry. This perspective is highly naive, because successful copying of nature is quite rare; the imitation of the wing of a bird, for instance, does not result in the ability to fly, as the history of aerodynamic techniques clearly exhibits [28]. Nevertheless, the concept of nature, which is presupposed in both the philosophical tradition and the literature on biomimicry, represents a technological model of nature. Thus, nature is understood in technological terms as productive and achievable [26]. According to the strong concept of biomimicry, the production of nature is already perfect and should be reproduced in biomimetic technologies; the technologies of the first industrial revolution are imperfect and should be ideal by the second industrial revolution by mimicking nature.

Freya Mathews, for instance, argues that "if human beings are to 'act from within nature', as the ethos of biomimicry implies we ought, then the requirement of reclusiveness applies to us, to our agency. It must allow our ends as well as our means, our designs, to be shaped by who and what is out there in our environment." [26] Mathew added: "under current interpretation nonetheless, biomimicry takes our ends as it finds them in our consumer society, and merely looks to 'nature' for the "design solutions" that will enable us to attain those ends with less rather than more disruption to life-systems of planet-earth" [26]. Mathew's argument is significant in the transformation of architecture to become biomimetic and contribute toward attaining SDGs.

The supplementary use of biomimetic technologies presupposes continuity between nature and technology, due to concerns about the perfection of nature. On the other hand, it presupposes a discontinuity between nature and technology because it concerns the supplementing of nature [26]. If biomimetic technology is essentially supplementary and cannot be reduced to nature, the ontological status of

Table 6.1 Comparison between strong and weak biomimicry [26]

Aspects of comparison	Mimesis	Nature	Technology	Ethics
Strong biomimicry	Naive conceptualization of mimicry	Presupposes perfection of nature	Cannot deal with the supplementary of biomimetic technology	Risk of committing the naturalistic fallacy
Weak biomimicry	Acknowledges supplementary of mimicry	Presupposes desistance of nature	Cannot deal with the supplementary of biomimetic technology	Avoids naturalistic fallacy

this supplementary technology is unclear. What is exactly clear that it is not natural in biomimetic technology? Table 6.1 lists the comparison between strong biomimicry and weak biomimicry due to mimesis, nature, technology, and ethics [26]. From a philosophical perspective, it is concluded that the strong concept of biomimicry is more problematic than the weak concept of biomimicry. In all four dimensions (mimesis, nature, technology, and ethics), the strong concept has major issues to resolve, whereas the weak concept has to deal with major issues in only two dimensions. Box 6.7 highlights the biomimicry and city sustainability linkage.

Box 6.7: Biomimicry and City's Sustainability Linkage

Even in a stable world, no city or urban region as presently configured could be sustainable on its own. (McGranahan et al. [32])

6.9 Future Directions

Biomimicry has an important role in greening cities in both the industrialized and developing worlds, but it remains an open question whether it is anywhere near capable of having a pronounced-enough effect to dispel the present environmental crisis [29]. This is scalable and depends on the green and sustainable policies and regulations in each country in the race toward achieving the Sustainable Development Goals (SDGs).

Based on the intellectual movement in design and innovation, biomimicry may indeed offer an essential means to progressively transform technological patterns; however, to be really effective, it may require more than a little regulation. In coming years, biomimicry specialists may well be open to learn from various traditions in design and innovation, including the following:

(a) Appropriate technology movement.
(b) Science movement.
(c) Values in design movement.
(d) Radical environmentalism [29].

If biomimicry incorporates the findings of the other social and intellectual movements, then it might be transformed into a more defined environmental practice. Accordingly, we may deem this eco-friendly type of biomimicry rather akin to "eco-mimicry" of the mimicking of nature for eco-friendly innovation, whereas biomimicry may not involve eco-friendly technological design, invented and regulated by experts without much open innovation inputs; nonetheless, "eco-mimicry" operates to the following:

- Inherent sustainable forms from an environmental and social point of view [29].
- Encouragement of decentralization and localism.
- Democratic when it comes to decision-making.
- Understood by all, not just by experts (collectively).
- Provide sensitivity to the need to disperse power rather than concentrate it.

All these suggestions work within the social dimension of innovation and sustainability and reflect some of the currents flowing through contemporary corporate social responsibility theory. Therefore, it is a theoretical possibility that may be woven into the regular practice of bio-inspired innovation sponsored or supported by corporate organizations [29]. Figure 6.36 presents this linkage while Fig. 6.37 shows the conclusion framework argument and criticism of the biomimicry concept.

6.10 From Bio-inspired Innovations to Eco-mimetic Cities

There is an interconnection between bioinspired innovations and eco-mimetic cities. The fundamental "principles and rules of life" could have their most valuable applications in transforming cities into life regenerative ecosystems and reconnecting the ecosystems to the broader natural ones. In this contest, the HO + K, the

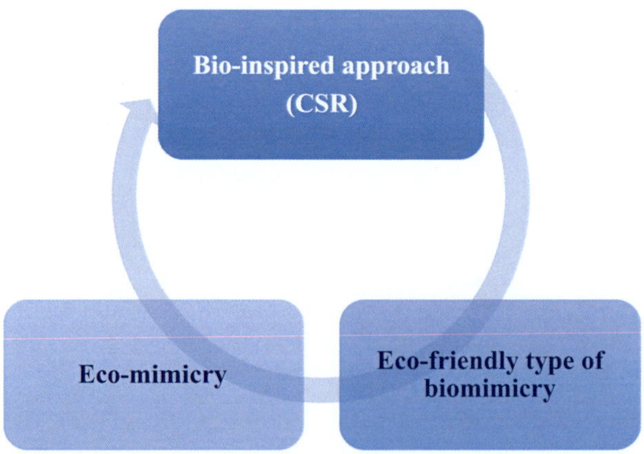

Fig. 6.36 Bio-inspired link to promote the concept of eco-mimicry and eco-friendly type of biomimicry. (Image source: Developed by authors)

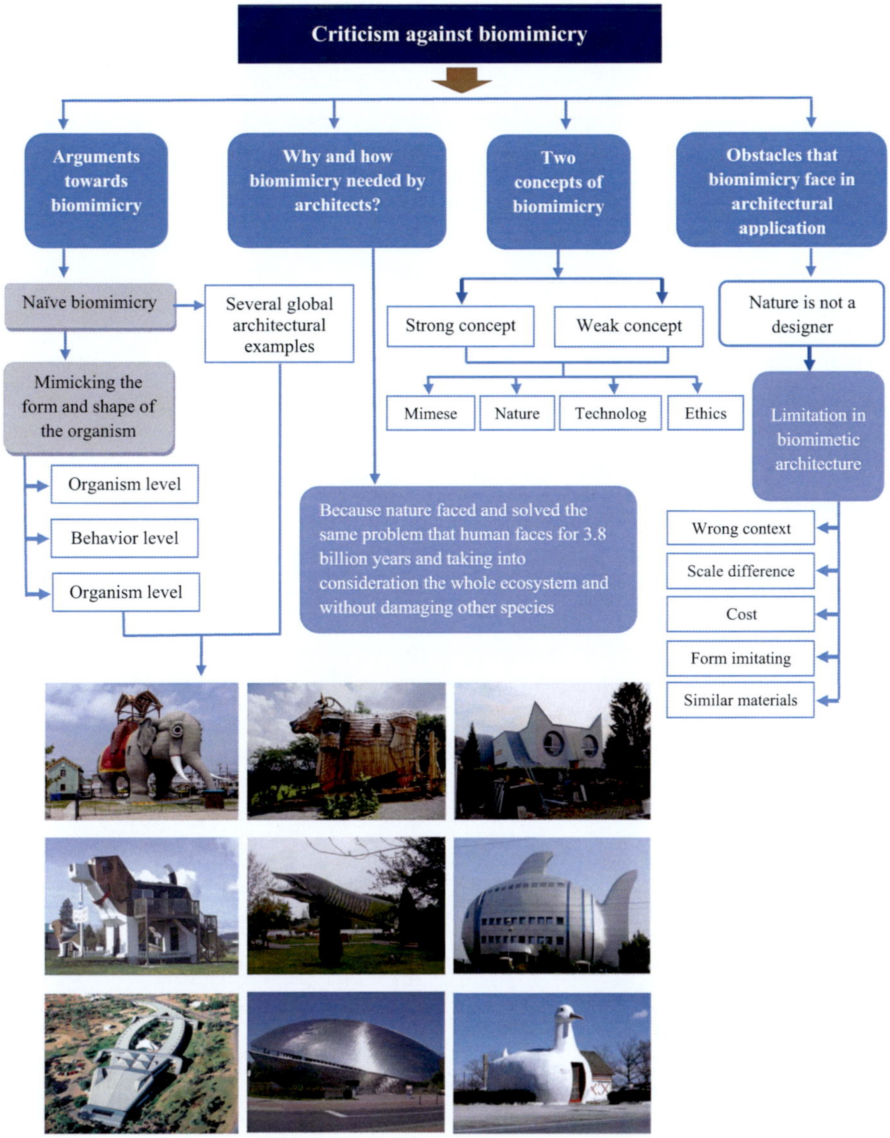

Fig. 6.37 Conclusion framework for the criticism against biomimicry. (Image source: Developed by authors)

global design, architecture, engineering, and planning firm, was founded in St. Louis in 1955, which started collaborating with the Biomimicry Guild (today Biomimicry 3.8) of Janine Benyus in 2001, before forging a formal alliance in 2008 [30]. Working together on the planning and design for the new city of Lavasa in India, they developed the fully integrated thinking (FIT) process for innovation, which assisted in integrating the life's principles into the full range of the design cycle. Taryn Mead, initiator of the FIT project at HO + K, and Dayna Baumeister from the Biomimicry Guild identified 40 different functions a city must perform, from collecting, storing, filtering, and distributing water to generating and distributing energy, and conducted extensive research into biological models for each of these functions. Figure 6.38 summarizes the main points of the city's different functions based on the Biomimicry Guild. The FIT assists in turning the design challenges into opportunities for better solutions that go beyond "doing less bad" to making a positive impact in multiple areas. Such 'life's principles encourage viewing the designs, as part of the complex and adaptive systems of a specific location. Nevertheless, the FIT matrix can simply reveal the latent potential of "place," including site selection, available ecological services, potential partners, and opportunities [30].

It is observed that the real potential of biomimicry for sustainable, regenerative, and resilient cities, hence we could call this "eco-mimicry" as it is based more on mimicking how local eco-systems work to transform the city itself into a "more natural" eco-system, and simultaneously releasing the pressure on "real" natural ecosystems, yet reaching a balance between the built and non-built environments [30].

There are no technical or regulatory barriers that we cannot overcome – as the very productive collaboration between the Bullitt Foundation/LBC teams and the Seattle City Regulators, the Seattle Department of Planning and Development, and the Washington Department of Health and Seattle Public Utilities demonstrates [30]. But nonetheless, there is a profound mind shift that has to be undertaken by the wider community of built environment professionals – from architects and

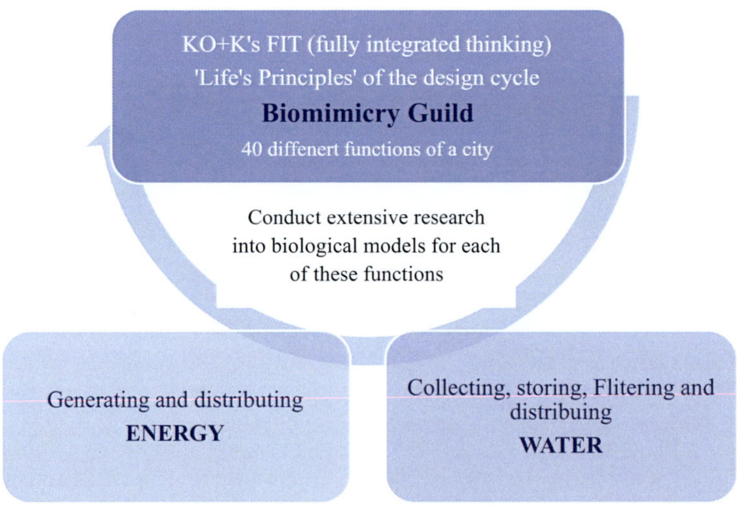

Fig. 6.38 Fully integrated thinking process for innovation – HO + K and Biomimicry Guild. (Image source: Developed by authors)

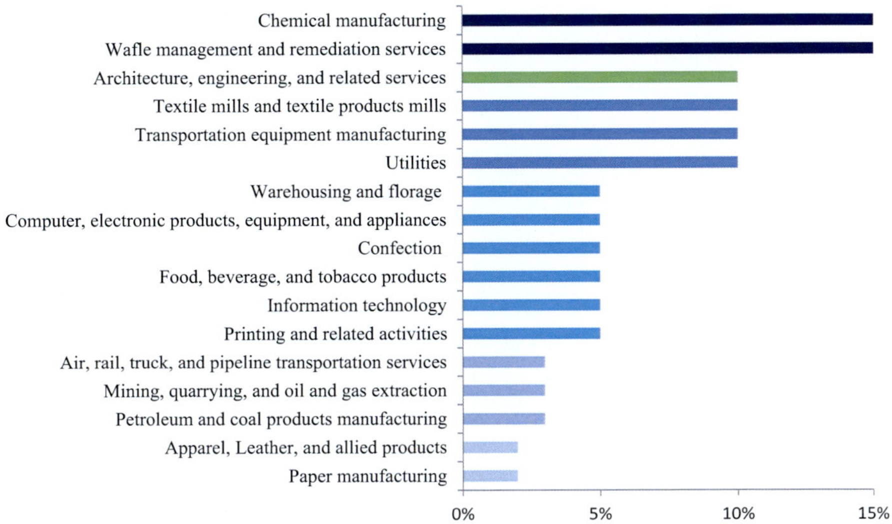

Fig. 6.39 Biomimicry's estimated market impact in 2025. (Image source: Developed by authors after https://www.google.com/url?sa=t&rct=j&q=&esrc=s&source=web&cd=&ved=2ahUKEwjplpLh3NjvAhXirnEKHbCTBE0QFjAJegQIBxAD&url=https%3A%2F%2Fhrcak.srce.hr%2Ffile%2F200540&usg=AOvVaw3-91MW-KlSrxsr3jb2oZZk(

engineers, to city planners and regulators, to real-estate developers and property owners – to design and rebuild cities that will be "socially just, culturally rich, and ecologically restorative." Biomimicry has just recently started to be perceived as a medium that should eventually lead to revolutionary changes in the economy. The transposition of biomimicry into business and commercial use could transform large portions of various industries in the coming years and could ultimately impact all segments of the economy, as illustrated in Fig. 6.39. Industries that could be particularly affected include utilities, transportation equipment, chemical manufacturing, storage and waste management, architecture, and engineering [31] (Box 6.8).

Box 6.8: The Outcome of a True Biomimetic

For a design to be truly Biomimetic, the outcome should not only be physically inspired by nature but also have sustainable features and improvements. (Janine Benyus)

6.11 Conclusion

Critics of biomimicry cling to a dangerous, dated cultural conception of humans as the ultimate life form. They argue that biomimicry diminishes the role of human designers, but it doesn't. Human beings are of central importance in biomimicry, since the approach depends on the designer's ability to distill what he/she learns about biology into a set of abstract design principles, which can be implemented to solve human design challenges. Critics also argue that we should not seek nature's advice when it comes to design, since nature designs incrementally. In this context, species are built from a binding template provided by their evolutionary ancestors. This fact doesn't mean the outcomes of natural selection have nothing to offer us, but nevertheless, we can learn, for instance, from a leaf how to harness energy without creating toxic byproducts and from a diatom how to effectively absorb mechanical stresses. Critics also state that the biomimicry mission of sustainability forces humans to repress their impulse to build. In addition, biomimicry recognizes that humans, and human developments, are in no way separate from the ecosystems in which they exist. Humans are part of nature, as biotic citizens. Biomimicry should continue to build, but our designs should support rather than suffocate life on Earth. Finally, some critics contend biomimetic architecture that references animal and plant forms that reflects a weakness of human meaning. This neglects a large body of evidence suggesting humans have a positive emotional response when their built environment contains naturally occurring geometries of any kind, including those of plants and animals. The incorporation of a general comprehension of science and biology in architectural design will be huge in the production of a built environment that adds to the well-being of human communities, while expanding positive integration with common carbon cycles.

References

1. Maglic, M. J. (2014). *Biomimicry: Using nature as a model for design*. University of Massachusetts Amherst. Retrieved 2 April 2018, from https://scholarworks.umass.edu/cgi/viewcontent.cgi?article=1984&context=theses
2. Dhakhwa, T. (2021). *Does nature hold the answer to sustainability? Biomimicry as ecological innovation*. Yale Environment Review. Retrieved 28 August 2021, from https://environment-review.yale.edu/does-nature-hold-answer-sustainability-biomimicry-ecological-innovation-0
3. Block, V., & Bart, G. (2016). Ecological innovation: Biomimicry as a new way of thinking and acting ecologically. *Journal of Agriculture and Environmental Ethics, 29*(2), 203–217. https://doi.org/10.1007/s10806-015-9596-1
4. Vogel, S. (2000). *Cat's paws and catapults: Mechanical Worlds of nature and people*. Norton. Retrieved 4 April 2021 from https://www.yumpu.com/en/document/read/64107060/pdf-online-cats-paws-and-catapults-mechanical-worlds-of-nature-and-people-pre-order
5. Ogochukwu, F. (2017). *Biomimicry and sustainable architecture: Reviews of existing literature*. Retrieved 4 April 2021, from https://www.researchgate.net/publication/331207896_Biomimicry_and_Sustainable_Architecture_A_Review_of_Existing_Literature
6. Oguntona, O., & Aigbavboa, C. (2019). *Barriers militating against the adoption of biomimicry as a sustainable construction practice*. MATEC Web of Conferences. Retrieved 4 April 2021 from https://www.matec-conferences.org/articles/matecconf/pdf/2019/15/matecconf_iconbee2019_03010.pdf
7. Singh, R. (2020). Biomimicry: Learning from nature. *Journal of Engineering Sciences, 11*(6), 533–547. Retrieved 29 August 2021, from https://jespublication.com/upload/2020-110680.pdf

8. Newton, J. (2016). *Buildings around the world made to look like all sorts of creatures*. Mail Online. Retrieved 10 April 2018, from https://www.dailymail.co.uk/news/article-3790822/Animal-house-Buildings-world-look-like-sorts-creatures-cats-koalas.html
9. Kim, J., Park, K., & Park, M. (2021). The design characteristics of nature-inspired ecofriendly buildin. *International Journal of Latest Trends in Engineering and Technology, 9*(4), 167–183. Retrieved 30 March 2021, from https://www.ijltet.org/journal/152030268027.%202140.pdf
10. This is not Biomimicry! Littlegreenseed. (2012). Retrieved 31 March 2021, from https://littlegreenseed.wordpress.com/2012/12/02/this-is-not-biomimicry/
11. Wojcicki, Z., Grosel, J., Belostotsky, A., Akimov, P., & Sidorov, V. (2017). OMA Research of Sky Tower in Wroclaw, Poland. *MATEC Web of Conferences, 117*, 00177. https://doi.org/10.1051/matecconf/201711700177
12. *Colombo Lotus Tower*. Lakpura™. (2021). Retrieved 3 September 2021, from https://lakpura.com/pages/colombo-lotus-tower
13. *Iconic Lotus Tower should open to more foreign investors: Sri Lankan President*. News.cgtn.com. (2021). Retrieved 3 September 2021, from https://news.cgtn.com/news/2020-06-17/Sri-Lanka-s-Lotus-Tower-project-should-be-open-to-more-investors-RodF92kbJe/index.html
14. *20 Fenchurch Street – Rafael Viñoly Architects*. Rafael Viñoly Architects. (2021). Retrieved 5 September 2021, from https://www.vinoly.com/works/20-fenchurch-street
15. Kennedy, E. (2017). *Biomimicry in industry: The philosophical and empirical rationale for Reimagining R&D (Ph.D)*. University of Akron. Retrieved 30 March 2021, from https://etd.ohiolink.edu/apexprod/rws_etd/send_file/send?accession=akron1491505696395819&disposition=inline
16. McDonough, W. (2005). The wisdom of designing Cradle to Cradle. *TED talks internet*. Retrieved 30 March 2021, from http://www.ted.com/index.php/talks/william_mcdonough_on_cradle_to_cradle_design.html
17. McDonough, W., & Braungart, M. (2008). *Cradle to Cradle: Remaking the way we make things* (p. 192). Vintage Publisher. Retrieved 30 March 2021, from https://books.google.com.eg/books/about/Cradle_to_Cradle.html?id=BCCUPwAACAAJ&redir_esc=y
18. Kaplinsky, J. (2006). Biomimicry versus humanism. Wiley online library. *Architectural Design, 76*(1), 66–71. Retrieved 30 March 2021, from https://onlinelibrary.wiley.com/doi/abs/10.1002/ad.212
19. Capra, F., & Luisi, P. (2014). *The systems view of life*. Cambridge University Press. Retrieved 30 March 2021, from https://www.cambridge.org/core/books/systems-view-of-life/35186BA5B12161E469C4224B6076ADFE
20. Wilson, A., Réale, D., Clements, M., Morrissey, M., Postma, E., Walling, C., et al. (2010). An ecologist's guide to the animal model. *Journal of Animal Ecology, 79*(1), 13–26. Retrieved 30 March 2021, from https://doi.org/10.1111/j.1365-2656.2009.01639.x
21. Booth, G. (2012). "Super-Bodies." Miracles of nature. *BBC*. Retrieved 30 March 2021, from https://flixi.com/tvshows/richard-hammonds-miracles-of-nature/season/1
22. Benyus, J. (2002). *Biomimicry innovation inspired by nature* [Ebook]. Harper Collins Publishers. Retrieved 18 April 2018, from https://www.academia.edu/38300413/Janine_M_Benyus_Biomimicry_Innovation_Inspired_by_Nature_2002_Harper_Perennial_1_
23. Mathews, F. (2011). Towards a deeper philosophy of biomimicry. *Organization & Environment, 24*(4), 364–387. https://doi.org/10.1177/1086026611425689
24. Buraczynski, K. (2013). *The limitations of biomimetic architecture*. Retrieved 31 March 2021, from https://scalar.usc.edu/works/the-limitations-of-biomimetic-architecture/index
25. Lee, M. (2017). Decision making approaches for complex adaptive systems in built environment. *International Journal of Management and Applied Research, 4*(4), 222–232. Retrieved 31 March 2021, from http://www.ijmar.org/v4n4/17-017.html
26. Blok, V., & Gremmen, B. (2016). Ecological innovation: Biomimicry as a new way of thinking and acting ecologically. *Journal of Agricultural and Environmental Ethics, 29*(2), 203–217. https://doi.org/10.1007/s10806-015-9596-1

27. TEDxBigApple. (2012). *Extreme biomimetics* [Video]. Retrieved 31 March 2021, from https://www.seas.harvard.edu/news/2012/03/joanna-aizenberg-talks-extreme-biomimetics-tedxbigapple
28. Vincent, J., & Mann, D. (2002). Systematic technology transfer from biology to engineering. *Philosophical Transactions of the Royal Society of London. Series A: Mathematical, Physical and Engineering Sciences, 360*(1791), 159–173. https://doi.org/10.1098/rsta.2001.0923
29. Marshall, A. (2013). Biomimicry. In S. O. Idowu, N. Capaldi, L. Zu, & A. D. Gupta (Eds.), *Encyclopedia of corporate social responsibility*. Springer. https://doi.org/10.1007/978-3-642-28036-8_273
30. Scheffer, O. (2016). From biomimicry to ecomimicry: Reconnecting cities—And ourselves—To Earth's balances–The Nature of Cities. *The Nature of Cities*. Retrieved 31 March 2021, from https://www.thenatureofcities.com/2016/11/30/from-biomimicry-to-ecomimicry-reconnecting-cities-and-ourselves-to-earths-balances/
31. Ivanić, K., Tadić, Z., & Omazić, M. (2014). *Biomimicry – An overview*. Retrieved 12 September 2021, from https://hrcak.srce.hr/136003
32. Mcgranahan, G., J. Marcotullio, P., Bai, X., & Balk, D. (2005). Urban Systems. In J. Eades & E. Ezcurra, Ecosystems and Human Well-being: Current State and Trends (p. 817). Retrieved 24 July 2022, from https://www.researchgate.net/publication/242451977_Urban_Systems.

Chapter 7
Biomimicry Innovation, Opportunities, and Post-COVID Regulations

> Viruses are all around us. In a square meter, 800 million viruses a day fall from the sky. They are a hugely important part of our world and our evolution. But these pandemics happen when the natural defences are broken. Meanwhile climate change is just putting everything under stress. So, we've got this natural system that is out of kilter. (Janine Benyus [4])
> The recovery from the COVID-19 crisis must lead to a different economy. (António Guterres – UN Secretary General, 31 March 2020)

7.1 Introduction

The new coronavirus disease (COVID-19) attacked every inch of our society at its core, claiming lives and people's livelihoods (Fig. 7.1). By April 2020, nearly half of the world's population was under some form of lockdown, with over 3.9 billion people in more than 90 countries or territories having been asked or ordered by governments to stay at home [1]. The potential longer-term effects on the global economy and those of individual countries are dreadful and terrible in all sectors. According to the UN report, "Shared responsibilities, global solidarity" highlights the urgent need to respond to the socioeconomic impacts of COVID-19 [2]. This idea begs the question of what this new economy is, whether it is a "green economy," a "resilient economy," or an "inclusive economy." This needs a long debate to reach the best for all.

The coronavirus (COVID-19) continues to devastate human lives and livelihoods across the world – striking the most vulnerable and hardest. This is specifically true for millions of people on the move [2]. In fact, the upheaval caused by COVID-19 is all around us and in every kilometer of space around the world, causing a severe health threat that has never been seen before in our lifetime [2]. Figure 7.2 shows the novel coronavirus SARS-CoV-2, while Fig. 7.3 presents the corona plant created by nature for inspiration, not for devastation.

M. Aboulnaga, S. E. Helmy, *Biomimetic Architecture and Its Role in Developing Sustainable, Regenerative, and Livable Cities*,
https://doi.org/10.1007/978-3-031-08292-4_7

Fig. 7.1 Streets in Cape Town, South Africa, during COVID-19 lockdown in 2020. (Image source: https://commons.m.wikimedia.org/wiki/File:Kloof_%26_Long_street_billboard_20020401_140037.jpg#)

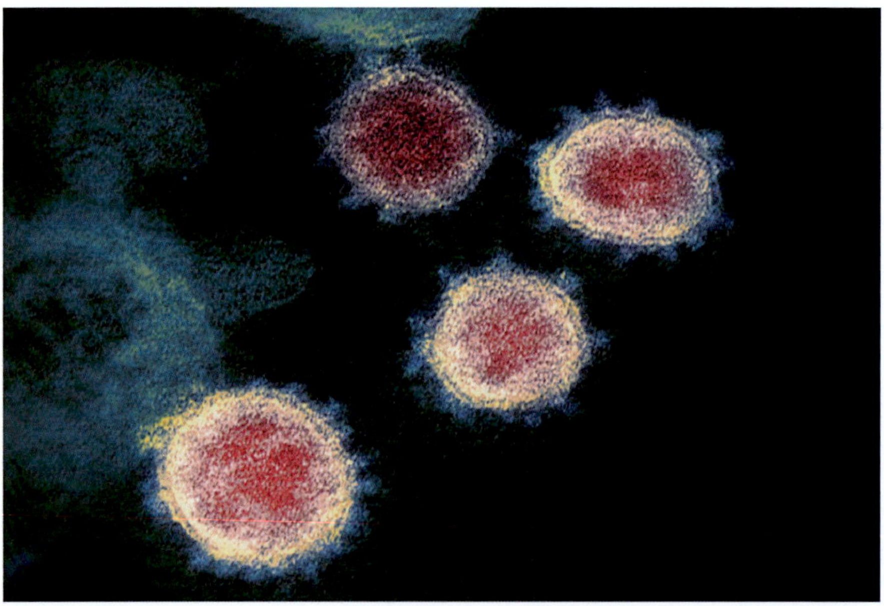

Fig. 7.2 Novel coronavirus SARS-CoV-2 that severely hit the world in 2020. (Image source: https://commons.wikimedia.org/wiki/File:Novel_Coronavirus_SARS-CoV-2.jpg)

Fig. 7.3 Nature created corona plant for inspiration not for devastation. (Image source: https://commons.wikimedia.org/wiki/File:Corona_plant.jpg#)

Fig. 7.4 Main roads are completely empty in Malaysia during the COVID-19 lockdown in 2020. (Image source: https://commons.wikipedia.org/wiki/COVID-19_lockdown#/media/File%3AMCO_covid-19_Penang_Malaysia_(49687038551).jpg)

(a) (b)

Fig. 7.5 Empty roads and highways during the COVID-19 lockdown in the UK and USA. (**a**) A digital road sign "STAY HOME, SAVE LIVES, KEEP YOUR DISTANCE" in Belfast, Northern Ireland, UK. (**b**) A digital highway sign "SAVE LIVES, STAY HOME, ESSENTIAL TRAVEL ONLY". (Images' source: (**a**) https://en.m.wikipedia.org/wiki/File:COVID19_lockdown#/media/File:Belfast_COVID19_Traffic_Management_Sign.jpg; (**b**) https://commons.m.wikimedia.org/wiki/File:2020-04_19_14_42_26_Variable_message_sign_reading_%22Save_lives_Stay_Home_Essential_Travel_Only%22_along_southbound_Interstate_95_just_south_of_Exit_29_in_Beltsville_Prince_George%27s_Country,_Maryland.jpg#)

During the coronavirus (COVID-19) pandemic, cities experienced prolonged lockdown in the year 2020 as well as the year 2021 as illustrated in Figs. 7.4, 7.5, and 7.6. Highways were empty except for emergencies, and roads in cities were empty of vehicles and residents as well, due to governments' confinement policies and lockdown across the globe. This has been extended to city markets in South Africa before and during lockdowns (Fig. 7.7a, b), urban spaces, and closed cafes (Fig. 7.8) as well as long lines waiting outside a supermarket in Italy (Fig. 7.9) and vacant shopping malls (Fig. 7.10). COVID-19 does not only affect people in cities but also those who are on the move and most vulnerable (Box 7.1).

After being in quarantine for several months in 2020 and part of 2021, it becomes a clear fact that our built environment, especially our indoor environment (homes) and cities (urban spaces), needs a makeover. Homes not only function as a shelter and a place to live in but also fulfill the new needs amidst COVID-19, including private space for working from home, a gym, shared space for family activities, and more connection with nature. These spaces should ensure fresh airflow through natural ventilation and cross ventilation and availability of sunlight and daylight. All of that should be integrated into our new approach of designing buildings, mainly residential buildings (homes) and indoor spaces in general. No matter if it is before the COVID-19 pandemic or after it, biomimicry is still relevant to the health of our planet and our well-being.

Box 7.1: The COVID-19 and Most Vulnerable

The COVID-19 crisis is an opportunity to re-imagine human mobility. (António Guterres – UN Secretary General, June 3, 2020)

Fig. 7.6 Empty streets in Milan Italy due to COVID-19 lockdown in 2020. (Image source: https://commons.m.wikimedia.org/wiki/File:Corso_Vittorio_Emanuele_II_Milnao.jpg#)

Figure 7.11 illustrates the coronavirus sculpture located in the Biomedical Research Institute of UNAM, Mexico [4], while Fig. 7.12 presents the world's total death per capita as per September 15, 2021, due to the coronavirus (COVID-19) pandemic. It is clear from Fig. 7.12 that the highest rates of death are reported in

(a)

(b)

Fig. 7.7 City markets in South Africa before and during COVID-19 lockdown 2020. (**a**) Greenmarket Square fully occupied before the COVID-19 pandemic – Cape Town. (**b**) Greenmarket Square fully empty during the COVID-19 lockdown – Cape Town. (Images' source: (**a**) https://commons.m.wikipedia.org/wiki/File:Greenmarket_Sqaure_20200320.jpg#, (b) https://commons.m.wikipedia.org/wiki/File:Greenmarket_Sqaure_20200327.jpg#)

Fig. 7.8 Urban spaces and closed cafes in Hanoi, Vietnam, amidst COVID-19 lockdown. (Image source: https://commons.m.wikimedia.org/wiki/File:Vọng_Ba_Lâu_2,_Hà_Nội_001.jpg#)

Fig. 7.9 A long queue at a Supermarket in Italy during COVID-19, 2020. (Image source: https://commons.m.wikimedia.org/wiki/File:Ian_Art_Photography_-_Italy_100_people_queuing_at_a_supermarket_during_coronavirus_outbreak_mar-2020.jpg#)

Fig. 7.10 ICONSIAM shopping mall closed during COVID-19 lockdown in 2020. (Image source: https://commons.m.wikimedia.org/wiki/File:ICONSIAM, remained _open_for_supermarket_ and_takeway_restaurant_during_the_COVID-19.jpg#)

Fig. 7.11 The coronavirus sculpture located in the Biomedical Research Institute of UNAM, Mexico, since before the COVID-19 pandemic. (Image source: https://commons.wikimedia.org/wiki/File:Escultura_de_Coronavirus.jpg)

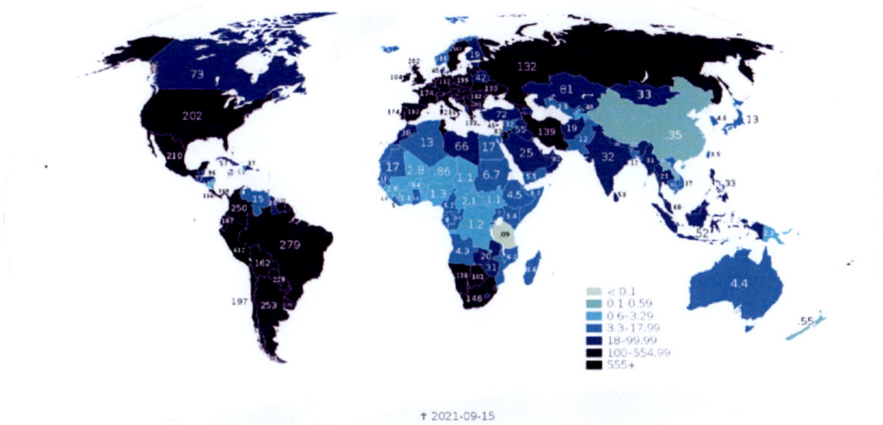

Fig. 7.12 World's total death per capita as per September 15, 2021, due to the coronavirus (COVID-19) pandemic. (Image source: https://commons.m.wikimedia.org/wiki/File:COVID-19_Outbreak_World_Map_Total_Deaths_per_Capita.svg#)

Fig. 7.13 Challenges facing healthcare sector. (Image source: Developed by authors)

North America and South America, Russia, Europe, Iran, and South Africa [3]. Due to the magnitude of the devastation caused by the coronavirus, a sculpture has been built in Mexico to depict such a horrific and terrible crisis.

Through many years of competition and adaptation, nature has produced conclusive medicines and designed solutions. Species such as the chemical compound found in the dogfish shark, which protects it from viral infections, may contribute to the response to the COVID-19 crisis. Viruses' detection is very difficult, and it is impossible in some cases. Though living cells have capabilities in detecting faint traces of molecules with their sensitive biochemical receptors. The instant detection of molecules associated with pathogens and other chemicals is allowed by mimicking the way receptors work and having them induce a macroscopically visible color change [4]. This offers insights into fighting the coronavirus (COVID-19).

One of the government sectors hit by the coronavirus is the healthcare sector, which is currently facing unprecedented challenges and severe shortages. Figure 7.13 illustrates the four main challenges facing such an important and prioritized sector (healthcare).

Looking for innovative materials to create lifesaving products is essential. Moreover, finding cheaper, easier, faster, and more efficient ways to manufacture them is critical. The brilliant idea is to mimic how nature produces materials, such as taking CO_2 from the air and turning it into biomass like trees do. Innovators have already tried this and created polymers that contain up to 50% CO_2 by mass, which also helps lock-up this harmful greenhouse gas (GHG) [4]. In order to reduce the negative feelings and anxiety, active participation is important. Therefore, biomimicry plays a vital and valuable role in finding solutions to the world's complex and painful challenges by turning to nature's designs for guidance and wisdom [5].

There is no doubt that the four challenges (Fig. 7.13) facing the healthcare sector in normal times are totally different from those of COVID-19. In medical waste, plastic is a main contributor to the plastic waste problem that the world has today. To tackle such a major problem, developing a biodegradable alternative to plastic is a must and key in eliminating waste. Alternative materials, for personal protective equipment and other uses like packaging, are also essential [4]. Insect cuticles could assist partially in solving this problem, by taking inspiration from those found in the rigid exoskeleton of houseflies for a "new plastic" to provide protection without adding weight. In addition, the coronavirus (COVID-19) can survive on surfaces for a specific period of time; hence, there is an urgent need to sanitize surfaces regularly. If scientists, biologists, and material experts can find a solution inspired by the red seaweed, which produces an antibacterial film to protect it from infection, this will help in developing similar antiviral films. One of the most important points that should not be neglected while looking for nature-inspired solutions, yet keeping in mind, is a race against us and the sixth mass loss of the world's biodiversity [4].

At the early beginning of the coronavirus in 2020, it puts business worldwide at a pause. This virus also shows that human beings can adapt and make different changes in behavior to reduce the human environmental impacts. It was noteworthy to notice the mitigation in carbon dioxide (CO_2) in 2020; it might be lower than any year in recent history [6]. Nitrogen dioxide (NO_2), a well-known pollutant of vehicles and industrial production, has suddenly disappeared from the water and air; it had suddenly cleared up from water and the air. The environmental damage of human activities is more visible now than ever. Will this give a chance for the rainforests, coral reefs, and other hotspots of biodiversity to offer the next "miracle drug"? Lifesaving treatments are more in demand than ever before [6].

> We have a unique opportunity to design and implement more inclusive and accessible societies. (António Guterres – UN Secretary General, 6 May 2020)

In 2020, the microscopic infection COVID-19 has reshaped our macro world in ways that we could never have predicted. So, never underrate the power and effect of non-visible small creatures. In the first half of 2020, the coronavirus (COVID-19) spread to almost all countries. According to WHO, there are 251,266,207 confirmed cases of COVID-19 globally, including 5,070,244 deaths as of 7.19 pm (CET), November 11, 2021 [7]; an increase of 6.21% in the confirmed cases; and 4.94% deaths from the rates on October 8, 2021.

The feelings of fear, distress, and anxiety are typical feelings and understandable for many during the COVID-19 crisis right now. Although nature has already found innovative solutions to problems and global challenges, humans still look for inspiration, but not all the solutions and ideas are applicable. Nature has shown the secrets to survival after billions of years of research and design through organisms such as animals, germs, and birds, which are outstanding engineers.

Biomimicry means exhibiting these secrets and replicating them to confront and solve the challenges facing humankind. Nature has many possible solutions to future pandemics and existing viruses. For example, bats as a type of species seem to be able to carry these viruses, and they do not relent to such diseases. One of the powerful keys to human health is examining these species and extracting the profound design principles of their abilities to be carriers. It is not useful to wipe them out through hunting, climate change, and habitat loss, but this could literally be translated into eliminating their potential to assist us in finding preventions, cures, and treatments.

Biodiversity and species' conservation are vital to human health. It is imperative to honor what other species have to teach and educate us. Therefore, protecting such species is essential in the time of COVID-19. Learning from specific species could lead to different guidelines, which enable different aspects in the future to redesign the context with a more open mentality for resilience in the context of good adaptation performance to the coronavirus (COVID-19) (Box 7.2).

Box 7.2: Biodiversity and Human Behavior

Biodiversity is a natural repository for more than half of the medicines that we develop. A large number of disease outbreaks are zoonotic, primarily caused by ecological damage. By changing human behavior, we can make a big difference in terms of handling future pandemics. (Dr. Carlos Zambrana-Torrelio—Associated Vice-President for Conservation and Health at EcoHealth Alliance and a Research Associate at the Bolivian National Herbarium, in a discussion with Dr. Sean O'Brien—President and CEO, NatureServe)

7.2 Can Biomimicry Innovation Assist in Limiting the Spread of COVID-19?

The "great pause," as Janine Benyus referred to the COVID-19, is occurring in numerous corners of the globe and introducing a great opportunity for reflection. In addition, it is an incredible chance to profoundly think and ask questions [8]. Thinking innovatively and inspiringly from nature should commence swiftly to assist cities in the current devastating crisis. It is important to be ready and reshape the future to have good identification and management of the pandemic. Nevertheless, the age of COVID-19 proved to offer an opportunity to turn to nature for solutions, counsel, and wisdom. The world needs to change and move forward for innovation. Thus, it is time for biomimicry to answer that call [5].

The coronavirus could be considered as a life's natural defense; it is as much of an immune system as a human has and that is to keep the ecosystem and everything balanced. COVID-19 is like Mother Nature sending everyone to their rooms [8].

As a change due to the coronavirus, biomimicry is being picked up as a foundation for cities, in order to reinterpret how cities should operate in a post-pandemic future. For example, can economies be reshaped to work as an ecosystem with the principles of strong circulation, so money acts like the blood that reaches every single cell, instead of facing the obstacles that currently happen [8]? The same struggle is in the social system, with the disconnected community networks that fail to support people that need help. There is no doubt that one of the most important pillars in our lives is currently our homes. According to such circumstances of COVID-19, it has become much more essential to think beyond net-zero energy buildings (nZEB) to "healthy buildings." Thus, it is vital to propose new ways of building new and existing structures that protect us from viruses spreading while connecting us with nature, that is, biophilic (Boxes 7.3 and 7.4).

Box 7.3: Biodiversity and Biomimicry

"You go out, and you say, 'Wow, look at that butterfly's wings; they are really bright in the sun and then [they become] iridescent.' Biomimicry is the step of asking, 'What are you trying to do?' It's erasing the false boundary between you and the other organism." The butterfly is working to survive, to mate or to communicate—all of the things us humans must do, too. (Janine Benyus [8])

Box 7.4: Human and Health of the Earth

Human health depends on the health of the planet. (Dr. Ganga Shreedhar)

According to the vastly changing environment, innovative thinking in solving problems by adopting environmentally friendly manners is important in order to bring about new products to the market, especially amidst COVID-19; but, it is not an easy task [9]. Biomimicry has never been more critical to the health of our planet than it is now, since the world is dynamically changing under the influence of the COVID-19 pandemic.

In this regard, there is a question to be raised? How can biomimicry contribute to a safer future for humankind? In fact, biomimicry, as a pool of thinkers, collaborates between technology, science, design, and art depending on the learning gained from living nature and applying the best solutions and innovation to human societies. Figure 7.14 illustrates the key points that help us during the pandemic and crisis.

In the field of virus science, every virus has a specific genetic material, either DNA or RNA; some specific tests can be created for detection and early diagnostics. The PCR (polymerase chain reaction) technology is the base of these tests to identify the presence or absence of a virus within the sample. The situation is urgent, and there is not much transparency on how specific the tests for the COVID-19 are in addition to the ratio of false-positive and false-negative results [10].

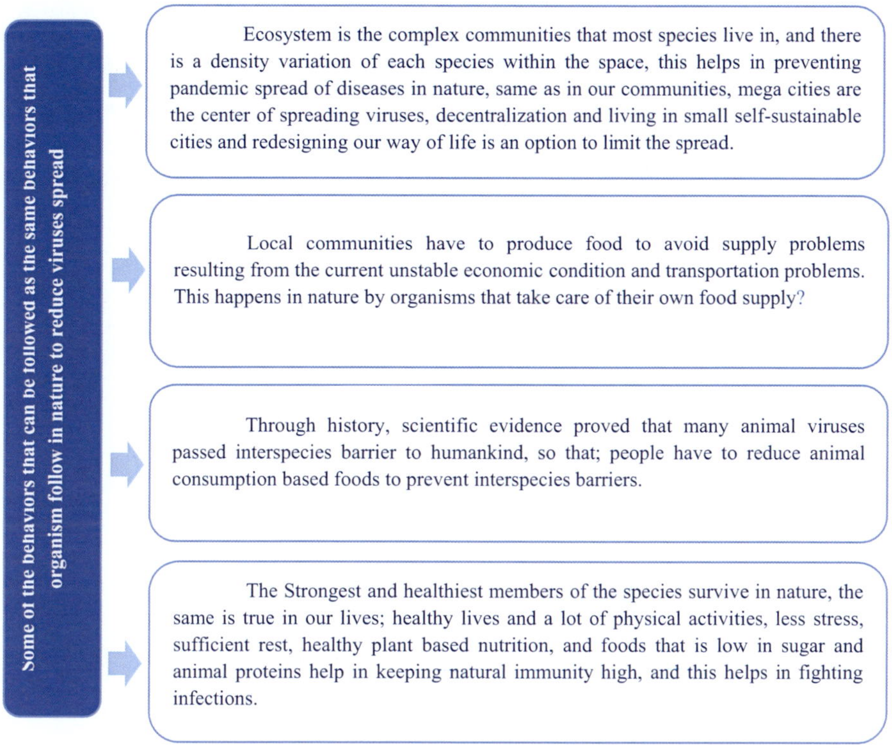

Fig. 7.14 Mimic nature behaviors reduce viruses' spread. (Image source: Developed by authors)

7.3 Scoping Challenge That the World Face and the Context of COVID-19

According to the novelty of the coronavirus (COVID-19), there is currently a sort of curing, medical treatment, and vaccines such as Pfizer, AstraZeneca, and Moderna as well as Johnson and Johnson, Sinovac, and Sputnik-V (under approval by the World Health Organization – WHO). It is clear that a novel virus explains why there are no absolute medical solutions yet.

Moreover, the specific detection of how long the virus lasts and survives on different types of surfaces (metal, plastic, cardboard, etc.) in buildings is not calculated. It can vary from one type to another. Building occupants should follow the instructions and do their best to protect their health, as they are always in direct contact with the virus indoors, especially in closed spaces [5]. It is necessary to issue a call for action to the entire biomimicry community to think, inspire, and develop innovative materials science and create novelty. Figure 7.15 presents the fields that should be realized amidst the COVID-19 crisis.

The target of official development assistance (ODA) investments is to have innovation that is increasingly and essentially needed to effectively respond to and

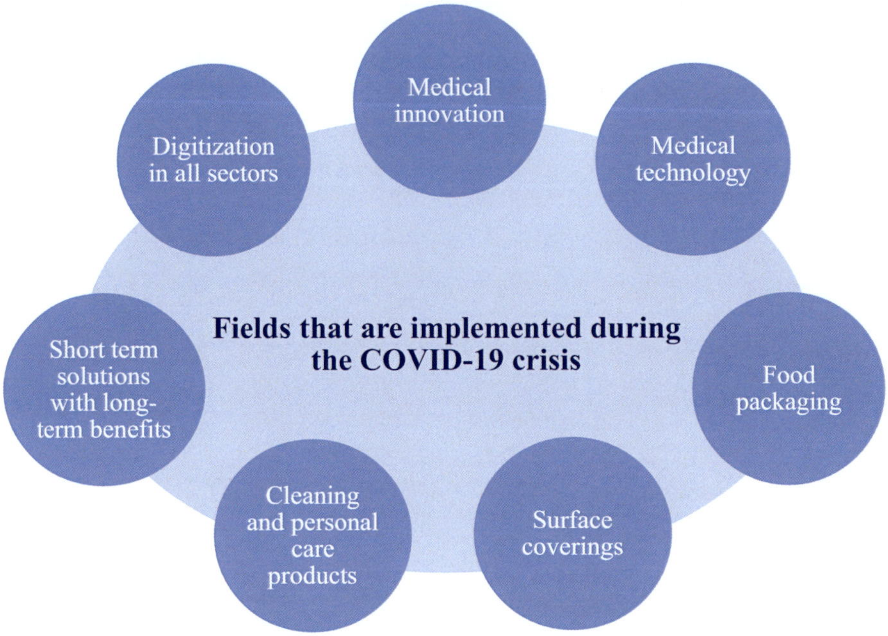

Fig. 7.15 Fields implemented in the age of COVID-19. (Image source: Developed by authors)

recover from the COVID-19 crisis in low- and middle-income countries [11]. Since COVID-19 continues to wreak havoc on the global economy more than a year after the pandemic began, it has also persisted in spurring significant innovation in various industries and companies [12].

Figures 7.16 and 7.17 show the story of COVID-19 and the role of social media during the pandemic, respectively. It is clear from Fig. 7.17 that social media participation and the continuation of its role during the pandemic are immense, especially using pages of international and local organizations as well as press institutions on the Facebook and Twitter platforms to update information about COVID-19 around the clock for users.

7.4 Innovation for COVID-19 and Beyond

The coronavirus (COVID-19) pandemic is considered a colossal challenge to societies around the globe, due to the fast infectious spread among people, especially the "Delta" type that leads to keep people having sufficient social distance between each other in 2021. People have been confined and self-isolated for several weeks to months to limit the spread of this virus. Social distancing measures have become a new way and new lifestyle for communicating with each other; they have both

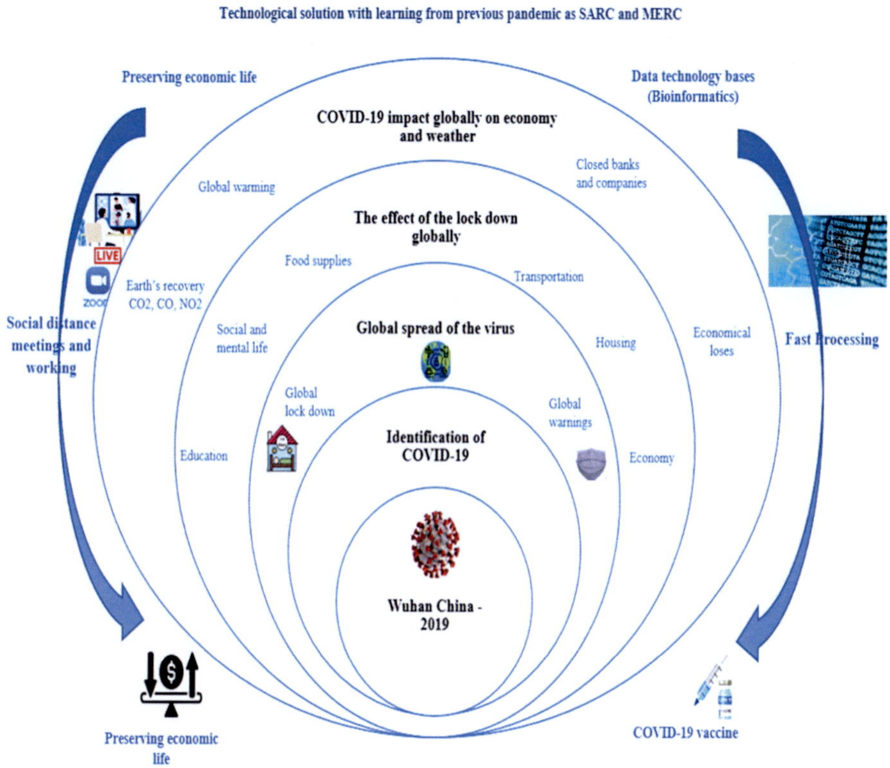

Fig. 7.16 Infographic story of the coronavirus (COVID-19) pandemic. (Image source: Developed by authors after https://www.mdpi.com/1660-4601/18/11/5645)

negative and positive impacts on different aspects, such as transportation, education, and economies.

As a result of the reduction in population movements and human activities, GHG emissions have been decreased, and the ozone layer has been improved. This had a positive impact on Earth, weather, and the environment. This means that the COVID-19 pandemic has negative effects on humans and their activities but positive impacts on nature [13] (Box 7.5).

Box 7.5: COVID-19 and Nature

COVID-19 offers 'Utopian Glimmer' of new world, but we must give back to nature. (Janine Benyus)

The ability of viruses to adapt to the changing environmental conditions and evolve with it is very fast. Therefore, strategies for accurate, rapid, and long-term detection as well as effective protection and regular updates of treatments are always

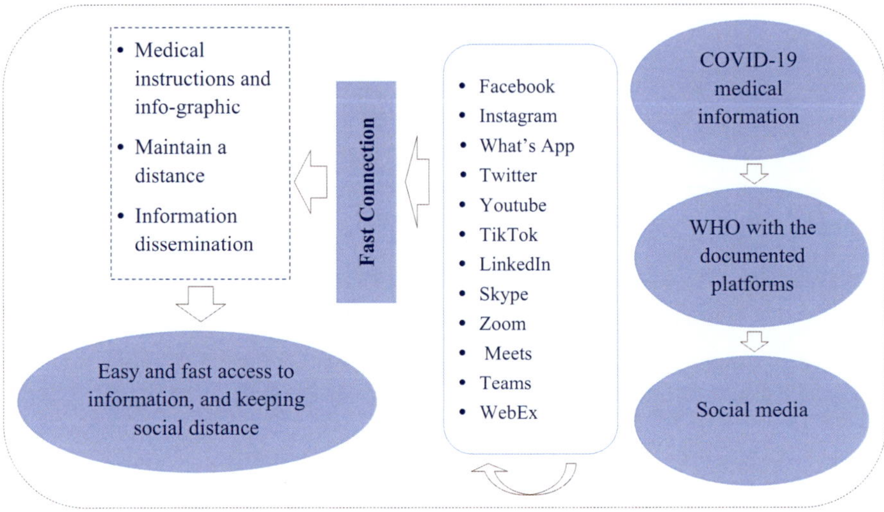

Fig. 7.17 Social media and its continuation role during the COVID-19 pandemic. (Image source: Developed by authors after https://www.mdpi.com/1660-4601/18/11/5645)

needed to control the spread of the virus, which requires multidisciplinary solutions. Can biomimicry in architecture and nature be one of the solutions? In fact, the role of materials science is very important in this regard since it helps in having a quick response to emerging viral diseases by developing tools to detect the viral protein structure, design care devices for accurate and rapid detection, and develop vaccination treatment and strategies [14].

7.4.1 Materials' Inspiration from Nature

Materials' inspiration from nature has been seen in different fields, including buildings and their façades. Over the past few decades, materials science and nanotechnology have been playing significant roles in this direction. Each is investigating particle structure: (a) nanotechnology studies particles with sizes ranging from 1 to 100 nm; and b) materials science studies the structure and function of materials. However, these two structures regularly go hand in hand in many applications in the built environment [15].

Nature spent millions of years evolving a playbook of engineering and material solutions. Scientists are searching for and putting forth effort in advanced characterization techniques in order to discover and unlock the secrets of natural materials. Moreover, scientists are using what they learned to design synthetic materials with more advanced properties. To mimic the bone, for instance, now there are nanocomposites, bringing together toughness and strength, but this is considered a very difficult and challenging process because it is not easy to create biomimetic materials that rely on a perfect assembly of organic and inorganic components at the microscale and nanoscale, with precise structural framing – tangible and intangible [15].

7.4.1.1 Natural Nanostructures' Mimicry

In the age of COVID-19, natural nanostructures' mimicry can play a major role in combating the spread of the coronavirus virus. To provide an extraordinary and unique property – seen in fauna and flora – nature relies on nanostructures. Organisms like gecko and chameleon have nano-fibers; the chameleon's under skin has nano-crystals that enable it to change its color reacting to external stimuli [15]. Nonetheless, the gecko has nano-fibers on its feet to allow it to hang itself from the ceilings and climb walls while resisting gravity. Hence, new technologies can be created by looking, identifying, extracting, and translating as well as mimicking natural nanostructures. For example, adhesive tapes that use nanostructures that mimic those found on gecko feet are now commercially available with potential applications in aerospace engineering, robots, prosthetics, wearable devices, and gloves [15].

7.4.1.2 Fighting the COVID-19 by Nano-materials and Nanotechnology

Can nano-materials and nanotechnology assist in containing the spread of COVID-19? Nanotechnology in the medical field played a vital role this year more than ever. It can be defined as a design and an application of many materials and devices – at least one dimension is less than 100 nanometers. In the field of medicine, the application of nanotechnology is called "nano-medicine," which includes the use of nano-materials for diagnosis, curing, and control of the disease. Nanotechnology has contributed to solutions for testing, vaccines, and preventing the spread of the COVID-19 pandemic [16]. Nevertheless, the virus has also highlighted scientific shortcomings in testing, such as the long delay in widely available testing and the test's inability to detect the virus for days after infection [15].

By exploiting nanotechnology, there are opportunities to develop strategies related to the prevention, diagnosis, and treatment of COVID-19 and other infections; nanotechnology helps in managing various aspects of COVID-19. The most vital roles for nanotechnology are the following:

- Develop safe personal protective equipment (PPE) and surface coatings that inactivate the virus in order to reduce the spread of the coronavirus
- Create a sensitive nano-based sensor for quick infection detection
- Develop new drugs with specific functions, such as lowering toxicity, and sustaining release, specified to work with the lungs
- Establish nano-based vaccination to improve immune responses [16]

Nano-based approaches seem to be the near future and the first choice for the development of the most effective therapies for a wide range of diseases, including the COVID-19.

7.4.2 COVID-19 Pandemic Changed the Face of Life

The past two years (2020 and 2021) witnessed a major change in the face of life globally. A worldwide interference occurred due to the coronavirus (COVID-19) pandemic, which has never been perceived before. World leaders have never experienced such a magnitude of a crisis before since World Wars I and II. Moreover, the narrow time window for inventing new pharmaceutical therapies by Research and Innovation (R&I) teams was also bounded by a very short period compared to the standard norms. Many countries across the world are exerting huge efforts to enact and implement strategies and control policies to avoid significant losses.

Despite such colossal efforts, death rate is still high according to the WHO dashboard, which was recorded at 4,887,600 people on 18 October 2021 at 10:38 CET [7]. Nevertheless, this pandemic has already imprinted itself upon several aspects of life as citizens are experiencing psychological and social anxiety in addition to disrupted national economies and growth as well as resilience in many sectors. This is a sort of threat, and the new source of stress cannot be understood by only scientific reductionism and specialized thinking, because it has many side effects on them, psychologically, socially, and economically [17].

The points that should be taken into consideration are how the current pandemic is linked to ecological sustainability and resilience. Furthermore, rethinking about the complex interactions between human and nature's health that led to the crisis and why the context in the unsustainable societies is crucial. As a result, research and inventions should now focus on finding transdisciplinary resilience links and dialogue on health issues and biosphere life-supporting processes in order to deal with the COVID-19 crisis.

The main lesson learned in particular is that the COVID-19 pandemic revealed that societies and city sectors are not resilient enough and not well prepared to anticipate and cope with amplified disease risks on the scale of COVID-19. In such a critical time of severe hazard (COVID-19), leaders and medical staff proved to be the champions of an ongoing journey that is not ending yet. The expected duties are problem identification, options' establishment, action plans' development, execution of the plans on time, and good communication through media to transfer the ideas to the public. Moreover, they need to understand the ongoing mechanisms for achieving clarity about the current situation and desired outcomes [17]. Although it is not the year 2022 yet, the COVID-19 pandemic is considered a message to us from nature. Such a message needs to be decoded and thought of profoundly to come up with city innovative solutions. Box 7.6 highlights such a notion.

The Earth is a self-regulating biosphere. Nature and human activities are interconnected and interdependent. Climate change (CC) is causing direct and indirect

Box 7.6: Climate Crisis and Pressure on Nature

The ongoing climate crises are due to the fact that humanity has placed too many pressures on the natural world with damaging consequences. Therefore, understanding the whole human-nature system and how the planet functioning is of primary importance. (United Nations [17])

hazards to human health due to CC disturbances to the Earth's physical systems. The self-regulating ecosystems have been fragmented due to human behavior, for example, deforestation, rapid population increase, and land use interference. These actions led to many consequences, one of which is the global COVID-19 pandemic that we are facing now. The Coronavirus (COVID-19) and other viruses were living in symbiosis with certain animal species and have now transferred from these species to humans and caused severe consequences to human health [17].

The lesson is that all should consider "how nature and biomimicry architecture can support the efforts to encounter and offset the spread of COVID-19?"

7.4.3 COVID-19 Positive Insight

There are several negative outcomes from the pandemic, such as illness, death, fear, and anxiety for people. Nevertheless, it highlights several important life lessons for all humans globally, at the national and local levels, as well as for societies and individuals. Figure 7.18 illustrates the nine positive insights gained from the COVID-19 crisis, however, more to be learned by days [17].

The current and unexpected crisis that attacked the whole world from the leaders to the poorest people equally taught us the lesson that there is nothing taken for granted and everything can be changed in one second; in addition, it let us know the importance of structural and personal consciousness, vigilance, and ethical responsibility to enhance the resilience of our cities' and communities' infrastructure to steadfastly cope with hazards. In addition, COVID-19 upgrades the level of hygiene of people and sanitizes their purchases and hands frequently. Furthermore, it influenced people's perspectives on their interactions with ecosystems and with others. However, more importantly, it shows the impacts of the "human-nature" dichotomy (division or being a part) [17]. Therefore, it sets a bill for personal, societal, and global changes and transformations toward nature.

Lately, biomimicry is being picked up as a foundation for cities, as one of the most important pillars in reimagining how they might operate in the future. Janine Benyus said that the biggest system we have is our homes. She spent a lot of time creating new ways of existing in tandem with nature, rather than in contrast or opposition. It is true that net-zero buildings (NZBs) are great, but can we go further? The target from this way of thinking is to heal the planet rather than to have a less negative impact [4]. It is clear that the time of the COVID-19 pandemic has contributed

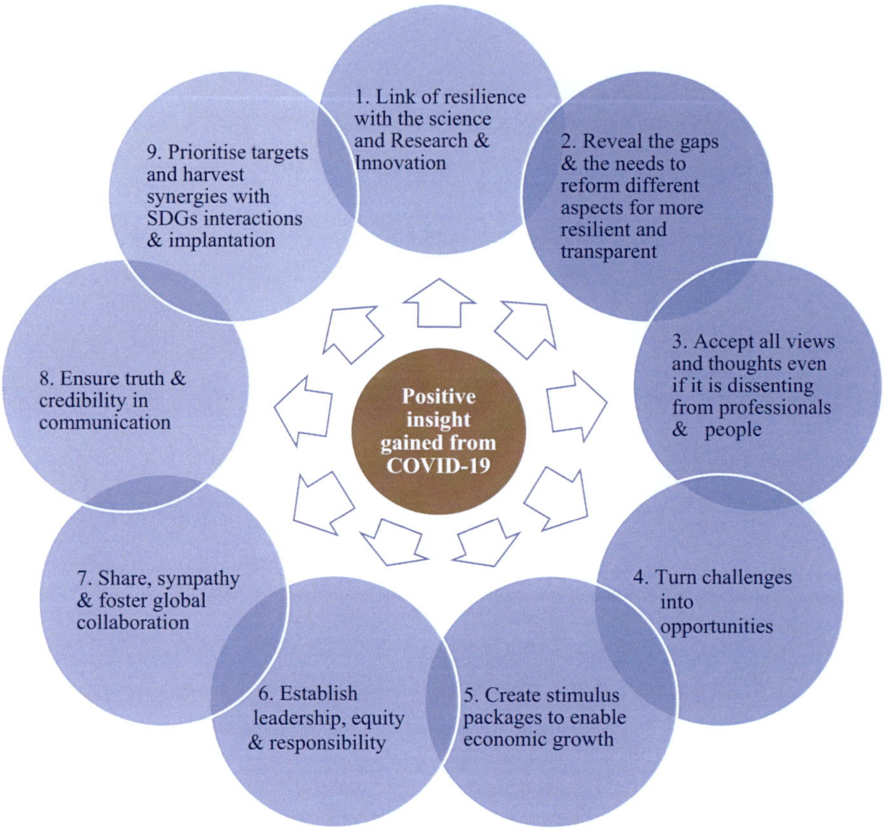

Fig. 7.18 Positive insights gained from COVID-19. (Image source: Developed by authors)

to resetting our natural world and people's mindsets – and Benyus could not be happier and set the notion as mentioned in Box 7.7 while Box 7.8 lists the connection between nature, organisms, and biomimicry.

Cities and their infrastructure play a key role in fitting and adapting to the impacts of COVID-19 on both the people and communities. Hence, it is essential to address the guidelines for healthier urban spaces' and buildings' indoors during the time of COVID-19. The next section portrays these guidelines for healthy urban areas.

Box 7.7: Change Behavior Towards Nature

You don't have to leave this planet, but you might just have to change your busyness imperative. (Janine Benyus [4])

Box 7.8: Nature, Organism, and Biomimicry

You go out, and you say, 'Wow, look at that butterfly's wings; they are really bright in the sun and then [they become] iridescent.' Biomimicry is the step of asking, 'What are you trying to do?' It's erasing the false boundary between you and the other organism. (Janine Benyus)

7.5 Guidelines for Healthier Urban Spaces' and Buildings' Indoor in COVID-19

The coronavirus (COVID-19) affects various sectors like health and the way we work, communicate, and live. It has been witnessed during the past two years that the level of the awareness and the importance of designing cities to be health-friendly and provide well-being for the people living in them have been slightly enhanced. Most people spend 90% of the year indoors, whether at home, work, school, or university or at leisure places depending on the day and age [18]. The COVID-19 pandemic comes as a chock to all different aspects of life. The lockdown period has changed the way that people and communities live, work, and interact to survive in such a horrible situation. Therefore, making the built environment resilient during and post-COVID is a must for most outdoor and indoor spaces (housing, public buildings, and workplaces) [18]. The main question is about the possibility of transforming cities and buildings to provide healthy living? In fact, it is a huge task to be carried away by everyone; however, many countries have failed in providing such an answer.

7.5.1 Urban Spaces

If the COVID-19 pandemic continues to stay longer, as all indicators disclose, it will and will live with us; it becomes more imperative to focus on how we live and grow in urban accumulation. During the COVID-19 pandemic, the majority of, if not all, urban spaces were vacant bylaw, besides; there are many issues that are concealed when it comes to cities' resilience. Thus, the question is "What are the cracks that COVID-19 has bared in city planning? How can we have a better future amidst or post-COVID regarding the urban planning system and its resilience? Redesigning and thinking about cities investigate the relationship between the pandemic and urban health infrastructure, climate resilience, informal settlements and work, social infrastructure, and commercial investments, with the goal of stocking a direction for urban planning to mitigate the risks of future health rises [19]. There are some elements in cities and urban areas (Fig. 7.19). All of these elements should be implemented in a new way, especially at the time of COVID-19.

The development of cities has evolved over the years! There are emerging types of cities such as green cities, smart cities, and sustainable cities. It is imperative, therefore, to present and discuss the relevant development of cities, including healthy cities, green cities, smart cities, and all in cities as well as sustainable cities, resilient cities, and rising cities as shown in Fig. 7.20.

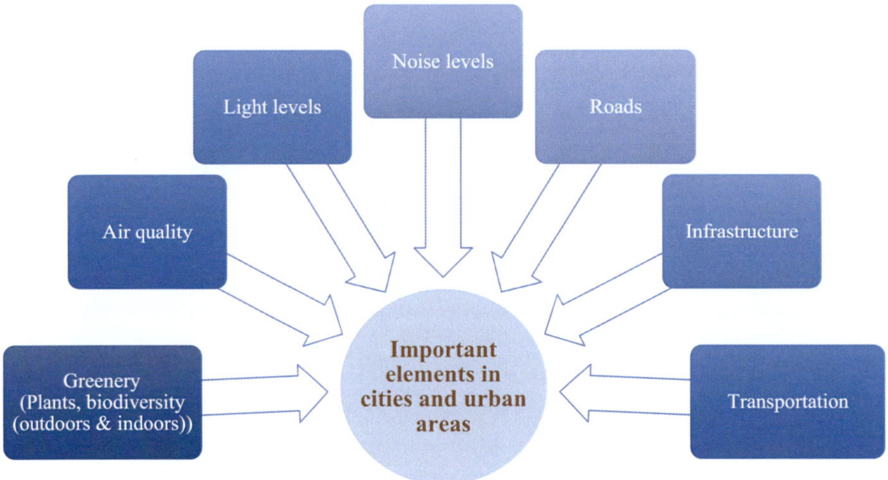

Fig. 7.19 Important elements to be considered in cities and urban area amidst the COVID-19 crisis. (Image source: Developed by authors)

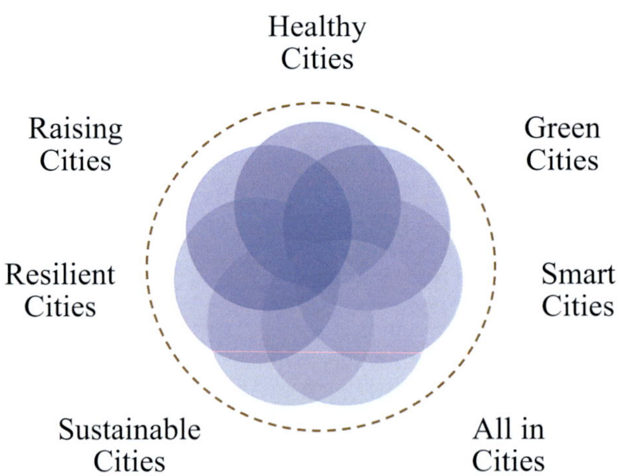

Fig. 7.20 Types of related cities in the age of COVID-19. (Image source: Developed by authors)

7.5.2 Healthy Cities

The spread of the COVID-19 pandemic globally has caused an urgent need to make healthy cities in all aspects and parameters. It is becoming a new and key parameter in developing such cities amid the current crisis. This includes the health sector where consolidation of the infrastructure of urban healthcare will lower the number of patients with new infections, and disruptions in hospitals and general healthcare services, particularly public health, such as immunization, will minimize urban populations' exposure to more infections and disease spread. There is an urgent need to strengthen urban primary health structures, physically and financially with effective and advanced monitoring, supervision, and responsibility. The administration and management at the city level should be seamlessly manifested at all levels of healthcare. They must work together in a distinctive and standard pandemic-related operating procedure [19]. Our cities and nature should be viewed as a whole, with neither being able to exist without the other in order to provide livable and healthy environments.

Several studies have shown that there is a close link between a person's proximity to nature and their mental health and well-being. The King Baudouin Park in Jette – Brussels, Belgium – provides a large variety of species and diverse habitats and enhances livability for people, as illustrated in Fig. 7.21.

Green areas have a positive impact on humankind's mental health and encourage physical activity and social interaction. In addition, green areas create spaces for both physical and mental restorations, which leads to a decrease in the chances of having cardiovascular disease, obesity, and depression, while improving visual acuity and productivity. Moreover, this helps countries to reduce the cost of treatments and the threats to the welfare of societies. Therefore, the creation of more green areas will not only contribute to healthy cities but also consequently bring financial benefits [18]. Other examples are shown in Figs. 7.22 and 7.23 (Box 7.9).

Box 7.9: Lessons to Protect Nature

We are saying, 'Let's make ourselves as small as possible and not harm anything around us.' But let's break that open and say, 'Let's put the wellness not just inside our homes, but outside the walls.

She added:

How can we go from meeting our own needs to doing what other organisms do, which is hailing goodness and benefit to everything around us? (Janine Benyus)

7.5.3 Green Cities

Some of the fractures introduced during the COVID-19 crisis are the lack of open spaces, ineffective resource management for the affected population, and disruptions in energy consumption and food supply. These are manifested due to the absence of appropriate urban planning to attain resilience and sustainability. The

Fig. 7.21 King Baudouin Park provides a large variety of species and diverse habitats – Jette – Ganshoren, Brussels, Belgium. (**a**) Water and greenery in Jette park. (**b**) Birds, water, greenery, and humans in Jette park. (**c**) Trees and plants in Jette park. (**d**) Flowers and greenery in Jette park. (Images' source: https://commons.m.wikimedia.org/wiki/Category:Parc_Roi_Baudouin,_Jette)

Fig. 7.22 Public parks and water provide wide green areas for walkability, a variety of species, and varied habitats. (**a**) A green park in Central Brussels (**b**) A government building surrounded by water in Ghent. (Images' credit: (**a**) and (**b**) Mohsen Aboulnaga)

Fig. 7.23 A public park in Songdo – Incheon, South Korea – provides wide green areas and variety of species and habitats. (Image credit and source: Mohsen Aboulnaga)

response to COVID-19 and climate change needs a special updated and recovery strategy, to avoid inauspicious impacts during pandemics. The priority and focus should be on green and clean urban mobility, circular and green economy, and green building frameworks (Fig. 7.24) [19]. Also, the priority is to accelerate investment in green infrastructure and renewable energies. Additionally, parallel to the mentioned transformations, the actions in the transportation sector are also needed to line up with the green agenda to have resilient and sustainable cities that are able to face future pandemics. The user behavior and transit preferences should be implemented in a balanced way with new decisions in the transportation sector to avoid congestion. For example, to achieve this balance, encourage people to use public transportation. To achieve that, it should be well ventilated with good pricing. Hence, creating well-ventilated public transport with congestion caps and pricing and encouraging the use of nonmotorized transport are some of the ways of how this balance can be achieved [19]. Moreover, a green city could be classified as the city that incorporates the five following parameters: (a) optimize energy supplies, (b) provide clean and potable water 24 hours seven days a week, (c) furnish bicycle and walking lanes, (d) establish recycling of water and waste systems, and (e) incorporate green technologies in buildings (Fig. 7.25).

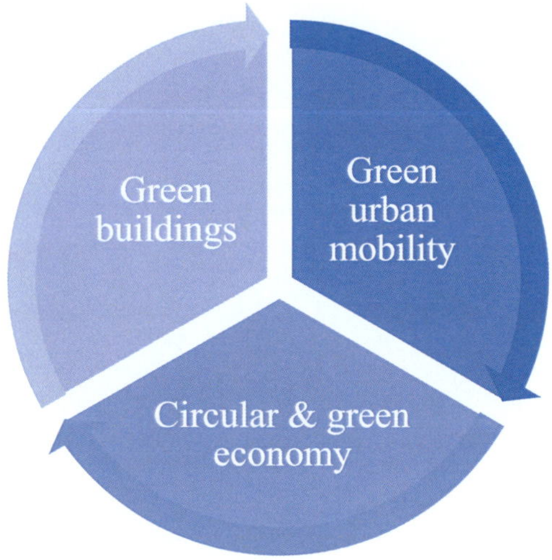

Fig. 7.24 Framework for green cities. (Image source: Developed by authors)

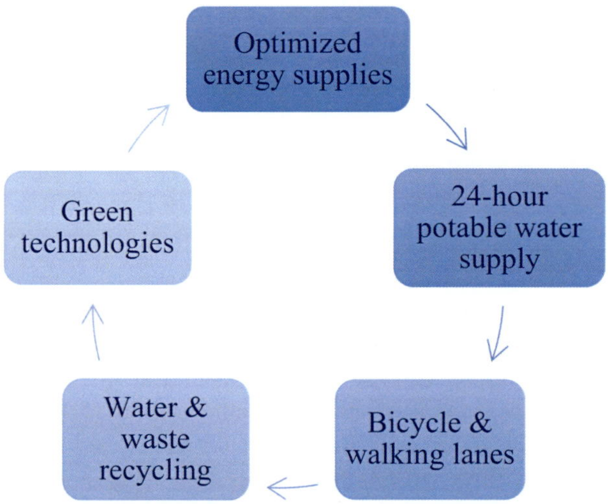

Fig. 7.25 Features of green cities. (Image source: Developed by authors)

7.5.4 Smart Cities

The concept of smart cities has been immensely used and witnessed globally through the coronavirus (COVID-19) pandemic, but mainly through smart applications and technologies, but not fully utilized in the city's infrastructures across the world, especially in developing countries.

During the COVID-19 crisis, the dependence on smart solutions and applications became at its highest capacity and daily use during the complete lockdown in 2020 and 2021, mainly in purchasing goods; paying bills; learning and education via online education platforms such as Zoom, Meet, Teams, and Blackboard; and even registering for the COVID vaccines in every country. The COVID-19 pandemic has triggered the urgent need to transform into a Digital Era. In this context, will smart citizens drive smart cities or will the latter transform citizens to be smart? In fact, the COVID-19 pandemic and conferment policies did it. Peoples at different ages have a major role to play in a smart city in all sectors. Nevertheless, Fig. 7.26 illustrates an example of the role people play in a smart city [20].

A smart city is defined as "the city that includes smart networks into which everything is hooked." It can also be described as "a city that acts more like a living organism in a smart way," and/or it can be characterized as "a city that can respond to your needs in a smart mean" [20, 21]. For instance, the City of Dubuque in Iowa, USA, has developed smart water meters to offer data to citizens via a community portal, so that residents can observe their water usage and even compare the readings online with those of their neighbors and even reported them online [20]. In Dublin, Ireland, the city has worked with the city council to open up the vast amounts of data it has, which led to clever little applications such as ParkYa, which uses traffic data to find people the best parking space in the city [20]. In Singapore, Stockholm, and California, IBM (International Business Machine) has been commissioned to gather traffic data and run it via algorithms to predict where a traffic jam will occur, an hour before it has happened [20]. In Rio de Janeiro City in Brazil, the city authority built a NASA-style control room where banks of screens collect data from sensors and cameras located around the city and analyze the traffic situation [20].

Another bold example of a smart city is "Fujisawa smart town" in Yokohama, Japan, which was built in 2018 and was due for full completion in 2020. The Fujisawa smart city exhibits the state-of-the-art technology in developing a smart city. The Fujisawa smart city uses an entire range of energy and smart solutions and IOT as well as intelligent electrical appliances and smart monitoring systems, such as solar cells, heat pumps, lithium-ion batteries, and LED lighting and sensors as well as energy-efficient washing machines and refrigerators, and smart building automation technology to monitor the city's daily performance [22].

The clean energy in Fujisawa smart city is generated by integrating renewable energy (solar PV) nature on top of a public social building in the city to generate

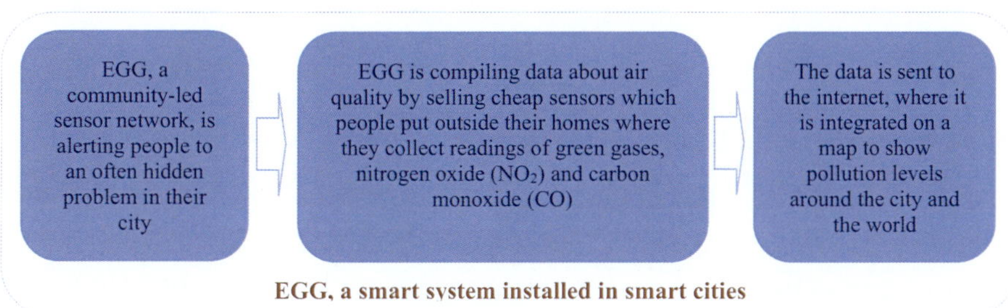

Fig. 7.26 People and smart cities – EGG system. (Image source: Developed by authors after [20])

(a) (b)

Fig. 7.27 Integration of nature with renewable energy (solar PV) on top of a social building in Fujisawa smart city, Yokohama, Japan. (**a**) Combined Solar PV panels with nature. (**b**) View of PV panels integrated with a green roof. (Images' credit and source: Mohsen Aboulnaga)

Fig. 7.28 Integrated solar PV panels on the roofs of residential buildings to produce clean energy from renewables in Fujisawa smart city, Japan. (Image credit and source: Mohsen Aboulnaga)

energy for the building as shown in Fig. 7.26. Each residential house has its own green energy and electricity generated from solar PV panels mounted on the house's roof, a heat pump to heat the water, and smart sensors to save energy (Fig. 7.27) [23]. Furthermore, solar PV panels – each generating 6 kWp – were installed on the Fujisawa smart city peripheries to generate electricity and sell it to the Yokohama power plant, ensuring economic viability and financial sustainability (Fig. 7.28). Moreover, a smart building in a smart city can be defined as listed in Boxes 7.10 and 7.11 (Fig. 7.29).

(a) (b)

Fig. 7.29 Installed solar PVs on the surroundings peripheries of the site of Fujisawa smart city, Japan, to generate electricity from renewables. (**a**) Installed solar PV panels on the city peripheries. (**b**) Renewable energy to attain economic viability. (Images' credit and source: Mohsen Aboulnaga)

Box 7.10: What Is Meant by a Smart Building?

A Smart building interacts with the power grid to optimize the energy flows between them.

A sophisticated operation of the building services represents a significant opportunity for energy efficiency, cost reduction, and stabilization of the smart grid. [24]

Box 7.11: Definitions of a Smart Building

Smart buildings are those that achieve significant energy savings by taking advantage of improved technology and materials in terms of structure, appliances, electrical systems, plumbing and HVAC.

A building that deals with the increased amount of information available from this wider range of sources will allow these systems to become adaptable, and enable a Smart Building to prepare itself for context and change over all timescales. [25]

7.5.5 All-In Cities (Inclusive)

Responding to pandemics should be improved through an urban planning system that includes collected data aggregated on the number of people, number of households, and areas of vulnerability, as well as the availability of sanitation and hygiene amenities. In order to have proper responses and preparations, the number of people living in vulnerable areas and information on the civic amenities has to be known. Also, it is so important to collect data on formal and informal settlements and labors in cities, to measure the required purse for sufficient access to food supplies, housing, sanitation, and financial services [19]. One of the most critical analyses for post-pandemic rehabilitation is the examination of the effect of the pandemic on job

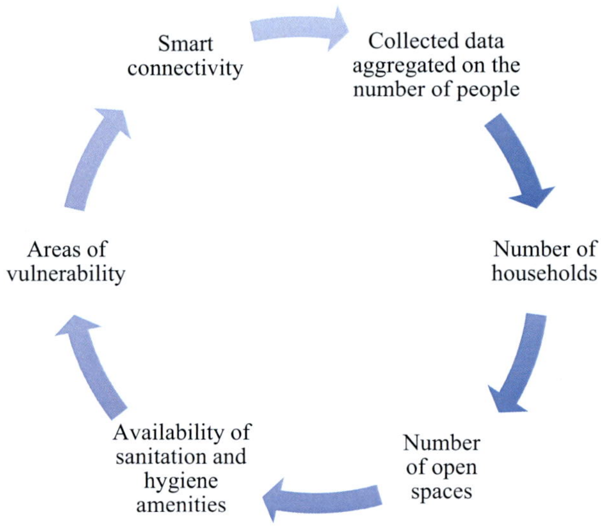

Fig. 7.30 Features of All-In Cities concept. (Image source: Developed by authors)

losses and employment. To limit the deadliest impact of future pandemics, it is a must to have better data management in urban planning [19]. Figure 7.30 summarizes the All-In Cities concept.

7.5.6 *Sustainable Cities*

Sustainable cities refer to "cities around the world that are actively making changes to become more sustainable by recognizing the interests, motivations, and models of these cities" [26]. According to Ban Ki-Moon, former Secretary-General of the United Nations, "Building sustainable cities – and a sustainable future – will need open dialogues among all branches of national, regional and local government. And it will need the engagement of all stakeholders" (Box 7.12) [27]. Sustainable cities can be described as the cities that can achieve the following 12 points shown in Fig. 7.31. Thus, policy makers, planners, and architects should consider the three pillars of sustainability in developing cities. These three pillars are (a) environmental, (b) economic, and (c) social that should be interwoven and overlapped in every process of creating the city not only that but also ensure that the other dimensions of sustainability (viability, livability, and equity) are coherently attained.

Box 7.12: Sustainable Cities

Building sustainable cities – for a sustainable future – will need an open dialogue among all branches of national, regional and local government – And will need the engagement of all stakeholders. (Ban Ki-Moon, former Secretary General of the United Nations)

Fig. 7.31 Features of sustainable cities. (Image source: Developed by authors)

In this context, to attain viability, economic development must abide by the supportive capacity of the ecosystems, and depletion of nonrenewable resources must be avoided; to achieve livability, this means correspondence of the environment to social needs, which can refer to the concept of quality of life; and finally, to ensure equity, an interaction between the economic and social dimensions should be in place to create affordability for all [28, 29]. However, having motioned to green cities, it is clearly imperative that a green city can be termed a sustainable city, if it first achieves the four main issues presented in Fig. 7.32 in addition to ensuring good governance and institutional improvement.

7.5.7 *Resilient Cities*

The coronavirus (COVID-19) has revealed that all cities globally are not resilient enough, specifically in the health and infrastructure sectors. Cities were unable to withstand the scale of the horrible shock of the COVID-19 and act in a resilient way, even some cities in the developed world failed to contain such a shock. There is more urgency than ever to enhance the resilience of cities to ensure that they can steadfastly confront future shocks and crises. It is imperative to define city resilience (Box 7.13) and the four areas that drive city resilience (Fig. 7.33).

Fig. 7.32 What turns a green city into a sustainable city? (Image source: Developed by authors)

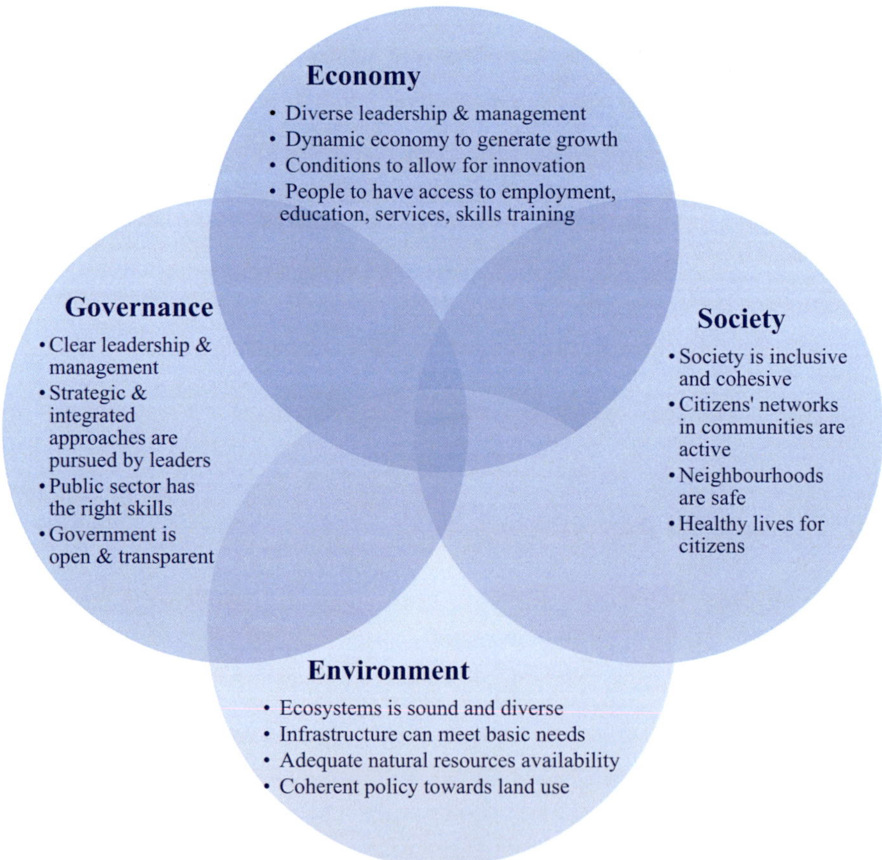

Fig. 7.33 Four areas to drive cities' resilience. (Image source: Developed by authors after OECD, https://www.oecd.org/cfe/regionaldevelopment/resilient-cities.htm)

Box 7.13: Resilient Cities

Resilient cities are cities that have the ability to absorb, recover and prepare for future shocks (economic, environmental, social and institutional). Resilient cities promote sustainable development, well-being and inclusive growth. (Organisation of Economic Cooperation and Development – OECD [30])

7.5.8 Rising Cities

As a result of the post-pandemic impact, the world is currently facing significant changes in many aspects like the physical, social, and financial infrastructures of cities. Safety is considered the most important factor in the future changes. Buildings' designs have to be transformed in accordance to the health protocols that have been followed during the crisis. The world is in a transitional period and is encountering a shrinking post-pandemic globalization. There may be a reconceptualization of cities as singular nodes of trade, commerce, and knowledge as well as food production.

One of the most essential steps is the restrictions on commodities and labor; the mobility of these will be restricted within specific governance zones. Cities may also develop as self-sufficient hubs [19] as shown in Fig. 7.34. The significant modifications in this type of cities, the administration, and the way of living will mainly happen as a result of the COVID-19 crisis. Our understanding of cities, as dense conglomerations of constant social interactions, might be overhauled. In this context, people should accept and be in harmony with these changes and verify that the system is going through changes that go along with this understanding, be it in mobility, housing, employment, or financial investments. It is difficult to carry out all these changes in an instant future; therefore, stakeholders must coordinate their efforts in making such changes that limit future pandemics and mitigate short-term consequences [19].

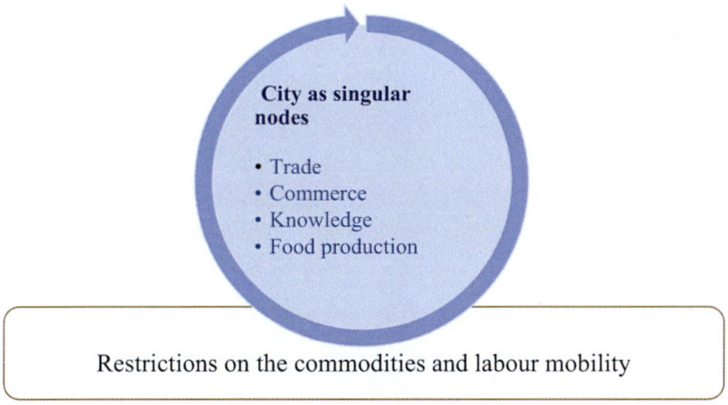

Fig. 7.34 Rising cities as singular nodes. (Image source: Developed by authors)

7.5.9 Urban Context and Pandemic Role

Cities of the future will have specific qualifications with the rapid growth of urban populations. The key elements of designing future cities are culture, local values, populations, and health [19]. As the pandemic continues, and the virus manifests itself in new ways, the number of cases has surpassed 243 million, with nearly 4.94 million deaths worldwide [31]. This is a strong reminder that urbanization has changed the way individuals and communities live, work, and interact and that it's urgent to make resilient systems and to limit the spread of infectious diseases today and tomorrow, and obviously, urbanization is achieving a success. Taking into consideration that 54% of the world's population is currently living in urban areas and it is forecasted that 70% of the population will live in urban areas by 2050, more green areas and open spaces are urgently needed to provide healthy spaces in the age of COVID [32, 33].

The image of the future city includes many pillars such as walkability, mixed use, mobility, and many other elements as presented in Fig. 7.35. It's a place that is scaled to people, providing safe pedestrian environments and open space for public life. Most importantly, it is a place that is grounded in the unique culture and values of its people while embracing modern technology and the global community. If you look back to history, you will discover that the future city idea engaged some city building processes and features from the past and added some modern approaches in transportation and environmentally friendly buildings' systems [34].

People's health and lives have become the most important factor in cities, especially at the time of COVID-19. Hence, the city of the future should value people and their daily experience as well. Therefore, architects, engineers, urban designers, and planners can set the foundations for healthy vibrant communities that are

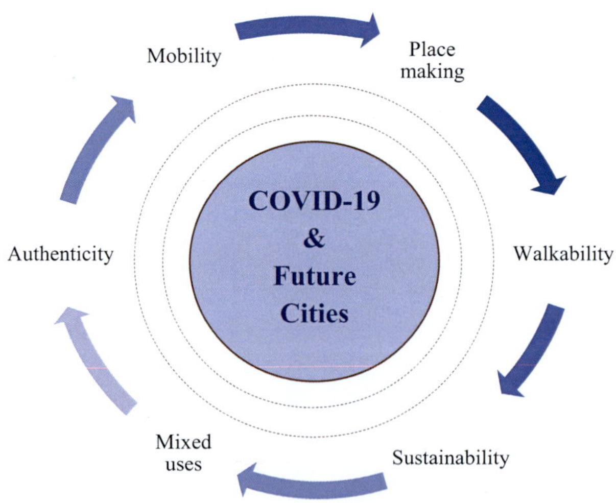

Fig. 7.35 Basic strategies for future cities. (Image source: Developed by authors)

lasting and adaptable over time by following key principles that put people at the center, yet providing better environmental indoor quality (EIQ). Successful cities are unique reflections of their climate, people's behavior, history, and economy as well as culture. It is not a one-size-fits-all solution and such principles are relevant anywhere in the world [35].

The COVID-19 pandemic has revealed the significance of green urban spaces to human health and well-being. The pandemic has seen an increase in people's use of green areas, such as local parks, public beaches, marinas, and national parks, which are safer than closed spaces. These areas provide important health benefits that can mitigate some of the stress associated with the pandemic. They can also provide a feeling of social cohesion as people continue to practice social distancing [36]. Box 7.14 illustrates the visits' rates to green areas during COVID-19.

It is imperative in the age of COVID-19, to think deeply about how the six elements of the future city (place making, walkability, sustainability, mixed uses, authenticity, and mobility) operate and are closely linked. Inspiration and innovative thoughts should be linked with nature to drive solutions to the current COVID crises (Fig. 7.36).

Box 7.14: Visits' Rates to Green Areas During COVID-19

If you live closer than 50 metres to a green area, you will visit a green area about 3–4 times a week, but if you live 1,000 metres away from a green area you will only visit it approximately once a week. Studies also show that if you have a green area close to your home, you are more likely to visit other green areas. (Daniel Hojniak and Camilla Julie Hvid)

Fig. 7.36 Inspiration and dynamic thinking for future cities amidst COVID-19. (Image credit and source: Mohsen Aboulnaga)

7.5.10 Indoor Buildings' Spaces

A new way of designing our buildings is critical to enhance people's health, because people spend from 90 to 95% of their time indoors [37, 38]. Designing the elevators with special specifications that meet the standards of social distancing to prevent the spread of COVID-19 is necessary as illustrated in Fig. 7.37. Buildings that include open areas, courts, and spaces for play or sport or that are well connected to outdoor spaces and are more likely to promote healthy living [36]. Poor air quality, noise, and an uncomfortable and polluted environment indoors or outdoors are not permitted and accepted anymore in the age of COVID-19.

Cities and their buildings are playing vital roles in stimulating health and well-being. Some of the examples depicting the impression of three key elements (place making, walkability, and sustainability), for designing buildings and cities to

Fig. 7.37 Physical and social distance inside and outside buildings. (**a**) A sign of procedures for riding an elevator in New York. (**b**) Arrangement of classrooms seating in one of UAE schools. (**c**) Arrangement of tables and seats in a café in Seattle, WA. (**d**) Dining outdoor in a Restaurant in Newburgh, New York. (**e**) Lecture's virtual setting for education in Portugal. (**f**) People waiting outside an office to finish business in Tamil Nadu. (Images' credit and source: (**a**) Shanluan, https://commons.wikimedia.org/wiki/File:Elevator_social_distancing_sign_New_York_City_COVID19_Pandemic.jpg; (**b**) Akbarali, https://commons.wikimedia.org/wiki/File:School_Classroom_seating_arrangements_during_the_time_of_covid_19_UAE.jpg#; (**c**) Carl Lender, https://commons.wikimedia.org/wiki/File:Hotwire_Cafeteria_Now_(50141063392).jpg#; (**d**) Daniel Case, https://commons.wikimedia.org/wiki/File:Dining_alfresco_at_TGI_Friday%27s,_Newburgh,_NY,_during_COVID-19_pandemic.jpg#; (**e**) Agéncia Lusa, https://commons.wikimedia.org/wiki/File:Estudo_em_Casa_(Principais_acontecimentos_que_marcaram_o_ano_2020,_Ag%C3%AAncia_Lusa).png#; (**f**) P Jeganathan, https://commons.wikimedia.org/wiki/File:Covid_19_Tiruppur_Tamil_Nadu_IMG_20200420_104009670_HDR.jpg#)

Fig. 7.38 Inspiration and dynamic thinking for greenery in urban spaces for future cities in the age of COVID-19. (Image credit and source: Mohsen Aboulnaga)

Fig. 7.39 Inspiration and dynamic thinking of water in urban spaces for future cities in the age of COVID-19. (Image credit and source: Mohsen Aboulnaga)

promote healthier urban living, are presented in Fig. 7.38 and healthy urban spaces with water are listed in Fig. 7.39. Details of these settings are depicted in Figs. 7.40, 7.41, 7.42, 7.43, 7.44, 7.45 and 7.46. Additionally, key elements to design buildings and cities for healthier urban living are shown in Fig. 7.47 (Box 7.15).

Box 7.15: Connect People to Site, History, and Knowledge

As architects and planners, we need to reconnect to the site, to place, to history, to knowledge—but also reconnect people to people, and people with buildings, all with modern techniques and advances. (Andy Cohen, Co-CEO of Gensler)

Building design in the time of COVID-19 should be different and innovative (future and healthy building) to meet the current challenges and ensure the provision of environmental indoor quality (EIQ), which could be costly in existing buildings due to the retrofitting process and new installations. Nonetheless, an innovative

Fig. 7.40 Inspiration of water, greenery, wood, and stones in urban spaces for future cities during COVID-19, London. (Image credit and source: Mohsen Aboulnaga)

example for the future and healthy building emerged in 2021, where long terraces were incorporated into its façades to furnish seating areas in open air (not closed spaces), allows for sunlight via sun breakers, daylight, and fresh ventilation that aid in virus killing, while also providing a nice visual comfort and relaxing view. External staircases were also created on the façades to enter the flats or exit from it to the outdoor, hence minimizing close contact inside the building using a standard staircase (Fig. 7.48). Figure 7.49 illustrates a close view of the building design overlooking a river to reduce social stress during the COVID-19 crisis and beyond. Table 7.1 also presents the indicators for healthy, safe, and sustainable housing at the preexisting and new constructions [37].

The COVID-19 pandemic ensures the importance of collaboration and strong interdisciplinarity between researchers and practitioners, with several backgrounds

Fig. 7.41 Inspiration of water, greenery, and stones in urban spaces for future cities amidst COVID-19, London. (Image credit and source: Mohsen Aboulnaga)

Fig. 7.42 Inspiration greenery in urban spaces for future cities in the age of COVID-19, Brussels. (Image credit and source: Mohsen Aboulnaga)

Fig. 7.43 Inspiration greenery in urban spaces for future cities during COVID-19, Milan. (Image credit and source: Mohsen Aboulnaga)

Fig. 7.44 Inspiration of water in urban spaces for future cities in the age of COVID-19, Paris. (Image credit and source: Mohsen Aboulnaga)

Fig. 7.45 Inspiration of water ad greenery in urban spaces for future cities in the time of COVID-19, Bressanone. (Image credit and source: Mohsen Aboulnaga)

Fig. 7.46 Inspiration of water and greenery in urban spaces for future cities in the age of COVID-19, Paris. (Image credit and source: Mohsen Aboulnaga)

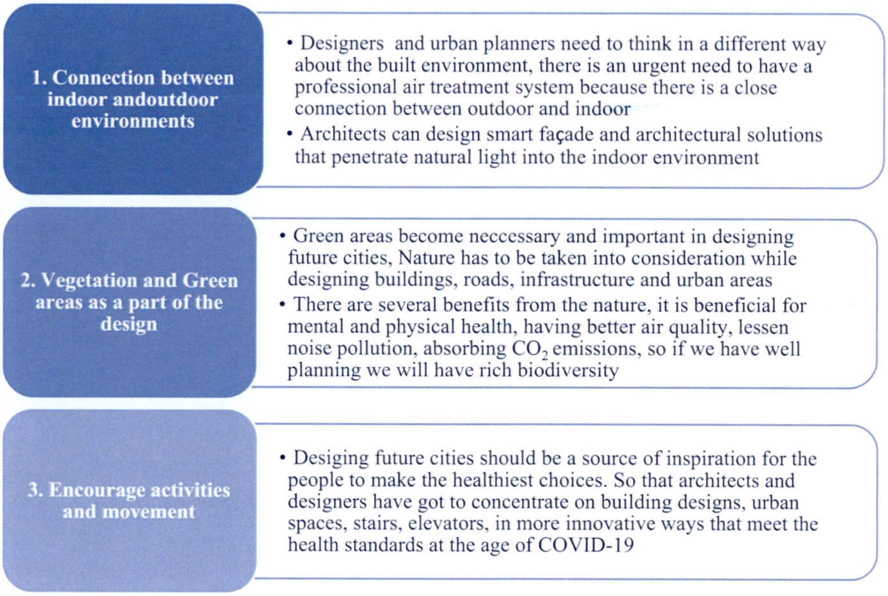

Fig. 7.47 Key elements for designing buildings and cities to promote healthier urban living. (Image source: Developed by authors)

in the health and technical field, to tackle the main public health issues related to the built environment and to the housing demands [37].

7.6 Post-COVID Pandemic

After passing through the extraordinary virus, the most important question would be what is the permanent impact of the COVID-19 pandemic? Through history, each crisis disappeared and left several structural changes behind it, for example, the 1918 Global Flu Pandemic helped topple empires. The post-COVID crisis will be a "new normal." How should leaders, CEOs, and boards plan and prepare for the post-COVID age, even as they respond to the immediate crisis ahead [37]? Responding to the pandemic needs special considerations, measures, and strategies; this should happen by now, next, and beyond as shown in Fig. 7.50. Many actions should also be taken into account at both immediate-, medium-, and long-term actions (Fig. 7.51).

Fig. 7.48 Future and Healthy Building, as a key contributor to healthy cities. (Image credit and source: Bruno Barral, https://commons.m.wikimedia.org/wiki/Category:Architecture#/media/File%3AAbreBlanc2.jpg)

Fig. 7.49 View of the future and healthy building showing the terraces experiencing fresh air, sunlight, daylighting, and accessibility from outdoor staircases. (Image credit and source: Bruno Barral, https://commons.m.wikimedia.org/wiki/File:AbreBlancDetail.jpg#)

Table 7.1 Indicators for a healthy, safe, and sustainable housing

Investigated areas	
1.Green materials, elements, and green spaces	
Preexisting constructions	**New constructions**
Renovate the buildings with green walls and green roofs if possible. Use sustainable materials for maintenance	Vegetation for shading Affirm the presence of terraces Greenery views from windows
2. Double using of spaces, flexibility, adaptability, and compliant functions located within the buildings	
Preexisting constructions	**New constructions**
Flexibility of the indoor spaces Compatibility between different functions in the building Quality of living spaces and keeping health standards and sanitization measurements	Using smart furniture with multipurpose use Design the buildings with multipurpose functions Taking into consideration the new health and social distance standards according to the pandemic
3. Reappropriation of the basic principles and archetypes of sustainable architecture, thermal comfort, and indoor air quality (IAQ)	
Preexisting constructions	**New constructions**
Appropriate orientation of the spaces Regular maintenance of ventilation. Cooling and/or heating systems, where present	Correct orientation of living spaces, according to the solar radiation, natural heat, and lighting supplies Using nano-materials Ventilation technologies that purify the air
4. Waste water management and water consumption	
Preexisting constructions	**New constructions**
Implementation of household water treatment technologies Installation of backflow valves on sprayers	Installation of several bathrooms taking into consideration the automatic sanitizing and cleaning process Implementation of technical plant engineering measures (i.e., well-maintained plumbing, scaled bathroom drains, backflow valves on sprayers) Using gray water for irrigation
5. Urban waste management	
Preexisting constructions	**New constructions**
Collecting garbage in a specific place to collect them Application of "returnable" practice to allow for an immediate and significant reduction in waste Having smart bins instead of regular bins	Availability of garbage spaces More than one access to the garbage room Innovative digital and smart devices and systems to facilitate waste collection Smart bins instead of regular bins Development of networks where possible
6. Automation in housing and electromagnetic fields	
Preexisting constructions	**New constructions**

(continued)

Table 7.1 (continued)

Investigated areas	
Install the new home automation systems Using smart lighting, sensors, detectors, and smart systems Monitoring the building	Installation of housing automation and home smart systems, by using detectors and sensors Automated windows and shutters Cameras linked with mobile applications Control and manage indoor air quality Having a security system for elderly and sick people
7. Indoor environment and finishing materials	
Preexisting constructions	**New constructions**
Normal cleaning methods Using appropriate cleaning products that kill viruses Automatic sanitizing devices everywhere If it is possible to replace the current finishing materials with other environmentally friendly and high-performance furnishing materials	The rise of hygienic standards and clean materials Purchasing air filtration system Using antibacterial finishing materials Having sufficient amount of openings for good ventilation to enhance indoor air quality Enhance indoor environment in several aspects such as thermal comfort, good illuminance, nontoxic materials, and ergonomic furniture

Source: Developed by authors after D'Alessandro et al., https://www.mattioli1885journals.com/index.php/actabiomedica/article/view/10115

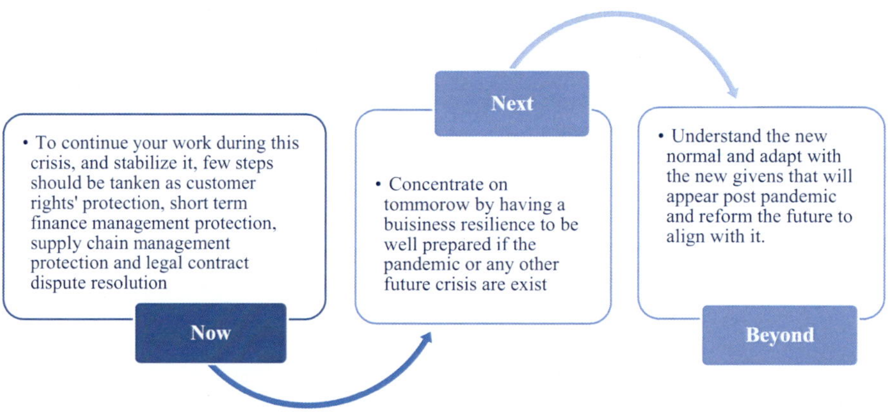

Fig. 7.50 Strategy that companies should consider across now, next, and beyond. (Image source: Developed by authors)

7.6.1 New Construction Considerations

The construction sector and the building industry should have tremendous changes and shifting priorities in some of the main specifications, in order to meet the standards of health protection at the time of COVID-19. For both existed buildings that are under renovation and buildings that are built from scratch, there are several

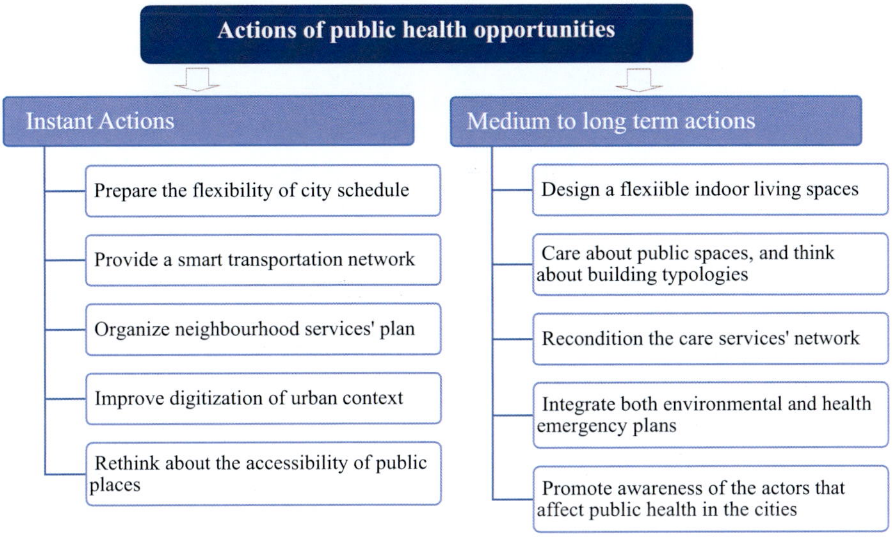

Fig. 7.51 Instant and medium- to long-term actions of public health opportunities. (Image source: Developed by authors)

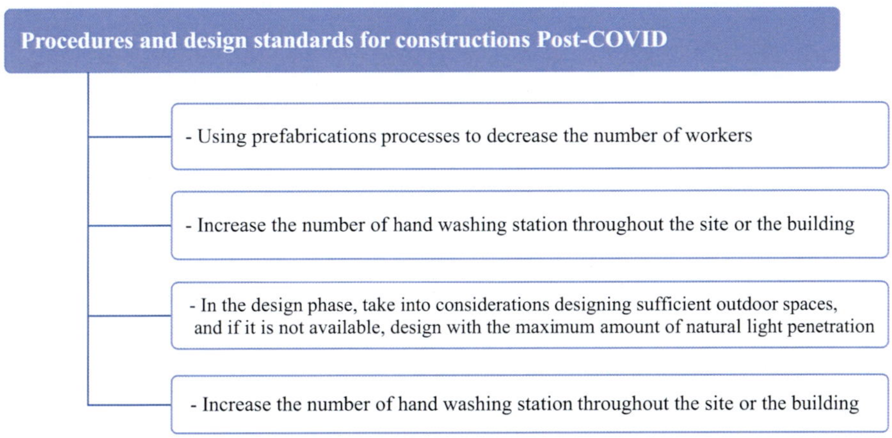

Fig. 7.52 Instant and medium- and long-term actions of public health opportunities. (Image source: Developed by authors)

building procedures and design standards that could be integrated to make them post-COVID-19 friendly; some of these procedures are listed in Fig. 7.52.

There are many design strategies and shared goals among health and global design initiatives. Aspects that should be taken into consideration while designing new buildings include (a) improvement of ergonomic, (b) safety, (c) indoor quality, (d) daylight and sunlight, and (e) fresh air and cross ventilation. This is not only for seeking healthy environment but also contributing to the creation of better spaces

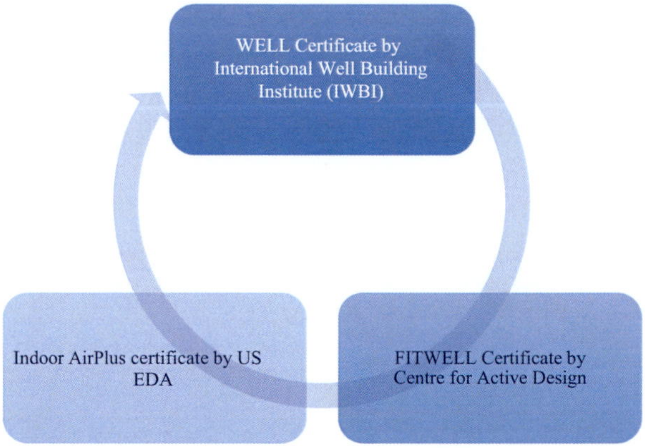

Fig. 7.53 Certifications of human health and wellness in the buildings. (Image source: Developed by authors after https://www.swinter.com/party-walls/the-great-indoors-creating-a-healthier-and-safer-built-environment/)

Key issues in the new normal (Post-COVID age)	
Globalization and trade	• Reorienting supply chains and manufacturing
Technology and innovation	• Technology reshaping the working world • Emerging technologies fighting COVID-19 • Innovating for the new normal
Societal impacts	• Strengthening safety nets • Reexamining companies' role in society • Preparing for the post-pandemeic generation
Behavioral shifts	• Transforming consumer psychology • Trust, polarization, and information

Fig. 7.54 Important issues in the age of COVID-19. (Image source: Developed by authors)

for all users [35]. Integrating health strategies has become a necessity, not prosperity, especially under the severe coronavirus pandemic. Many certifications are related to people's wellness in buildings, as shown in Fig. 7.53. The WELL certificate is created by the International WELL Building Institute (IWBI). This certificate program includes many aspects related to human health, such as mental wellness and community engagement. The recent version has ten concept categories such as thermal comfort, sound, materials, air, water, light, mind, nourishment, movement, and community. The IWBI created these categories to incorporate all areas where the users interact with a building and the ways where their health could be improved [35].

Among other certifications is the Fitwel certificate, which is a certification system for healthy buildings and communities. The Fitwel certificate was developed by the Center for Active Design (CAD). This rating system has been upgraded to address health issues relevant to the COVID-19 pandemic. It is not only focuses on indoor quality and human comfort but also addresses the connection between physical inactivity and mental health. Moreover, the Indoor airPLUS certificate, which was created by the US EDA, tackles some chemical elements that affect human health in general and people that suffer from respiratory conditions in particular. This certificate emphasizes additional safeguards for moisture control, improved HVAC system, combustion pollutant reduction, low-emission materials, radon resistance, and pest prevention [35]. Hence, the goal of Indoor airPLUS is to have qualified homes with improved air quality, low pollutant level, and protection from combustion pollutants.

COVID-19 witnessed an expedited examination of many automated transmission approaches in manufacturing, such as three-dimensional printing, which has the potential to deliver significant benefits in terms of speed, cost, precision, and materials. This may lead to new business models and reshape the current business, not just in manufacturing but also in related sectors such as logistics. Figure 7.54 illustrates some key issues in the post-COVID age [37]. Also, the medium-term risk mitigation during the age of COVID-19 is illustrated in Box 7.16.

In conclusion, as WHO stated, building healthy housing is a complex issue and a multi-sectoral responsibility that could be achievable, only if a contribution is made by all relevant players and stakeholders. The COVID-19 pandemic has placed all countries' profound housing crisis under the spotlight and further highlighted the need to address it in a systematic and innovative way [36].

Box 7.16: Medium-Term Risk Mitigation

In the medium term, the pandemic will drive risk-mitigating procedures designed to track employee health, reduce human-to-human interactions, and upgrade ventilation and physical barriers. (EY Building a better working world [35])

7.7 Conclusion

There is no doubt that the pandemic resets the whole world, and it has contributed to reforming the face of the world and reorganizing our mindsets. Human healing and planet healing are not detached processes, but the healing of people's fear, anxiety, and illness is a part of the basis for healing the Earth.

What has been illustrated above expresses how biomimicry can contribute as an approach during the COVID-19 crisis. The amazing capability of biomimicry, as an inspiration and innovative tool, assists in limiting the spread of COVID-19 through the manufacturing of nano-materials, especially clean surfaces – expect easy-to-clean materials and fixtures – and becomes the standard in hygienic design. Natural nanostructures' mimicry has also been discussed and the way people can fight the pandemic by nanotechnology engaged with nano-materials.

The changes that occurred to the whole world have been discussed. In addition, some of the guidelines for healthier urban spaces and building indoors at the age of COVID-19 have been highlighted. Moreover, it has been referred to how life will be post-COVID age and how it will have a great effect on the different sides of our normal life, yet it has to have new norms that we should adapt to and meet its new standards to survive.

Biomimicry, as a way of thinking, should be the ideal approach of now and the future. Searching for the link between this field of science and our daily life is a must; how it can contribute in mitigating this virus and other challenges is an open question to be addressed.

References

1. Alasdair, S. (2020). Coronavirus: Half of humanity on lockdown in 90 countries. *Euronews*. Retrieved 7 October 2021, from https://www.euronews.com/2020/04/02/coronavirus-in-europe-spain-s-death-toll-hits-10-000-after-record-950-new-deaths-in-24-hou
2. United Nations. (2020). *SHARED RESPONSIBILITY, GLOBAL SOLIDARITY: Responding to the socio-economic impacts of COVID-19*. Retrieved from https://unsdg.un.org/sites/default/files/2020-03/SG-Report-Socio-Economic-Impact-of-Covid19.pdf
3. De Best, R. (2021). *COVID-19 deaths per capita by country | Statista*. Statista. Retrieved 20 October 2021, from https://www.statista.com/statistics/1104709/coronavirus-deaths-worldwide-per-million-inhabitants/.
4. Udovyk. O., (2020). *Biomimicry and COVID-19 response. UNDP Ukraine*. Retrieved 7 May 2021, from https://www.ua.undp.org/content/ukraine/en/home/blog/2020/biomimicry-and-covid-19-response.html .
5. Avampato, C. (2020). *Biomimicry Institute*. Biomimicry.org. Retrieved 30 May 2021, from https://biomimicry.org/a-quiet-war-using-biomimicry-in-the-time-of-coronavirus/
6. Seckler, I. (2020). *Biomimicry Institute*. Biomimicry.org. Retrieved 30 May 2021, fromhttps://biomimicry.org/asking-nature-a-day-keeps-the-doctor-away/
7. *WHO Coronavirus (COVID-19) Dashboard*. Covid19.who.int. (2021). Retrieved 8 October 2021, from https://covid19.who.int
8. IDLE, T. (2020). Benyus: COVID-19 offers 'Utopian Glimmer' of new world, but we must give back to nature. *Sustainable Brands*. Retrieved 26 September 2021, from https://sustainablebrands.com/read/defining-the-next-economy/benyus-covid-19-offers-utopian-glimmer-of-new-world-but-we-must-give-back-to-nature
9. Lipholt, N. (2019). *Biomimicry – Where nature is changing innovation*. The Netherlands: University of Twente. Retrieved from https://essay.utwente.nl/78549/1/LIPHOLT_BA_bms.pdf
10. Terzin, T. (2020). *Biomimicry can be used to decrease or prevent the impact of future outbreaks of viral diseases, says Augustana biology professor*. Ualberta.ca. Retrieved 15 September 2021, from https://www.ualberta.ca/augustana/news/2020/04/biomimicry-can-be-used-to-decrease-or-prevent-the-impact-of-future-outbreaks-of-viral-diseases-says-augustana-biology-professor.html

11. Prabhu, J. (2020). *Innovation, development and COVID-19: Challenges, opportunities and ways forward.* OECD. Retrieved 16 September 2021, from https://www.oecd.org/coronavirus/policy-responses/innovation-development-and-covid-19-challenges-opportunities-and-ways-forward-0c976158/
12. Cromwell, J., & Kotelly, B. (2021). A Framework for Innovation in the COVID-19 Era and Beyond. *MIT Sloan Management Review*. Retrieved 18 September 2021, from https://sloanreview.mit.edu/article/a-framework-for-innovation-in-the-covid-19-era-and-beyond/
13. Khalifa, S., Swilam, M., El-Wahed, A., Du, M., El-Seedi, H., Kai, G., et al. (2021). Beyond the pandemic: COVID-19 pandemic changed the face of life. *International Journal of Environmental Research and Public Health, 18*(11), 5645. https://doi.org/10.3390/ijerph18115645
14. Tang, Z., Kong, N., Zhang, X., et al. (2020). A materials-science perspective on tackling COVID-19. *Nature Reviews Materials, 5*, 847–860. https://doi.org/10.1038/s41578-020-00247-y
15. Unlocking the Next Generation of Science with Nature-Inspired Materials and Biomimicry. News-Medical.net. (2021). Retrieved 17 September 2021, from https://www.news-medical.net/whitepaper/20210302/Unlocking-the-Next-Generation-of-Science-with-Nature-Inspired-Materials-and-Biomimicry.aspx
16. Campos, E.V.R., Pereira, A.E.S., de Oliveira, J. L., et al. (2020). How can nanotechnology help to combat COVID-19? Opportunities and urgent need. *Journal of Nanobiotechnology, 18*, 125. Retrieved 27 September 2021, from https://doi.org/10.1186/s12951-020-00685-4
17. Zabaniotou, A. (2021). *A systemic approach to resilience and ecological sustainability during the COVID-19 pandemic: Human, societal, and ecological health as a system-wide emergent property in the Anthropocene* (pp. 116–126). Retrieved from https://doi.org/10.1016/j.glt.2020.06.002
18. Hojniak, D., & Hvid, C. (2021). *Healthy buildings, cities and you – How to design future living environments – Urban Insight.* Urban Insight. Retrieved 18 September 2021, from https://www.swecourbaninsight.com/healthy-buildings-cities-and-you-how-to-design-future-living-environments/
19. Ratho, A., & John, P. L. (Eds.). (2020). *Rethinking Cities in a Post-COVID-19 World.* ORF and Global Policy Journal. From https://www.orfonline.org/wp-content/uploads/2020/06/Rethinking-Cities-ORF1.pdf
20. Wakefield, J. (2013). Tomorrow's cities: Do you want to live in a smart city? *BBC News.* Retrieved 24 October 2021, from https://www.bbc.com/news/technology-22338561
21. *What is a Smart City? – Centre for Cities.* Centre for Cities. (2021). Retrieved 24 October 2021, from https://www.centreforcities.org/reader/smart-cities/what-is-a-smart-city
22. Euklidiadas, M. (2021). *Fujisawa Sustainable and Smart Town: planning for the next 100 years. Tomorrow City.* Retrieved 25 October 2021, from https://www.tomorrow-city/a/fujisawa-sustainable-smart-town
23. Balch, O. (2011). Japans' Fujisawa Sustainable and Smart Town – A model for 'smart life'? *The Guardian.* Retrieved 25 October 2021, from https://www.theguardian.com/sustainable-business/fujisawa-sustainable-smart-town-future-model
24. Ghaffarianhoseini, A., Berardi, U., AlWaer, H., Chang, S., Halawa, E., Ghaffarianhoseini, A., & Clements-Croome, D. (2015). What is an intelligent building? Analysis of recent interoperations from an international perspective. *Architecture Science Review, 59*(5), 338–357. https://doi.org/10.1080/00038628.2015.1079164
25. *What is a Smart Building?*. Tripelpundit.com. (2013). Retrieved 24 October 2021, from https://www.triplepundit.com/story/2013/what-smart-building/54101
26. *Sustainable Cities.* Sustainable Cities Network. (2021). Retrieved 24 October 2021, from https://www.sustainablecitiesnet.com
27. Ki-moon, B. (2013). *Remarks at "Sustainable Cities Days" | United Nations Secretary-General. Un.org.* Retrieved 24 October 2021, from https://www.un.org/sg/en/content/sg/speeches/2013-12-12/remarks-sustainable-cities-days

28. Aboulnaga, M. M., & Abdullah, S. (2012). *Sustainable cities: Strategy and indicators for healthy living environments*. World Renewable Energy Congress – Sweden 8–12 May 2021, Linköping University, Sweden. https://doi.org/10.3384/ecp110573018
29. Tanguay, G. A., Rajaonson, J., Lefebvre, J. F., & Lanoie, P. (2010). Measuring the sustainability of cities: An analysis of the use of local indicators. *Ecological Indicators, 10*, 407–418.
30. *Resilient Cities – OECD*. Oecd.org. (2021) Retrieved 24 October 2021, from https://www.oecd.org/cfe/regionaldevelopment/resilient-cities.htm
31. *WHO Coronavirus (COVID-19) Dashboard. Covid19*. who.int. (2021). Retrieved 21 October 2021, from https://covid19.who.int/
32. *Designing the Cities of the Future | Gensler*. Gensler. (2021). Retrieved 17 September 2021, from https://www.gensler.com/designing-cities-of-the-future
33. Department of Economic and Social Affairs Un.org. (2018). Retrieved 24 October 2021, from https://www.un.org/development/desa/en/news/population/2018-revision-of-world-urbanization-prospects.html
34. Capolongo, S., Rebecchi, A., Buffoli, M., Appolloni, L., Signorelli, C., Fara, G. M., & D'Alessandro, D. (2020). COVID-19 and Cities: from Urban Health strategies to the pandemic challenge. A Decalogue of Public Health opportunities. *Acta Biomedica Atenei Parmensis, 91*(2), 13–22. https://doi.org/10.23750/abm.v91i2.9615
35. Alvarez, J. (2020). *The Great Indoors: Creating a Healthier and Safer Built Environment*. Steven Winter Associates, Inc. Retrieved 26 September 2021, from https://www.swinter.com/party-walls/the-great-indoors-creating-a-healthier-and-safer-built-environment/
36. EY building a better working world. (2020). *Beyond COVID-19 what will define the "new normal"*. Retrieved 25 September 2021, From https://assets.ey.com/content/dam/ey-sites/ey-com/fi_fi/pdf/beyond-covid-19-what-will-define-the-new-normal.pdf
37. D'Alessandro, D., Gola, M., Appolloni, L., Dettori, M., Fara, G., Rebecchi, A., et al. (2020). COVID-19 and Living Spaces challenge. Well-being and Public Health recommendations for a healthy, safe, and sustainable housing. *Acta Bio-Medica, 91*(9-S), 61–75. https://doi.org/10.23750/abm.v91i9-S.10115
38. Roberts, T. (2016). *We Spend 90% of Our Time Indoors. Says Who?* BuildingGreen. Retrieved 24 October 2021, from https://www.buildinggreen.com/blog/we-spend-90-our-time-indoors-says-who

Chapter 8
Can Nature and Biomimicry Be the Solution for Net-Zero Cities? Discussions and Recommendations for the Future of Biomimetic Architecture

> Making peace with nature is the defining task of the 21st century. (Antonio Guterres, Secretary-General of the United Nations)

8.1 Introduction

Nature is an essential element in achieving the global goal of net zero by 2050. According to Antonio Guterres, the Secretary-General of the United Nations, nature needs a bailout. In conquering the COVID-19 pandemic, we can also avert climate cataclysm and restore our planet [1]. Nature can be the driver for biomimetic architecture that is considered one of the solutions for carbon neutrality (Fig. 8.1). He added that a green transition is needed to transform the world economy in the age of COVID as illustrated in Box 8.1.

Box 8.1: Nature and Green Shift of the World Economy
It is time to flick the "green switch" We have a chance to not simply reset the world economy but to transform it. (Antonio Guterres, Secretary-General of the United Nations)

Frank Gehry has stated unequivocally that time and place are critical in the creation of architecture [2]. In this context, is architecture that was built addresses time and place and yet integrates nature to offset the climate change crisis and contribute to achieving carbon neutrality? The answer is basically not yet, apart from a few examples worldwide, but this is not enough. Figures 8.2, 8.3, 8.4, 8.5 and 8.6 show some of these examples that address decarbonization and are inspired from nature.

> Architecture should speaks of its time and place, but yearn for timelessness. (Frank Gehry)

M. Aboulnaga, S. E. Helmy, *Biomimetic Architecture and Its Role in Developing Sustainable, Regenerative, and Livable Cities*,
https://doi.org/10.1007/978-3-031-08292-4_8

Fig. 8.1 Nature as a driver for biomimetic architecture, one of the solutions for carbon neutrality. (Image credit and source: Harloh Massey, https://commons.m.wikimedia.org/wiki/File:Temple_flowers_wide.jpg#)

Fig. 8.2 Recent iconic example of biomimetic architecture that addresses its time and place. (Image credit and source: Mohsen Aboulnaga and Amina El Haggan)

Fig. 8.3 Architecture that addresses its time and place through clean energy and renewables. (Image credit and source: Mohsen Aboulnaga Mohsen Aboulnaga and Aya Ghobashy)

Fig. 8.4 The Terra – The Sustainability Pavilion Expo 2020, a unique example of biomimetic architecture mimicking nature. (Image credit and source: Mohsen Aboulnaga and Amina El-Haggan)

8.2 Can Nature and Biomimicry Contribute to Achieving Net-Zero Cities?

Biomimetics is projected to be a promising approach, since it supports a paradigm shift toward sustainable and resilient cities and livable societies. The question of "whether nature and biomimicry can be the solution for net zero cities" is considered a turning point toward his path.

By chiming in the previous narration of this book, it is clear that the term "biomimicry" has become an approach for creating sustainable architectural design and buildings, yet it contributes to developing sustainable, regenerative, and livable cities to reach the main goal of COP 26, net zero by 2050 (Figs. 8.2 and 8.3). In this example illustrated in Figs. 8.2 and 8.3, it can be noticed that the famous architect Santiago Calatrava represents a model of biomimetic architecture that is inspired from nature. Such a model of biomimetic architecture reflects its time and place, being built in a desert climate, and generates its energy from clean and renewable sources. Indeed, it is a unique model that inspires architects and designers and paves the way for them to address the concept of net zero.

In this context, another unique example of biomimetic architecture that speaks for its time and place is Grimshaw's innovative architecture model, which is inspired from nature. It is Terra – The Sustainability Pavilion, where 19 energy trees (known

Fig. 8.5 The dynamic 18 "Energy-trees" Solar PV array canopies tracking the sun throughout the day, The Terra – The Sustainability Pavilion at Expo 2020. (Image credit and source: Mohsen Aboulnaga and Aya Ghobashy)

Fig. 8.6 The ripped and flowing façade mimicking wave movements inspired by nature – The Mobility Pavilion at Expo 2020. (Image credit and source: Mohsen Aboulnaga & Aya Ghobashy)

as E-trees) were created to generate 4 GWh of renewable energy as well as produce water from harvested rainwater and recycled water [3]. Every small E-tree of the 18 E-tree canopies mimicked the sunflower that was inspired from the dragon tree found in the Deseret (Figs. 8.4 and 8.5). These E-trees revolve 180 degrees to enhance the efficiency of the solar PV cells installed on top of the 130 canopies as shown in Fig. 8.5. Such innovative urban solar complex would also inspire architects to develop models that address the net-zero buildings in hot-arid climate zones.

It is obvious from Figs. 8.4 and 8.5 that such a tendency of biomimetic architecture, The Terra – Sustainability Pavilion, is one of the most vivid models to reach the goal of net-zero cities, if it is widely implemented in various buildings' forms of urban development in cities. In addition, the Mobility Pavilion – designed and developed by Sir Norman Foster and Partners – is an indicative and iconic model of biomimitic architecture which is inspired from nature, and it depicts sustainable mobility (Fig. 8.6); this could inspire architects as well.

8.3 Buildings Created Away from Nature and Not Biomimetics

On the contrary, most urban communities and architects who created models of architecture and buildings' examples in the past appear to isolate themselves from nature, except for a few. For example, office buildings in Barcelona, Spain (Fig. 8.7), may be similar somehow to those in Seoul and Incheon, South Korea (Fig. 8.8); an office building in Vienna, Austria (Fig. 8.9); and the dancing office building in Prague, Czech Republic (Fig. 8.10); modern office buildings in Berlin, Germany (Fig. 8.11); as well as some office buildings in Dan Hugae and Amsterdam in the Netherlands (Fig. 8.12); an office and hotel building in Yokohama, Greater Tokyo, Japan (Fig. 8.13); and the flaming office towers in Baku, Azerbaijan (Fig. 8.14) or even in main cities of the Middle East and North Africa (MENA) region such as that in Beirut, Lebanon (Fig. 8.15). These modern buildings were developed and built regardless of the differences in culture, topography, and climatic conditions of each region as shown in Figs. 8.7, 8.8, 8.9, 8.10, 8.11, 8.12, 8.13, 8.14 and 8.15. These buildings' examples are not addressing time and place or even integrating nature within their fabric or design concept.

The increasing level of attention toward sustainability worldwide in the past 5 years and the growing tendency of biomimicry amidst the deterioration of biodiversity and ecosystems, as well as climate and energy crises and COVID-19 crisis,

Fig. 8.7 Typical examples of office buildings in Barcelona, Spain. (**a**) Sabadell Bank office building. (**b**) La Caixa Bank office building. (**c**) A glazed façade office building. (Images' credit and source: Mohsen Aboulnaga)

(a) (b) (c)

Fig. 8.8 Examples of office buildings in Seoul and Incheon, South Korea. (**a**) Typical office building in Seoul. (**b**) A modern office building in Seoul. (**c**) An office and hotel building in Incheon. (Images' credit and source: Mohsen Aboulnaga)

have led to an urgent call for nature to be at the center stage of urban and city development.

In the hope for the future of architecture, buildings, and cities, adaptation to climate change and addressing COVID-19 measures in a better way to their context, culturally and environmentally, should be strategically developed and implemented. Nevertheless, this will never be achieved by imitating the form of flowers or trees in buildings alone but by imitating their behaviors in innovative and rethinking models that are more robust and dynamic and address time and place. Imperatively, this means that buildings should adapt to the context and environment they exist within. For instance, if the building is constructed in a hot-arid climate zone, architects should focus on the means by which the building will conserve energy for cooling and harvesting rainwater and recycling it as well as integrating solar PV panels for generating clean energy, whereas if the building is constructed in a cold or temperate climate zone, architects should concentrate on finding ways to conserve energy for heating and harvest rainwater and recycles as well and geothermal using heat pumps.

In contrast, nature, on the other hand, does everything in a proper and accurate mechanism on time, with no waste, more efficiency, and less energy and material use; this is what all organisms seek and how they behave. Additionally, nature is a flawless sample of sustainable, creative, and smart engineering. In this milieu, the environmental impact could consequently be decreased by following nature and biomimicry approach as well as applying and integrating

Fig. 8.9 An office building in the business center of Vienna, Austria. (Image credit and source: Mohsen Aboulnaga)

these principles in architecture from the early stage of conceptual designs and projects' development.

Therefore, it is vital to highlight the biomimicry and biomimetic architecture definitions and fully digest the term "biomimicry" before creating architectural

Fig. 8.10 The dancing office building in Prague, Czech Republic. (Image credit and source: Mohsen Aboulnaga)

design concepts. The root of the word biomimicry originates from the Greek *bios* meaning “life” and *mimesis* means “imitate” [4, 5]. Based on the book coverage, biomimicry presents immense potential for urban planning and cities following the analogy of city lungs (parks including trees) and human lungs that demonstrate such an ability, while the basic principles of sustainability appear to be “returning back to the roots of nature.”

(a) (b)

Fig. 8.11 Typical office buildings in Berlin, Germany. (**a**) Deutsche Bahn office building in the center Berlin. (**b**) A glazed façade of an office building in Berlin. (Images' credit and source: Mohsen Aboulnaga)

(a) (b)

Fig. 8.12 Office buildings in Dan Hugae and Amsterdam, the Netherlands. (**a**) Glazed facades of two office buildings in Dan Hugae. (**b**) A glazed façade of Akzo Nobel Headquarters in Amsterdam. (Images' credit and source: (**a**) Mohsen Aboulnaga, (**b**) Ceescamel, https://commons.m.wikimedia.org/wiki/File:Akzo_Nobel_Headquarters_(1).jpg)

Fig. 8.13 An office and a hotel building in Yokohama, Greater Tokyo, Japan. (Image credit and source: Mohsen Aboulnaga)

Fig. 8.14 The flaming office towers in Baku, Azerbaijan. (Image credit and source: Mohsen Aboulnaga)

8.4 Biophilic Architecture

Biophilic architecture represents an approach to create lungs in the city and provide livability. Figure 8.16 presents an example of biophilic architecture in one of the streets of the Royal Borough of Kensington and Chelsea in inner London Borough, UK, where the building skin acts as a lung for the city. While another example of a Japanese model of biophilic office building is The Pasona O2 office building, a former bank building in the business district of Otemachi in Tokyo, portrayed in Fig. 8.17. This building acts as an urban farm, a lung in the city, and factually sprouting leaves to achieve livability [6]. Moreover, Seoul City Hall building in South Korea is a unique example of biomimicry, where nature is integrated into the municipal office building to provide thermal comfort by adapting green walls and other solutions in the interior to reduce energy and maximize cooling by passive design [7] as shown in Figs. 8.18, 8.19, and 8.20.

8.4.1 Seoul City Hall Building, South Korea

In analyzing Seoul City Hall building from the biomimicry viewpoint, it can be easily said that it was inspired from nature since the building's form as presented in Figs. 8.18, 8.19, and 8.20, and its front façade mimics tulip flower leaves,

Fig. 8.15 An office building in Beirut, Lebanon. (Image credit and source: Mohsen Aboulnaga)

Tradescantia pallida leaves, and Cyclamen "'Stirling' flower as shown in Figs. 8.21, 8.22, and 8.23. The curved envelope shades the whole glass façade throughout the day in addition to the fact that the design is based on passive design means to reduce energy consumption and integrate nature inside the building in the form of green walls as shown in Figs. 8.19 and 8.20a–f.

Fig. 8.16 A biophilic example in one of the streets of the Royal Borough of Kensington and Chelsea in inner London Borough, the UK. (Image credit and source: Mohsen Aboulnaga)

Fig. 8.17 Pasona O2 office building acting as an urban farm and a lung in the city of Tokyo, Japan. (Image credit and source: 钉钉, https://commons.m.wikimedia.org/wiki/ File: Pasona_ building. JPG)

Fig. 8.18 Seoul City Hall building in South Korea – an example inspired from nature to provide thermal comfort. (Image credit and source: Mohsen Aboulnaga)

Fig. 8.19 The façade of Seoul City Hall building mimicking nature in the form of leave shading itself. (Image credit and source: Mohsen Aboulnaga)

(a)

(b)

(c)

(d)

Fig. 8.20 Interior of the Seoul City Hall building bringing nature in the indoor spaces, South Korea. (**a**) The glazed façade from inside the building. (**b**) Daylight impinges on the green wall. (**c**) Visitors enjoying the green walls that achieve passive design and comfortable interior. (**d**) Green walls covering the inside skeleton to provide livable spaces and thermal comfort. (**e**) One of the green walls covering the inside structure of the building. (**f**) Close-up view of the green walls inside Seoul City Hall building. (Images' credit and source: (**a**) Smiley.toerist, https://commons.m.wikimedia.org/wiki/File:Seoul_City_Hall_green_wall_1.JPG#. (**b**) Michael Sean Gallagher, https://commons.m.wikimedia.org/wiki/File:Seoul_City_Hall_-_October_2013_(10488420895)_jpg. (**c**) Smiley.toerist, https://commons.m.wikimedia.org/wiki/File:Seoul_City_Hall_green_wall_5.JPG#. (**d**) Smiley.toerist, https://commons.m.wikimedia.org/wiki/File:Seoul_City_Hall_green_wall_4.JPG#/. (**e**) Smiley.toerist, https://commons.m.wikimedia.org/wiki/File:Seoul_City_Hall_green_wall_2.JPG#. (**f**) Michael Sean Gallagher, https://commons.m.wikimedia.org/wiki/File:Seoul,_Korea_October_2013_(10488435674).jpg#)

(e) (f)

Fig. 8.20 (continued)

8.4.2 Resources Tower in Nanshan, Shenzhen: Guangdong, China

One of the iconic biomimetic architecture examples is the Resources Tower, colloquially referred to as "Spring Bamboo Skyscraper" completed in 2018, Nanshan, Shenzhen – Guangdong, China (Fig. 8.24). The 400-meter high skyscraper covering floor areas of 210,000 m^2 is inspired from nature and mimicked a spring bamboo [8] as illustrated in Figs. 8.25, 8.26, and 8.27. It can be seen from Fig. 8.25 that the iconic tower mimics the spring bamboo shoots and big bamboo sprouts in its conical-shaped structure; hence, it represents a recent and innovative model of biomimicry in architecture. The iconic 67-floor building is designed by Kohn Pederson Fox (KPF), and the stainless façade structure content is 80% recycled steel (Fig. 8.26); and the energy strategy through the unique building form contributes to reduce energy consumption and CO_2 emissions [9].

8.5 Narrative on Biomimicry, Biomimetic Architecture, and Lessons Learned

This book narrates many important topics concerning biomimicry and biomimetic architecture in urban areas and cities. These subjects were comprehensively addressed in the previous seven chapters.

Chapter 1 started with an introduction and background of the book and the reason why it is focusing on biomimicry and biomimetic architecture in the age of the COVID-19 pandemic and climate crisis. Also, it provided a review of biomimicry and biomimetic architecture's definitions as well as the biomimicry' significance in

Fig. 8.21 Seoul City Hall building design inspired from nature and mimicking tulip flowers' leaves. (Image credit and source: Lee Elnaga and Mohsen Aboulnaga)

the age of COVID-19, including attitudes, connectivity to nature and pro-environmental behaviors, and learning through neighborhood's access to green spaces amidst COVID-19 since nature is a great source of inspiration to biomimicry. Chapter 1 also highlighted the need for new approaches to attain livability and sustainable development with an emphasis on biomimicry and energy efficiency and circularity and biomimicry in the built environment as well as the SDGs. In addition, it attempted to address an answer to the thoughts about can biomimicry aid in developing urban spaces and creating healthier buildings' indoors? Portraying the role of biomimicry in climate change adaptation and highlighting biomimicry and post-COVID regulations was an essential part of this book. Moreover, the chapter

Fig. 8.22 Close-up view of the tulip flowers' leaves that inspire Seoul City Hall's main façade. (Image credit and source: Lee Elnaga and Mohsen Aboulnaga)

presented models of innovation in biomimicry and recent global examples of biomimetic architecture built in Expo 2020 Dubai. At the end of this chapter, readers would have a full comprehension of the book. Finally, the chapter highlights the importance of biomimetic architecture and its role in developing sustainable, regenerative, and livable cities.

Chapter 2 concentrated on a full review of biomimicry architecture and its inspiration to comprehend the biomimicry approach through background and history. It also highlighted how defining biomimicry is essential to understand the origins of biomimicry architecture and its meaning and how to integrate this approach to become a way of thinking for future architecture. The chapter also highlighted the

Fig. 8.23 Tulips, *Tradescantia pallida*, and Cyclamen leave inspiring Seoul City Hall building design and form. (**a**) The curved leaves of the tulip flowers. (**b**) *Tradescantia pallida* leaves providing shade. (**c**) Cyclamen "Stirling" flower leaves provide shade, Royal Botanical Gardens in Tasmanian, Australia. (Images' credit and source (**a**) Ks aka 98, https://commons.m.wikimedia.org/wiki/File:Tulips_(Black_Horse_hybrid_cultivar).jpg (**b**) Atamari, https://es.m.wikipedia.org/wiki/Monocotyledoneae#/media/Archivo: Tradescantia_ pallida_2000.jpg (**c**) JJ Harrison, https://commons.m.wikimedia.org/wiki/File: Cyclamen_ %27 Stirling%27.jpg)

evolution of mankind and its relationship through nature biomimicry and early conceptual design to learn the overlap layers between architecture and nature, in this context, taking nature as a model, measure, and mentor of the biomimicry design. In addition, the Helix model of biomimicry and its relationship with the environment and sustainable architecture as well as integrating the biomimicry approach as a way of thinking is also vital to be portrayed in biomimetic architecture. Moreover, the importance of biomimicry in architecture amidst the coronavirus (COVID-19) pandemic is crucially needed to be addressed in both buildings and urban areas. Furthermore, the biomimicry design methods and related processes, to bring about dynamic and vivid biomimetic architecture, were reviewed to draw lessons. Finally, architects and designers should have a full comprehension of biomimicry through different aspects to achieve sustainability and avoid being detached from nature.

According to the meaning, the father and creator of biomimicry and the way of integrating biomimicry approach to become a way of thinking have been explored. Additionally, the examination of biomimicry and early conceptual design and

Fig. 8.24 China Resources Headquarters inspired from nature – Shenzhen, Guangdong. (Image credit and source (**a**) Charlie Fong, https://commons.m.wikimedia.org/wiki/File: China_Resources_Headquarters_by_the_side_of_shenzhen_bay2020.jpg)

reverting to nature in the field of architecture are essential in the design process. Moreover, the overlap between architecture and nature and the role of nature as a model, measure, and mentor is also vital in the biomimetic design method and processes – known as the "biomimicry design spiral" to get the steps that help architects and urban designers to apply biomimicry way of thinking.

After knowing and having a deep understanding of biomimicry meaning, history, background, and nature as well as design methods and design spirals and the relationship between the environment and sustainable architecture, it was imperative to highlight and discuss biomimetic design strategies and goals. Furthermore, the

(a) (b)

Fig. 8.25 The Spring Bamboo inspiring the Resources Tower in Shenzhen, Guangdong, China. (**a**) *Phyllostachys edulis* new bamboo shoots in spring. (**b**) Close up view of a big bamboo sprouts. (Images' credit and source (**a**) Thayne Tuason, https://commons.m.wikimedia.org/wiki/File:Phyllostachys_edulis_new_bamboo_shoots_in_spring.jpg (**b**) Joi Ito, https://commons.m.wikimedia.org/wiki/File:Big_Bamboo_shoot_(Joi_Ito).jpg)

importance of biomimicry in architecture during the age of COVID-19 and post-COVID was illustrated.

Chapter 3 presented a review of the needed information to comprehend the influence of biomimicry on architectural design and the built environment through several points. It also highlighted and discussed the causes of biomimicry thinking, biomimicry in architecture, approaches and levels of biomimicry (organism, behavioral, and ecosystem), and the levels of biomimetic information and the application types of biomimicry in architecture. Chapter 3 also focused on the inspiration levels, whether visual inspiration from nature, conceptual inspiration, and computational inspiration from nature. In addition, Chap. 3 showcased the iconic global exemplary buildings to learn from and draw lessons. In this context, it was also vital to address biomimicry design lens and nature's design principles and how biomimicry attains sustainability. Moreover, Chap. 3 emphasized the future of biomimicry and biomimicry conditions in Egypt. Finally, it drew conclusions and lessons learned for a successful design process.

Having narrated the above, biomimicry in architecture and the cause of biomimicry thinking are imperative in the biomimetic design process. The two different approaches of biomimicry are the following:

Approach 1 (design looking to biology (a problem-based approach)). This approach looking to the living world for solutions requires designers and biologists to set out exactly their design problems and then link these to creatures in nature that have solved similar problems; this approach is an effective way led by designers setting initial goals and parameters for the design.

Fig. 8.26 The Resource Headquarters "Spring Bamboo" Tall building in Shenzhen Bay, Guangdong, China. (Image credit and source: Charlie Fog, https://commons.m.wikimedia.wiki/File:China_Resources_Headquarters_in_Dec_2020.jpg)

Approach 2 (biology to design (a design solution-based approach)). This approach depends on the previous knowledge of biological research and solutions and does not require searching for solutions in nature; then apply this knowledge on the design problem you already have. Identifying specific features or behaviors in an organism or ecosystem and following by using them as rules and guidelines for developing industrial or architectural designs.

Fig. 8.27 View of China Resources Headquarters from Shenzhen Bay at sunset, Guangdong. (Image credit and source: Milkomède https://commons.m.wikimedia.org/wiki/File:China_Resources_Headquarters_03-10-2018.jpg)

The three levels of biomimicry (organism, behavior, and ecosystem) were presented and discussed. Each level includes five possible levels of the mimicry; organism level refers to bio-mimicking a particular organism as plants or animals and may involve mimicking them partially or the whole organism's form, material, function, and process; behavior level mimics the attitude or the act that the organism does to survive or replicates in usual and how it functions. This may include a

Fig. 8.28 Nature inspiration for biomimicry in architecture – Young cones of *Picea pungens* from Colorado blue spruce. (Image credit and source: JJ Harrison, https://commons.m.wikimedia.org/wiki/File:Picea_Pungens_Young_Cones.jpg)

translation of the mechanism and how it behaves or relates to a larger context. The ecosystem level of biomimicry intends to create and mimic an entire ecosystem and the common basics and principles that allow them to function in a successful way, which combines the other two levels to obtain a sustainable environment. Hence, a profound understanding of nature, environment, ecology, and regenerative process is needed. Finally, the application types of biomimicry in architecture portrayed selected architectural designs and projects that obtained inspiration, ideas, and solutions from nature (Figs. 8.28 and 8.29).

The source of inspiration from nature encompasses three types, (a) visual inspiration, (b) conceptual inspiration, and (c) computational inspiration; each kind plays a different role. The biomimicry design lens and nature's design principles were also discussed to grasp a full understanding of nature's singular features, qualities, and principles that can be applied to and contribute to developing biomimetic design

Fig. 8.29 Nature inspiration for biomimicry in architecture – a young female cone cultivated tree in Keila, Estonia. (Image credit and source: Ivar Leidus, https://en.m.wikimedia.org/wiki/File:Pseudotsuga_menziesii_var._glauca_young_female_cone_Keila.jpg)

Fig. 8.30 Nature inspiration for biomimicry in architecture – unopened bunch of onion flowers. (Image credit and source: Vishalsh521, https://commons.m.wikimedia.org/wiki/File:Onion_flower_unopened_buds.jpg)

before formulating solutions that encompass nature principles. Also, biomimicry design lenses were created to furnish an understanding of the relationship between this concept and life's principles. It can be concluded that biomimicry for sustainability often qualifies as a tool to make human-designed products, materials, and the built environment. In addition, it can be stated that sustainable architecture and ecology are analogous, forming a synergy, and identical in one essential way; however, both of them are the arts of space forming and creating a relationship between man, nature, and related surroundings (Figs. 8.30 and 8.31).

Fig. 8.31 Learn from nature systems for biomimicry in architecture. (Image credit and source: Nevit Dilmen (talk), https://commons.m.wikimedia.org/wiki/File:Onion-flower-07966. jpg)

Eco architecture reevaluates the relationship between nature and culture in architecture while being at the same time the factor of reestablishing our relationship with the natural world. The future of biomimicry was also discussed. It could lead to a bright future, if we actively connect to nature and acquire inspiration from it to craft and create biomimetic architecture. However, this will require learning from natural systems, which serve as a guide for rational design of target functionalities (Fig. 8.32). By and large, it can be said that there are no architectural examples that mimic nature in Egypt, but nevertheless, there is a group of students who used biomimicry in an irrigation system in Fayoum. This suggested a sustainable solution to the irrigation problems to enhance lives in Fayoum south of Cairo. Such a system's inspiration was created from the giraffe and the dromedary camel. However, there may be some but not yet published till the date of publishing this book. At the end of Chap. 3, the future of biomimicry was highlighted, especially during COVID-19.

Chapter 4 addressed the sustainability and global applications of biomimicry approaches and levels in architecture. Chapter 4 also reviewed a set of selected global examples to derive lessons learned from portraying these biomimicry approaches. The chapter also presented a review of biomimicry applications in line with sustainability as well as depicted many examples of biomimetic buildings

Fig. 8.32 Nature inspires biomimicry in architecture. (**a**) Plantaginaceae, *Hebe speciosa*, pink cone flower. (**b**) Orchid primrose flower. (**c**) *Celosia spicata* flower and foliage. (**d**) *Echium candicans* flower. (Image credit and source: (**a**) Lisa, https://commons.m.wikimedia.org/wiki/File:Pink_Cone_Flower.jpg; (**b**) Schnobby, https://commons.m.wikimedia.org/wiki/File:Orchid_primrose.jpg; (**c**) Muhammed Mahdi Karim, https://commons.m.wikimedia.org/wiki/File:Celosia_spicata.jpg#; (**d**) Jean-Pol Grandmont, https://commons.m.wikimedia.org/wiki/File:Echium_candicans_JPG01.jpg)

worldwide along with different levels of application, which were investigated. Although biomimicry concepts are not widely recognized in the field of architecture, nevertheless many countries across the globe follow this design approach. As the country's awareness regarding the connection between environment, nature, and sustainability increases, the biomimetic buildings' design improves and flourishes. In addition, Chap. 4 presented the ideas formed from the formal or structural functional viewpoint. It can be noticed that a high level of engineering knowledge is essential in order to imitate the designs of nature and apply them to architectural design. Moreover, Chap. 4 reviewed, assessed, and evaluated different global examples of biomimetic architecture as well as it conducted and discussed a comparative analysis and evaluation of these global buildings' examples.

Specifically, Chap. 4 tackled biomimicry application in line with sustainability and presented the structural efficiency, ICD/ITKE research pavilion 2016/2017, including the material manufacturing, ICD/ITKE research pavilion 2012; waste management model; and the water management – Sahara forest project – seawater-cooled greenhouse. The chapter also highlighted the thermal control-self-regulating homeostatic facade system, including the Hispasat innovative building, and ends with renewable energies – "straw-scraper" building. Architectural design examples including 30 global buildings in Europe, the USA, Asia, Australia, and Africa were covered and discussed. In addition, examples of rainwater collectors and low-carbon buildings were highlighted to draw learning lessons from this path. Moreover, the role of robotic strategies in biomimicry in architecture and global biomimetic architectural examples were summarized. Sustainability comparative analyses and evaluation of these selected best examples were conducted and presented with emphasis on a comparative evaluation of biomimicry architecture and sustainability. Finally,

Chap. 4 puts forward a question – why do we need sustainable architecture? The reasons and benefits of developing sustainable architecture were also outlined.

Chapter 5 offered a full review of climate change through several mitigation and adaptation actions. It specifically addressed climate change in a global context and highlighted its vital signs. It also presented the causes and effects of climate change. Chapter 5 also highlighted the role of biomimicry in climate change mitigation and adaptation and emphasized climate change in Egypt, including the Egyptian position related to climate change, and the needs for adaptation and adaptation and mitigation activities – the Egyptian view on climate change issue. From Egypt, Chap. 5 moved to highlight climate change in Italy, particularly the environmental action plan and adaptation strategies in Italy, implementation means, monitoring, reporting, and evaluation and sectors and actions as well as the role of engaging stakeholders in governance, knowledge, and international dimensions, and ended with climate change assessments. Chapter 5 depicted also climate change in Germany, including policy and legal framework, adaptation strategies in regard to implementation means and monitoring, reporting, and evaluation as well as assessments' means and moreover presented Germany's Climate Action Plan 2050 (development plan, goals, and content, as well as sectoral targets). Chapter 5 ended by highlighting biomimicry and climate change in two folds: (a) how could biomimetic design respond efficiently to climate change and (b) biomimicry to mitigate greenhouse gas emissions in the built environment in terms of biomimicry for energy efficiency and the biomimetics' segregation and carbon storage. Finally, it presented the CH2 building in Australia as the best biomimicry in architecture case study in this regard. Finally, it highlighted the benefits of a biomimetic approach in architecture amidst climate crisis.

Linking the most successful sustainable and environmentally friendly biomimetic architecture case study of the building CH2 in Australia to climate change, does this building contribute to having a lower negative impact on the environment and nature? The benefits of a biomimetic approach to architecture over time have been demonstrated through different kinds of biomimicry forms, its short-, medium-, and long-term responses to climate change. In addition, to mitigate the carbon footprint for the built environment, other significant social and economic benefits of biomimicry exist. In this context, several authors investigated the relationship between biomimetic design and amended psychological and physical health.

Chapter 6 presented biomimicry, as one of the biologically sustainable approaches that assist in creating environmentally friendly cities. There are many critical thinking and arguments toward imitating nature's ideas, concepts, and behaviors from its organisms. On the other hand, there are many defenses toward this approach and many proofs that it is an excellent solution in different aspects. Chapter 6 also discussed the arguments toward biomimicry and naïve biomimetic architectural design examples. Additionally, Chap. 6 presented the defense of biomimicry and debates that biomimicry is not constrained by evolutionary instrumentalism. Moreover, the importance of biomimicry, as an aid and which does not limit the human ability to build innovative designs, was highlighted; plus the limitations in biomimetic architecture are presented and discussed. Based on the comparison of the strength versus weakness of biomimicry, the strong concept of biomimicry and the weak concept of biomimicry were depicted and discussed with future directions toward sustainability. Finally, Chap. 6 portrayed the notion "from bioinspired innovations to eco-mimetic cities" and the conclusion's framework for the criticism against biomimicry.

It can be clearly said that not all the examples in the architectural field that mimic nature, especially at the organism level or form level, should be environmentally friendly or sustainable. On one hand, thinking ecologically and looking to nature for inspiration should be conducted with a reason and not only imitates the form of the organisms but also interacts with nature to survive and sustain and applies the same concept to the buildings' design (Fig. 8.33). On the other hand, there was a defense of biomimicry; the biomimicry philosophy encourages humility in the face of our natural limits.

Humans are not ranked above other species with regard to eco-dependencies, but we cannot absolve ourselves of our reliance on nature. However, humility is not self-deprecating; William McDonough rejects the traditional "cradle-to-grave"

Fig. 8.33 *Echium candicans* flowers from nature to inspire biomimetic architecture. (Image credit and source: Velela, https://commons.m.wikimedia.org/wiki/ File:Echium_ candicans1311.jpg)

manufacturing model, in which up to 90% of materials used in production are cast off as industrial waste. Instead, McDonough advocates a "cradle-to-cradle" model. The "cradle-to-cradle" manufacture emulates nature's closed-loop use of materials. In nature, an organism's waste is cycled through the ecosystem and becomes nutrient for other living things. In "cradle-to cradle" manufacturing, waste from one production line becomes feedstock for another; and products at the end of their lifecycle safely reenter the environment and decompose into five biological nutrients or are disassembled into "technical nutrients" and upcycled, which means they are recycled to form higher-grade products.

Moreover, critics of biomimicry adhere to a dangerous, dated cultural conception of humans as the ultimate life form; and they argue that biomimicry diminishes the role of the human designer. Nonetheless, it does not, since human beings are a core importance in biomimicry, since the approach depends on the designer's ability to extract and refine what he/she learns about biology into a set of abstract design principles from nature that can be implemented to solve human design problems. Thus, in Chap. 6, both opinions of those who criticize biomimicry or support it, those who are for or against it, are presented. Furthermore, architects who adopt biomimicry as an approach to sustainable buildings will provoke a shift in thinking with their biomimetic design (Fig. 8.34). Finally, the future direction is presented.

Chapter 7 presented an overview of the coronavirus (COVID-19) pandemic and addressed how biomimicry innovation and post-COVID regulations can curb the devastating impacts of such a crisis. Chapter 7 also highlighted an important question for all – "can biomimicry innovation assist in limiting the spread of COVID-19?" And it attempted to furnish some answers to this question. Additionally, Chap. 7 sheds light on the scoping challenge that the world faces and the context of COVID-19 and illustrated the innovation for COVID-19 and beyond to fight the COVID-19 through the exploitation of materials' inspiration from nature including natural nano-structures' mimicry and nano-materials and nanotechnology.

The change in the face of life and the COVID-19 positive insights and opportunities were addressed. Moreover, Chap. 7 presented the guidelines for healthier urban spaces and buildings' indoors in COVID-19, including eight folds: (a) urban spaces, (b) healthy cities, (c) green cities, (d) smart cities, (e) All-In cities (inclusiveness), (f) sustainable cities, (g) resilient cities, and (h) rising cities. Furthermore, Chap. 7 focused on the urban context and pandemic role as well as indoor building's spaces, and finally, this chapter ended by depicting a discussion on the post-COVID age.

8.6 Concluding Remarks

Too many essential points can be drawn out in this book, especially in regard to climate, energy, food, and COVID-19 crises, but most importantly, we can deduce from this book the following conclusions:

Fig. 8.34 Nature for inspiring regenerative and green cities. (Image credit and source: Shareen Elnaga and Mohsen Aboulnaga)

- Biomimicry has existed since billions of years ago, and it was adapted to architectural design in architects' works.
- Biomimicry created a role as an inspiration to design and also nature as an origin of biomimetic inspiration – role of "model," "measure," and "mentor."
- Evaluation of the ideas of nature by the process of "distill," "translate," "discover," "emulate," and "evaluate," and it is regarded that the nature is perfect and the origin of design principles.
- Evaluation of the concepts of "structure," "function," and "system" in the context of perfect nature.
- Biomimicry as an architectural approach is still not largely widespread worldwide, few examples of buildings are existing in several countries,
- Revivalism of the importance of looking to nature and taking into consideration the environment and the other creatures that share our planet in order to keep the ecosystem and more environmentally friendly built environment.

- Biomimicry approach could play a role in reducing GHG as the building sector is the largest sector that affects climate change, so engaging biomimicry thinking from the early conceptual design stage is necessary.
- Biomimetic architecture creates lungs in cities for better livability.
- Biomimetic strategies are urgently needed to contribute to mitigating 45% of carbon by 2030 and net zero by 2050.
- Recent biomimetic architecture models proved that net-zero building can be successfully achieved.
- Biomimetic architecture can be the paradigm shift toward attaining the net-zero goals by 2050.

8.6.1 Theoretical and Methodological Framework

The theoretical and methodological frameworks to enable architects and urban designers can be summarized as follows:

- Develop an architectural design that addresses the current ecological, materialization, and special requirements and in this way particularly customizes its area and conditions.
- Create a cutting-edge sustainable and biomimicry architecture.
- Design biomimetic architecture is a gate that can interface architecture and design professions on a course to connecting design, ecological, and environmental issues in a sustainable solution and product.
- Design biomimetic architecture develops the way of thinking and designing that engages architecture and industrial design in a process of environmental and biological focus on more responsive, comfortable, and safer buildings.
- Identification of using biomimicry in architecture as a problem-solving methodology is vital.

Human activities in the built environment are progressively responsible for worldwide ecological and social issues or problems with tremendous extents of waste, material, and energy consumption impact on nature in addition to high-carbon emissions. Biomimicry is a multidisciplinary inventive tool, including a wide assorted variety of such domains as electronics, science, chemistry, materials science, engineering, and buildings, thinking about nature and copying its innovative capacities and functions, forms, and eco frameworks utilizing advanced innovation to solve human issues in reconciliation with nature (Fig. 8.35).

It is also possible to shift the future of design into mimicking nature's design principles and developing it into the built environment, which is able to gather and use natural resources efficiently, diversify, and cooperate to fully use the habitat as well as be capable of functioning as sustainable and regenerative (Fig. 8.35).

In addition, it is obvious that there are numerous parallels to be drawn among nature and architecture, some of which have been considered for quite a long time and others, which are just getting to be pertinent as looking for a cur; this stressed

Fig. 8.35 Integrated buildings with nature for a regenerative city. (Image credit and source: Mohsen Aboulnaga and Nada Nidal)

for the connection between the built environment and the natural environment, the latter is quite forgotten. While traditional approaches to sustainability emphasize energy conservation and reduced consumption, designing with full engagement and integration with nature results in a positive and regenerative design and results. Biomimicry, as opposed to being utilized as a logical strategy for imitating nature in the built form, however, is a system ought to be connected in a full encompassing sense, where architects and urban designers recognize the complex collaborations that occur inside the common world more essentially and comprehend the current situation and challenges within.

Not exclusively is nature – a promptly accessible source of motivation and inspiration that it is available in each particle around us. However, natural forms have also evolved within the same confines as humanity, using just the material and energy resources that are easily accessible on planet Earth. As human species continue to advance embracing our potential for future improvement, while respecting the collective wisdom of our predecessors, this should be pursued in a rational way. Thus, the possibility of producing a sustainable future, our successors will wind up with both achievable and rewarding aspirations. It is imperative to state that the building sector is in charge of half 48% of all energy consumption and greenhouse gas (GHG) emissions yearly; universally, the rate is considerably more prominent (US energy information administration); 76% of all power plant created power is utilized to operate buildings. Consequently, there is desperation for activity to urgently secure the environment, where biomimicry could contribute to mitigating climate change due to reducing GHG emissions in the building sector. Therefore, it also helps in reducing energy consumption and decarbonization; hence, nature is the solution (Fig. 8.36).

(a) (b)

Fig. 8.36 Nature inspiring architecture. (**a**) Tulip flowers from nature to inspire architecture. (**b**) Nature as a source of inspiration. (Images' credit and source: (**a, b**) Lee Elnaga and Mohsen Aboulnaga)

8.6.2 Main Recommendations

Based on the narrative of this book, the following main recommendations can be deduced to derive and foster biomimetic architecture:

- Facilitate the integration of biomimicry in academic curriculums of architecture and engineering, as a futuristic method of reducing the impact on the environment and attaining sustainability, since biomimicry is considered the advising and guiding principle for the paradigm shift to creative and adaptive education towards sustainability.
- Collaborate different disciplines of engineering, applied science, and biology to help designers reach the deepest form of sustainability in biomimicry on the ecosystem level, aiming for a different outcome, as well as graduate students who are not architects, biologists, or urban planners but biomimetic professionals.
- In order to achieve a sustainable environment, incorporate biomimicry as an evaluative tool from the beginning of the design process.
- Incorporate biomimicry as an evaluative tool from the early stage of design processes to attain sustainable environment.
- Change the mindset and raise awareness, and learn that biomimicry will not only assist in significantly reducing the environmental impact of buildings, but also it

has the potential to help in defining a whole new sustainable standards for architecture profession.
- Educate students and researchers on how to develop their design to serve societal, environmental, and economical needs, thus insuring a well-being and sustainable approach in design to achieve sustainable and livable communities and resilient cities.
- Teach biomimicry, as a design philosophy, and learn to model from the genius creations in nature and develop an increasing sense of gratitude and a passionate desire toward sustainability and protect the genius that surrounds us.

In this context, nature enlightens our live, and therefore, architecture should learn from nature and integrate within all processes of design and development (Figs. 8.37, 8.38, 8.39 and 8.40).

Fig. 8.37 Combine nature with architecture for a sustainable built environment. (Image credit and source: Lee Elnaga and Mohsen Aboulnaga)

Fig. 8.38 Integrate nature with architecture for sustainable cities. (Image credit and source: Lee Elnaga and Mohsen Aboulnaga)

8.7 Most Recent Innovative, Smart and Biomimetic Architecture in 2022

In short, it is clear that biomimetic architecture has reached a new height in recent years. The new smart, net-zero building by Zaha Hadid and Zaha Hadid architects was her last project before her death in 2016 [10]. The innovative building is BEEAH Headquarters. "The office of the Future" is built in Al Sajaa desert, Sharjah, UAE; and its concept is inspired from nature and mimics sand dunes in the vast nature desert of the UAE, while responding to the hot and dry climate as shown in Figs. 8.41 and 8.42. The building lines were carefully complied to reflect nature in

Fig. 8.39 Nature to inspire architecture for sustainable communities. (Image credit and source: Shareen Elnaga and Mohsen Aboulnaga)

Fig. 8.40 Nature inspiring architecture for a sustainable built environment. (Image credit and source: Shareen Elnaga and Mohsen Aboulnaga)

Fig. 8.41 A bird's-eye view of BEEAH Smart Headquarters inspired from desert sand dunes – an innovative net-zero building. (Image credit and source: Fearandloathingindubai, https://commons.m.wikimedia.org/wiki/File:BEEAH_headquarters.jpg)

(a) (b)

Fig. 8.42 View of the desert sand dunes inspiring the concept of BEEAH Headquarters' building by Zaha Hadid. (**a**) The edges of sand dunes inspiring the innovative building's design. (**b**) Created concave dunes and edges of desert's sand dunes. (Image credit and source: (**a**) JSPhotography2016, https://commons.m.wikimedia.org/wiki/File:Sunset_in_Al_Ain_Desert.jpg#, (**b**) Ravi Kumar, https://commons.m.wikimedia.org/wiki/File:Desert_(143403787).jpeg)

Fig. 8.43 View of BEEAH Smart Headquarters inspired from desert sand dunes in its flowing shaped building. (Image credit and source: Fearandloathingindubai, https://commons.m.wikimedia.org/wiki/File:BEEAH_headquarters1.jpg)

a smooth and flowing shape presenting the interconnecting "dunes" and resonating with surrounding landscape of sand dunes formed by prevailing winds into curved dunes and edges, which become convex when they intersect [11] as illustrated in Figs. 8.41, 8.42, and 8.43.

The building is also powered by solar energy array and equipped with next-generation technologies. The BEEAH Headquarters combined both sustainability and digitalization strategies, including six key industries: (a) clean energy, (b) environmental consulting, (c) green mobility, (d) waste management, (e) recycling, and (f) education [10, 11]. This innovative, smart, and iconic building was officially opened on March 30, 2022, and it represents the first smart, artificial intelligence and Internet of Things (IoT) building, which is also considered a new model of biomimetic architecture [10].

Finally, the current world challenges are posing tremendous pressure on all governments, local authorities, and planners as well as architects to act to achieve decarbonization and climate neutrality by 2050; if we want to achieve the set targets of the Glasgow Climate Pact (GCP) – "mitigate 45% of carbon by 2030 and reach net zero by 2050" [12] – nature and renewables should be integral parts in developing urban areas and cities as well as buildings. But if not, then authors don't anticipate reaching the GCP targets.

References

1. Nature's Future, Our Future – The World Speaks. (2021). Retrieved April 15, 2022, from https://www.iucn.org/news/protected-areas/202102/natures-future-our-future-world-speaks
2. Gehry, F. (2022). *Frank Gehry quote: Architecture should speak of its time and place, but yearn for timelessness*. Inspiring Quotes. Retrieved April 18, 2022, from https://www.inspiringquotes.us/quotes/gk6c_84MoQnb6
3. Architects, G. (2021). *The Terra – The sustainability pavilion expo 2020 Dubai*. GRIMSHAW Global. Retrieved April 15, 2022, from https://grimshaw.global/projects/dubai-expo-2020-sustainability-pavilion/
4. Beermann, K., & Chen Austin, M. (2021). An inspection of the life cycle of sustainable construction projects: Towards a biomimicry-based road map integrating circular economy. *Biomimetics, 6*(4), 1–20. https://doi.org/10.3390/biomimetics6040067
5. *WHA IS BIOMIMICRY? – Nordic Biomimicry*. Nordic biomimicry. (2022). Retrieved April 18, 2022, from https://www.nordicbiomimicry.org/what-is-biomimicry/
6. Andrews, K. (2013). *Pasona Urban Farm by Kono Designs*. Retrieved April 15, 2022, from https://www.dezeen.com/2013/09/12/pasona-urban-farm-by-kono-designs/
7. Wang, L., & Wang, L. (2013). *Lush Curvaceous Green Wall Grows in iArc Architects' New Seoul City Hall*. Retrieved April 15, 2022, from https://inhabitat.com/lush-curvaceous-green-wall-grows-in-iarc-architects-new-seoul-city-hall/
8. *Shenzhen China Resources Headquarters Tower*. WSPglobal. (2022). Retrieved April 18, 2022, from https://www.wsp.com/en-OM/projects/shenzhen-china-resources-headquarters-tower
9. China Resources Tower by Kohn Pedersen Fox (KPF). Kpf.com. (2022). Retrieved April 18, 2022, from https://www.kpf.com/projects/china-resources-tower
10. Hadid, Z. (2022). *BEEAH Group Officially Opens Ground-breaking New Headquarters, Unveiling the 'Office of the Future' – Zaha Hadid Architects*. Zaha-hadid.com. Retrieved April 18, 2022, from https://www.zaha-hadid.com/2022/03/31/beeah-group-officially-opens-ground-breaking-new-headquarters-unveiling-the-office-of-the-future
11. Sabith, S. (2022). *BEEAH HQ by Zaha Hadid is a stunning blend of design, technology and sustainability*. Retrieved April 18, 2022, from https://www.middleeastarchitect.com/projects/beeah-zaha-hadid?utm_medium=email
12. Unfccc.int. Decision -/CP.26. Glasgow Climate Pact. (2022). Retrieved April 19, 2022, from https://unfccc.int/sites/default/files/resource/cop26_auv_2f_cover_decision.pdf

Index

M. Aboulnaga, S. E. Helmy, *Biomimetic Architecture and Its Role in Developing Sustainable, Regenerative, and Livable Cities*,
https://doi.org/10.1007/978-3-031-08292-4

The manufacturer's authorised representative in the EU is Springer Nature Customer Service Centre GmbH, Europaplatz 3, 69115 Heidelberg, Germany. If you have any concerns regarding our products, please contact ProductSafety@springernature.com

Printed and bound by CPI Group (UK) Ltd, Croydon, CR0 4YY
30/09/2026
02224059-0012